The Best of AMATEUR TELESCOPE MAKING JOURNAL

Volume 1

Edited by William J. Cook

Published by

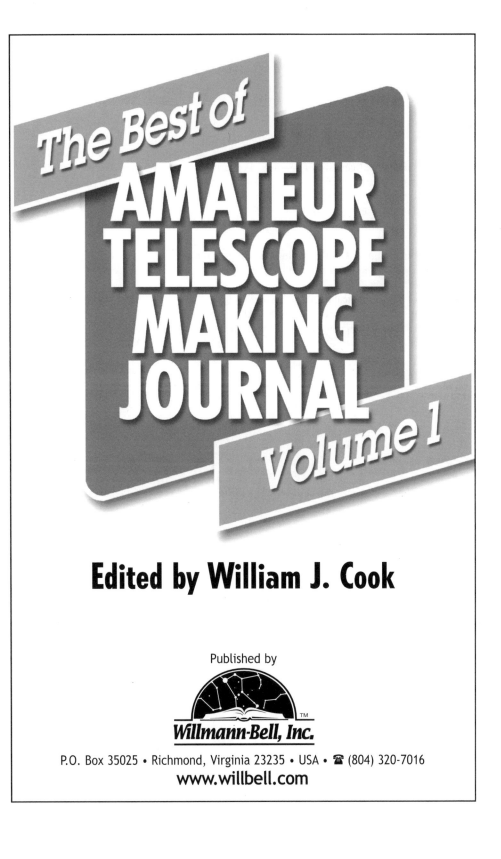

Willmann-Bell, Inc.

P.O. Box 35025 • Richmond, Virginia 23235 • USA • ☎ (804) 320-7016
www.willbell.com

Printed in the United States of America
First Printing 2003

Library of Congress Cataloging in Publication Data
The Best of Amateur telescope making journal / edited by William J. Cook
    p. cm.
  Includes bibliographical references and indexes.
  ISBN 0-943396-77-8 (v. 1) -- ISBN 0-943396-78-6 (v. 2)
    1. Telescopes--Design and construction. 2. Telescopes--Design and construction--Amateurs' manuals. I. Cook, William J., 1951-II. ATM journal

QB88.B47 2003
681'.412--dc21   200304190

03 04 05 06 07 08 08 10 9 8 7 6 5 4 3 2 1

Sonotube, Teflon, Kydex, Sorbathane, Pyrex, Formica, Ebony Star, Permatex are trademarks.

Table of Contents

Foreword

Albert Ingalls was first to give amateur telescope making wide exposure in the United States through his articles in *Scientific American*. Growing more enamored with the craft himself—and seeing a rising interest in his readers—Ingalls began to collect and publish material for the amateur telescope maker in book form. From a very thin book titled *Amateur Telescope Making,* published in 1926, *Scientific American* supported Ingalls in an effort that was to culminate in 1953 with three massive volumes—*Amateur Telescope Making I, II* and *III*—which have come to enjoy an almost scriptural status among telescope makers.

The hobby enjoyed increased momentum in the 1940s and 1950s. By then, articles and information on telescope makers and telescope making had 20 years in which to pique the interest of the more adventurous members of astronomical societies all over the world. And why not? It had been clearly demonstrated that with supplies readily available in every hardware store, and many home workshops, telescopes could be crafted that would rival many of the best professionally made instruments of the day.

In 1947, Allyn J. Thompson published *Making Your Own Telescope*. This little book, while not as ambitious and broad as the *ATM* series, placed in the hands of its readers a thorough but succinct description of what it takes to build a fine first telescope. It first appeared in serial form beginning with the November 1944 issue of *Sky & Telescope* magazine, which throughout its history has been eager to cover this facet of amateur astronomy and in which to this day ground-breaking articles frequently appear. In fact, my interest was spurred when I first came across John Gregory's classic article on the design and construction of the Cassegrainian-Maksutov in *S&T's* March 1957 issue.

That same year saw Jean Texereau's *How to Make a Telescope* make its first appearance in the English language, and it has since come to be recognized as *the* book on telescope mirror making. Neale Howard's *Standard Handbook for Telescope Making* made its debut in 1959 and Sam Brown's *All About Telescopes*—now in its 14th printing—hit the streets in 1967. And who could have done more than Richard Berry with the 14-year run of his publication, *Telescope Making* magazine?

More recently, Willmann-Bell, which has published a number of new ATM books, has also begun updating some of these classics. They have revamped the venerable *ATM* series—arranging articles in a more logical fashion—and have turned Allan Mackintosh's *Maksutov Circulars* into *Advanced Telescope Making Techniques, Vols. 1 and 2*. And fortunately for amateur telescope makers, the list now includes a new title, *The Best of Amateur Telescope Making Journal*, a body of work that consumed most of my spare time for the past 10 years.

I would like to take this opportunity to acknowledge those who helped *ATMJ* get started and provided valuable counsel along the way. First and foremost among these are longtime friends: Dick Buchroeder, John Gregory, Harrie Rutten, Martin van Venrooij, Diane Lucas, and the late Robert E. Cox. All of whom, with the exception of Bob, were Associate Editors.

Also offering valuable assistance was Australian ATM and optics technician, Roger Davis. For his efforts to solicit articles and promote *ATMJ* "down under," Roger was asked to become an Associate Editor in 1998.

In 1999, my daytime employer, Captain's Nautical Supplies, lightened the administrative load on my shoulders by taking over *ATMJ* and hiring M. Barlow Pepin, who brought new ideas to the effort, solicited articles, and acted as managing editor during *ATMJ's* last three years in publication.

I must also thank my wife Debbie and my children—Bill, Sean and Noelle. Their support was invaluable and their sacrifices for *ATMJ*, immeasurable.

Finally, I would like to thank the most critical component of any publication, those who supported *ATM Journal* with their articles. As I thumb through these two volumes, totaling more than 800 pages, I am truly impressed by the scope, depth, and quality of their contributions to the body of ATM literature, and pleased that I had the good fortune to be associated with their efforts.

Newspapers, journals and magazines are by their very nature transitory. Today's hot publishing vehicle, the WEB, has yet to prove that content posted today will be there well into the future—already, elaborately constructed sites have many links that no longer work! I have no doubt about the staying power of the book and I am personally gratified that the collective efforts of the many people who created *ATMJ* will be available for a long, long time.

William J. Cook
Seattle, Washington
January 2003

Issue 1

1.1 The Houghton Telescope: An Optimum Compromise?

By Harrie G. J. Rutten & Martin van Venrooij

A question often asked by amateur telescope makers is: "Which telescope is best?" Everybody who delves into this question will find out soon that each type of telescope has its own set of advantages and disadvantages. These disadvantages may, for instance, pertain to image quality, width of field, focal ratio, weight, etc. Consequently, it is impossible to get an unambiguous answer to the question, "Which telescope is best?".

The situation is more favorable for those who specialize in only one field; for instance, the observation of planets or deep-sky photography. For these tasks, specialized instruments are available with which optimum results can be achieved.

However, most amateurs want a universal instrument that can be used in different fields of observation and photography with a favorable combination of properties and only minimal disadvantages. This type of instrument may be called an "optimum compromise."

The authors have looked for such a compromise and the following discussion presents their findings.

1.1.1 Requirements for an Optimum Compromise

We believe that an instrument designed to meet the needs of most casual observers or astrophotographers could be achieved with an instrument possessing the following characteristics:

1. Compact design
2. Aperture 200–250 mm
3. Suitable for visual and photographic application
4. Minimum central obstruction (30–35% of the aperture at most)
5. High photographic speed, perhaps $f/4$.

The first requirement is, in our view, the most important. A compact telescope can be handled and transported easily and may also find a place in the car during vacation trips. The second demand is self-evident. A large aperture has many advantages. The third requirement is the most difficult to achieve, for three

1

reasons:

1. The image sharpness should be good not only in the middle of the field, but at the edges as well. Unfortunately, most amateur telescopes do not provide good off-axis imagery.

2. The focal surface should be flat, or nearly so. This is another area in which most amateur instruments fall short.

3. For astrophotography, the instrument must have a large spectral range, providing a high degree of color correction from red to violet.

The fourth demand refers to the need to keep the central obstruction as small as possible. Large obstructions cause diminished performance on low contrast objects. The last requirement is necessary for deep-sky photography where a relatively wide field and short exposure times are desirable.

1.1.2   Possible Instruments

A check on the existing instruments reveals that most of them do not satisfy the above-mentioned criteria simultaneously.

• A refractor with a large aperture and short focal length will have strong chromatic aberration.

• A short-focus Newtonian suffers from coma which degrades off-axis image sharpness.

• A Cassegrain, when designed as a short-focus instrument with a flat field, has a large secondary mirror, which may approach 55% of the aperture! This is also the case with various catadioptric designs such as the Schmidt-Cassegrain and the Maksutov.

The solution can be found only in the group of catadioptric derivatives of the Newtonian telescope; i.e. systems with a single mirror combined with a corrector to suppress the off-axis aberrations.[1]

An example of an instrument which meets most of our prescribed criteria is the Rutten Newtonian with a small corrector lens just inside the focus. However, this is an $f/15$ lunar and planetary telescope and would not allow us to achieve our requirements for a "fast" wide-field instrument. The first instrument we will consider consists of a spherical mirror and a Schmidt corrector. This type of instrument is called a Schmidt-Newtonian and should not be confused with the Schmidt camera. In the latter the corrector is placed at twice the focal length, and the design is frequently referred to as a "concentric" system. Image quality in this system is exceptionally good. However, this instrument may only be used for photography since the focal plane lies inside the instrument.

In the Schmidt-Newtonian shown in Figure 1.1.1, the corrector is placed inside the focus of the mirror. While the image plane is accessible for visual use, this design lacks the performance of the Schmidt camera in that off-axis aberrations

---

[1] Rutten, Harrie and Martin van Venrooij. "The Rutten Newtonian," *Telescope Making* 35, 1988/89, pp. 46-50.

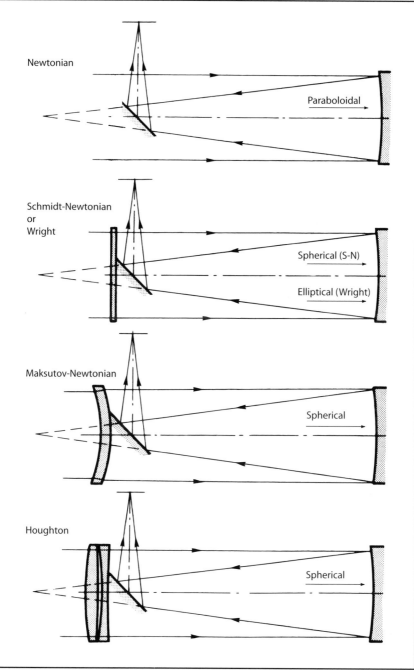

Newtonian

Paraboloidal

Schmidt-Newtonian
or
Wright

Spherical (S-N)

Elliptical (Wright)

Maksutov-Newtonian

Spherical

Houghton

Spherical

Fig. 1.1.1 *The Newtonian and some catadioptric derivations.*

such as coma and astigmatism are not fully corrected.

A better optical system is the Wright design, in which the mirror has been deformed into an ellipsoid. In this design, coma can be fully removed. However, this system still suffers from astigmatism and making an ellipsoidal mirror is not an easy task.

The next system uses a Maksutov corrector to remove the spherical aberration produced by the spherical primary mirror. The corrector is much closer to the mirror than in a Maksutov camera; therefore, coma cannot be completely corrected.

1.1.3 Enter the Houghton

The Houghton consists of a spherical primary mirror and a two-element corrector of zero optical power.

Thus, the corrector can remove spherical aberration without introducing the chromatic aberration normally associated with large doublet lenses. The particular two-lens design we analyzed was made by Lurie in 1979.[2] It is but one example of the various types of correctors Houghton proposed in 1944.[3]

Design methods for this type of instrument may be found in *Telescope Optics, Evaluation and Design.*[4]

1.1.4 Optical Performance

In Figure 1.1.2, we see the optical performances of four catadioptric systems compared to that of a Newtonian telescope. We show the images of a star on the best-fitting focal surface on the optical axis and at 10 mm and 20 mm from the axis. The last value corresponds approximately with the corner of a 24 x 36 mm negative. Compare the star images with the Airy disk for visual use of the telescope and a disk of 0.025 mm for photographic applications.

All instruments provide sharp images on-axis because spherical aberration has been completely eliminated. However, off-axis, the systems differ considerably. The Newtonian is not suitable for critical photographic applications because of its strong coma.

Coma is still present in the Schmidt-Newtonian (although to a lesser degree), and the Wright suffers from astigmatism. Note that while the Maksutov derivation is better than the Schmidt, only the Houghton is capable of rendering excellent image quality at the edge of the field of view.

Thanks to the long radius of curvature of the focal surface (nearly three meters!), we need no field flattener, and nearly all of the full image sharpness may be captured on a flat film plane.

[2] Turco, Edward. "Gleanings for ATM's—Making an Aplanatic 4-inch Telescope," *Sky and Telescope*, Nov., 1979, page 473.

[3] Houghton, J. L. U.S. Patent 2,350,112. May 30, 1944.

[4] Rutten, H.G.J. and M.A.M. van Venrooij. *Telescope Optics, Evaluation and Design*. Richmond, VA: Willmann-Bell, Inc., 1988.

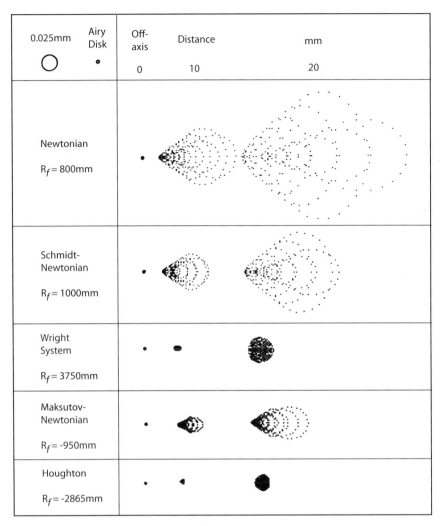

Fig. 1.1.2 *Spot diagrams for a 200 mm f/4 Newtonian and other catadioptric derivations.*

Furthermore, the color correction for the whole range from red to violet is excellent thanks to the fact that the corrector is a "zero power" element.

1.1.5   Diagonal Mirror Size

A disadvantage of the fast systems described is the presence of a rather large diagonal mirror. In order to receive all incoming parallel rays, the secondary mirror must create an obstruction no less than 29% of the full aperture. For off-axis bundles the size must be enlarged; and for the full illumination of a 24 x 36 mm negative, the size of the mirror (minor axis on a 200 mm instrument) would be approximately 46% of the aperture.

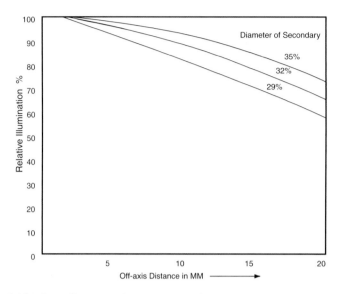

Fig. 1.1.3 *Relative illumination for instruments with secondary mirrors of various sizes.*

Table 1.1.1 Construction of a 200 mm $f/4$ Houghton telescope (data in mm)		
Radii of curvature:	Glass Type:	Thicknesses:
R1 = 1286.81	Bk7	D1 =16
R2 = −4810.34		D2 =3
R3 = −1286.81	Bk7	D3 =12
R4 = 4810.34		D4 =617.67
R5 = −1590.97		D5 =−793.84

The distance from the primary mirror to the center of the diagonal is 575 mm and the diagonal mirror has a minor axis of 64 mm.

This figure is much too large to allow the instrument to be considered good for visual work. However, photographic applications often allow a light drop-off at the edge of the field of view approaching 40%.

This would indicate that one could use a secondary considerably smaller than the 46% obstruction just mentioned. Figure 1.1.3 shows the off-axis illumination relative to the center of the field for diagonals having a minor axis of 29%, 32% and 35% of the aperture for a 200 mm telescope.

We find the mirror with a 32% obstruction acceptable and tolerate a light reduction of 34% at the edge of the field.

Table 1.1.1 gives design data for the 200 mm instrument derived from Footnote 2.

1.1.6   Conclusion

The authors believe that the Houghton telescope is very close to a universal telescope for the demanding amateur. The advantage of the Houghton may be described as follows:

- Compact design
- Closed system
- Good image quality
- Fast system, suitable for visual and photographic application
- All surfaces may be left spherical
- Common glass types may be used
- No spider since the diagonal is attached to the corrector
- Relatively simple to construct
- Radii of curvature in complementing pairs allow the lenses to be ground and tested against each other.

Because of these favorable properties, it is surprising that this type of instrument is rarely built by amateurs, and that it is not offered commercially. The authors hope that this article will contribute to a broader knowledge of this fine instrument among amateurs.

1.2   The World of Unobstructed Reflecting Telescopes
By José Sasián

The quest for the best images in amateur astronomy has stimulated the development of unobstructed reflecting telescopes. These telescopes have evolved as a way to obtain high contrast images from instruments that are relatively inexpensive and easy to fabricate.

The central obstruction in conventional reflecting telescopes like the Newtonian reduces image contrast, and the supporting spider introduces diffraction artifacts. Such effects are clearly visible in many of the photos of stars and planets published in *Sky & Telescope* and *Astronomy* magazines. These problems can be avoided with optical designs employing tilted mirrors, which permit the light path to remain unobstructed. The story of the modern unobstructed reflecting telescope starts in Germany, where Anton Kutter popularized the Schiefspiegler. His writings about this "oblique" telescope include articles, a book, and *Sky & Telescope Bulletin A*. He labored for twenty-five years to bring diffraction-limited reflecting optics within the means of the ATM.

Thanks to the efforts of the late Robert (Bob) Cox, the Schiefspiegler was introduced in the United States. Shortly thereafter, telescope makers like Oscar Knab and Al Woods began work on such instruments. Early on, Bob made a Schiefspiegler for himself and encouraged Richard Buchroeder, an optical engineer with strong ATM roots, to examine other configurations of the tilted component telescope—or TCT. Buchroeder analyzed and designed several unobstructed models, and his work culminated in the classic *Technical Report 68* and his well-known Tri-Schiefspiegler. In the meantime, Arthur Leonard was also working on "giving the astronomical mirror its theoretical definition" and designed and demonstrated the Yolo. Leonard contributed to developing the theory of TCTs and designed other telescope configurations of interest like the Solano.

Strongly influenced by Richard Buchroeder's *Technical Report 68*, I further developed the theories associated with unobstructed telescopes that Buchroeder had initiated at the University of Arizona. Combining optical shop experience and an understanding of the theory of unobstructed optics, I was able to refine the Yolo and design and make the unobstructed Newtonian. In this article, I present a glimpse of how unobstructed reflectors are designed and discuss some of the best-known telescopes. Two unobstructed designs that may be of interest to ATMs are presented as well.

1.2.1   Telescope Design

The optical design of an unobstructed reflecting telescope involves the layout of a mirror configuration, aberration correction, and an overall evaluation of the design. The aberration correction drives the design process because there is usually not enough design freedom to obtain a good image quality and a practical telescope configuration. Thus, image quality is usually achieved at the expense of telescope simplicity and practicality.

The Main Aberrations

The field of view of a telescope is the angular extent of the object or scene it can image. At the focal plane, the field of view is represented by the diameter of the image. Every point in the object emits a wavefront, or a bundle of light rays, that will, if our design is good, form a corresponding image point at the focal plane. To have a perfect image, these wavefronts must converge to a true image point. In practice, due to the inherent geometry of telescopes and to mirror fabrication errors, the wavefronts are deformed, and the rays do not converge to a single point. Instead they form a "spot" or "blur" pattern as illustrated by the familiar spot diagrams. The compilation of all these spots of light forms the telescope image.

Image defects caused by the geometry of telescopes are called aberrations. Those that have the greatest impact on image quality are spherical aberration, coma, and astigmatism. The wavefront deformation and through-focus spot diagrams corresponding to these aberrations are illustrated in Figure 1.2.1.

Aberrations are measured from the vertex to the highest point of the deformation; this distance is measured along the direction of light propagation and is usually expressed in wavelengths. When the amount of aberration is small (about $\frac{1}{10}$ wavelength of light) the image is essentially perfect within the limits imposed by the phenomenon of light diffraction. When the amount of aberration is about $\frac{1}{4}$ wavelength, the image is acceptable; and when there are several waves of aberration, the image is poor.

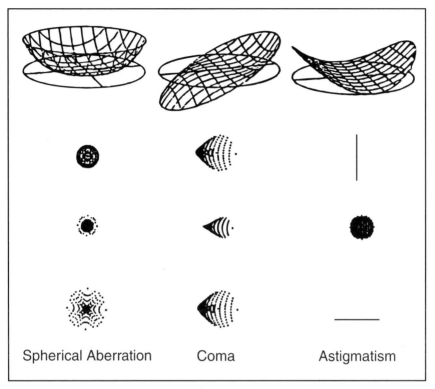

Fig. 1.2.1 *Wavefront aberrations and through-focus spot diagrams for spherical aberration, coma, and astigmatism Courtesy of R. Shack.*

In part, the task of the telescope designer is to correct or minimize the image aberrations. This is mainly done by causing the aberrations of one mirror to cancel the aberrations of the other, or by aspherizing one or more surfaces. If a mirror is polished with an asphericity similar to the wavefront deformation for spherical aberration, coma, or astigmatism, one can induce or correct such aberrations.

The design of unobstructed telescopes appears to be an obscure matter. This is because when the axial symmetry, typical in traditional telescope design, is broken by tilting the mirrors, the behavior of image aberrations becomes less simple.

The Newtonian and Cassegrainian designs can have only one aberration at the center of the field of view—spherical aberration due to their inherent axial symmetry. At off-axis field points, coma and astigmatism are generated. Coma mainly limits the extent of the usable field of view, and it grows linearly with the distance off-axis. For small fields astigmatism is usually negligible, and it varies with the square of the distance off-axis.

In contrast, in a tilted component telescope spherical aberration, coma, and astigmatism can all be present at the same time at the center of the field of view. The absence of axial symmetry around the field center results in this behavior. At off-axis field points, anamorphic distortion, coma and astigmatism (both varying linearly with the distance off-axis) and image plane tilt are the aberrations that affect the image.

Design Process

On one hand, the design of a known telescope configuration is simple because it is already understood. The design task becomes the adjustment of the various parameters to suit the particular requirements. Once in a while articles are published describing this process. For example, in *TM* #1, Buchroeder illustrates how to design Cassegrainian systems; Anton Kutter, in *Sky & Telescope Bulletin A*, describes the design of Schiefspieglers; and in *TM* #37, I describe how to design a Yolo telescope.

On the other hand, in the process of designing a new telescope, a novel mirror arrangement is proposed. The primary mirror, usually the largest one, is tilted so as to place the secondary outside the incoming light beam. The position of the secondary, its radius of curvature, and its tilt are chosen to yield a configuration physically possible and practical, and to correct aberrations. If more than two mirrors are involved, they are also used to accomplish such functions. The main concern in the design is to correct or minimize astigmatism, coma, and spherical aberration on-axis and linear astigmatism off-axis.

With designs employing only a few mirrors, there are not too many variables to correct for all the aberrations and obtain a practical telescope. Therefore, when the main aberrations have been corrected or minimized, then the image aberrations, the image anamorphism, image plane tilt, and the practicality of the configuration are evaluated. If these characteristics are acceptable, the main phase of the design process is completed. Other phases include checking mirror sizes and shapes, the design of testing configurations for the mirrors, tolerancing and light baffling.

Most amateur unobstructed telescopes have been designed with algebraic formulas to calculate the focal length f/ratio, and aberrations. Ray-tracing software has been only used to verify a theoretical prediction and optimize it. This design process implies a good understanding of telescope requirements—the "how?" and "why?" of previous designs and of optical design principles.

While the emphasis in designing and making unobstructed telescopes is here put toward obtaining a diffraction-limited image, there is no reason why these telescopes could not be designed with other goals in mind.

Design Considerations

The number of mirrors involved and their sizes are important considerations. The optimum number of surfaces is two since use of only one mirror leads to an uncomfortable viewing position, and three are too many—two can do the job. In addition, the light lost due to reflection is reduced when fewer mirrors are used. It is desirable to have small secondaries and tertiaries to reduce the telescope's volume and weight, and increase its portability.

The roughness of mirror surfaces may also scatter light that reduces the im-

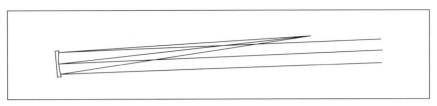

Fig. 1.2.2 *The Herschelian single-mirror telescope.*

age contrast. Aside from the atmospheric effects, the light scattered by rough or dirty surfaces is the limiting factor to obtaining perfect images in well-corrected unobstructed telescopes. In theory, and for the same amount of roughness, light scattering by mirrors is 16 times stronger than for refracting surfaces and about ½ as much as for a Mangin mirror surface. These facts indicate that the comparatively small amount of light scattered inherent in refracting telescope designs can be compromised if additional reflections are introduced to achieve comfortable viewing positions.

1.2.2   Telescope Designs

The following telescopes exhibit the best-known designs involving only mirrors and further illustrate how the design process is carried out.

One-Mirror Telescopes

A. The Herschelian

Figure 1.2.2. It is generally accepted that Herschel used this construction in his "front view telescope" to avoid the loss of light (about 33%) that a secondary mirror of speculum metal would have caused. This design was suggested by Lemaire in 1732 and tried by Herschel in his 20-foot telescope. Herschel seems to have been more concerned about light-gathering power than about optical quality. However, his decision to make a 40-foot front view telescope could have also been influenced by the simpler configuration. The mirror tilt in his instruments introduced about 20 waves of coma and 20 waves of astigmatism. It is surprising that, with this amount of aberration, he could observe details in the rings of Saturn along with certain spots and belts on the planets surface, as he reported in *Philosophical Transactions* articles between 1790 and 1806.

Tilting a concave spherical mirror introduces mainly coma and astigmatism over the field of view. The amount of these aberrations depends on the tilt angle, and therefore in any single-mirror telescope this effect is minimized. Spherical aberration can also be present; but if the focal ratio of the mirror is large enough (f/10 or greater), it becomes negligible. In order to minimize coma and astigmatism to a fraction of a wavelength, the f/number has to be increased to about 24 for the case of a 4-inch mirror. This results in a long and vibration-prone tube.

The astigmatism at the center of the field in a tilted spherical mirror can be corrected by making a double curvature surface. However, coma still dominates,

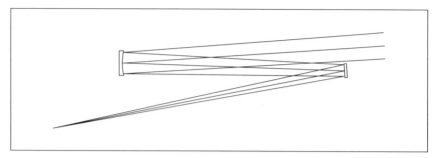

Fig. 1.2.3 *Anton Kutter's Schiefspiegler.*

and little speed is gained in trying to reduce the tube length or in improving image quality.

An off-axis paraboloid mirror is free of astigmatism, coma, and spherical aberration at the field center and could be used as an objective. Unfortunately the manufacture of this mirror is not an easy project for the amateur. Making a full paraboloid and cutting an off-axis section is not an attractive solution either.

The single mirror objective has the great advantage of simplicity, but it leads to an inconvenient viewing position, a long telescope tube, possible degradation of the incoming light beam by heat from the observer's head, and lower image quality due to aberrations. In spite of these facts, some amateurs still make unobstructed telescopes with a tilted, long-focus spherical mirror.

Two-Mirror Telescopes

A. The Schiefspiegler

The first-high performance unobstructed reflecting telescope was probably Anton Kutter's Schiefspiegler or oblique telescope[5] as it is illustrated in Figure 1.2.3. This design uses two spherical mirrors; the primary is concave and the secondary, convex. In this design astigmatism is corrected by making the aberration of one mirror cancel that of the other by the proper choice of secondary mirror tilt. The main residual aberration is coma, and it is minimized to ¼-wavelength by reducing the overall telescope speed to $f/26$ for a 4-inch primary.

A feature of Kutter design is that both mirrors have the same but opposite radius of curvature. This makes it possible to use the primary grinding tool for the secondary, and to test the convex surface by light interference against the concave mirror, which itself can be easily tested using the Foucault or Ronchi tests. However, additional work is required to polish the back surface of the secondary and in preparing the interference test.

B. The Yolo

Another unusual telescope design is Arthur Leonard's Yolo, named after a Cali-

---

[5] Kutter, A. *Der Schiefspiegler*. Biberach and der Riss, 1953.

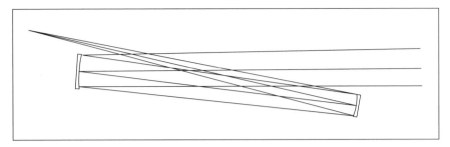

Fig. 1.2.4 *Arthur Leonard's Yolo.*

fornia county dear to the inventor. In this configuration he introduces a "warping harness"[6], and employs two shallow concave mirrors as shown in Figure 1.2.4.

This design corrects three aberrations at the center of the field: spherical aberration, astigmatism, and coma. Spherical aberration is corrected by hyperbolizing the primary, coma by the choice of the secondary mirror tilt, and astigmatism by mechanically deforming the secondary to induce a double curvature. The off-axis aberrations are minimized by increasing the focal length. This degree of correction enables the Yolo to have a faster speed and a larger aperture than is possible with a number of other TCTs. For example, Leonard has designed a 12.5-inch system operating at $f/15$.

One disadvantage of this configuration is that the secondary mirror size typically ranges from ⅔ to ¾ that of the primary. In designing Yolos, Leonard recommended minimizing the mirror tilt angles.

José Sasián introduced a refinement of the Leonard design in his article, "A Practical Yolo Telescope"[7]. This instrument features no warping harness and has better aberration correction. This design not only corrects spherical aberration, on-axis astigmatism, and on-axis coma; but linear astigmatism and linear coma as well. The key is to use the secondary mirror tilt to correct linear astigmatism and to further aspherize both mirrors. This refinement makes the Yolo "aplanatic" and provides a full degree of diffraction-limited performance for a 5-inch aperture working at $f/8.6$. The penalty for this performance is the difficulty of making the special aspheric secondary mirror. The secondary asphericity includes the three types of wavefront deformation discussed above.

The making of the secondary mirror for this aplanatic design showed that it is easy to make very smooth double-curvature surfaces with the degree of precision needed. The fact that double-curvature surfaces are easy to produce is of importance in telescope making because these surfaces allow one to correct both astigmatism and image anamorphism, thereby making feasible configurations that otherwise could not be pursued.

---

[6] See example in Allan Mackintosh's *Advanced Telescope Making Techniques*, Willmann-Bell, 1986.

[7] Sasián, José M. "Gleanings for ATM's—A Practical Yolo Telescope," *Sky & Telescope*. August 1988, p. 198.

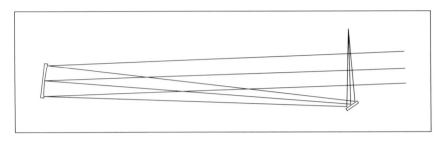

Fig. 1.2.5 *José Sasián's unobstructed Newtonian.*

C. The Unobstructed Newtonian

An example of the design flexibility that is gained by using double curvature surfaces is the unobstructed Newtonian illustrated in Figure 1.2.5.[8] This design pairs a standard long focus paraboloid primary mirror with a very strong double curvature secondary. The radius of curvature in one principal meridian of the secondary is twice as long as in the other principal meridian. This asphericity is used to correct the astigmatism contributed by both mirrors. In spite of the large secondary angle, coma, linear astigmatism, and image plane tilt can be minimized to obtain diffraction-limited images around the field center. This design has the usual advantages of the Newtonian—a comfortable viewing position at most elevations (even without an extra reflection), and the potential for satisfactory use with even a simple mounting. In addition the secondary can be made oversized to avoid a possible turned-down edge, and can be left circular.

Three-Mirror Telescopes

A. The Tri-Schiefspiegler

The image quality of the Schiefspiegler is limited when spherical surfaces are used, and the original Yolo required an undesirable warping harness. To solve the problem of obtaining a well-corrected image at the center of the field of view without a special aspheric or warping harness, Richard Buchroeder designed and demonstrated a three-mirror telescope that has been named Tri-Schiefspiegler,[9] see Figure 1.2.6. Like the Schiefspiegler, his design requires the primary and secondary mirrors to have the same, but opposite, radii of curvature and the tertiary to be a very long-radius concave mirror. The primary is a standard ellipsoid to correct spherical aberration, the secondary mirror mainly corrects coma, and the tertiary corrects astigmatism. These latter two are spherical and are ½ and ⅓ the size of the primary, respectively. The design has the advantage of being compact; a 12.5-inch unobstructed aperture working at *f*/20 can be assembled in a tube less

---

[8] Sasián, José M. "Gleanings for ATM's—An Unobstructed Newtonian Telescope," *Sky & Telescope.* March 1991, p. 320.

[9] Buchroeder, Richard A. "Gleanings for ATM's—A New Three-Mirror Off-Axis Amateur Telescope," *Sky & Telescope.* December 1969, p. 418.

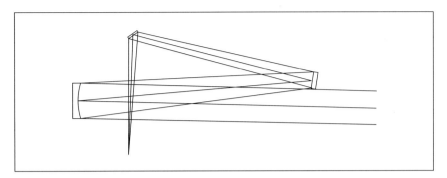

Fig. 1.2.6 *Richard Buchroeder's Tri-Schiefspiegler.*

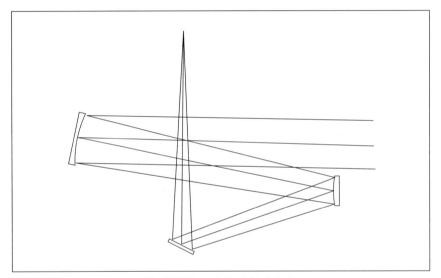

Fig. 1.2.7 *Kutter's Tri-Schiefspiegler.*

than 90 inches long. Makers of Buchroeder's design often praise the superb images that it provides.

Kutter's Tri-Schiefspiegler,[10] Figure 1.2.7, and Art Leonard's Solano,[11] Figure 1.2.8 are other examples of three-mirror telescopes corrected at the center of the field of view. As in Buchroeder's design, these require spherical surfaces for the secondary and tertiary mirrors, and under- or over-correction of the primary to control spherical aberration. They, too, have advantages and disadvantages.

B. Corrected Paraboloid

---

[10] Kutter, Anton. "Gleanings for ATM's—A New Three-Mirror Unobstructed Reflector," *Sky & Telescope.* January 1975, p. 46.

[11] Mackintosh, Allan. *Advanced Telescope Making Techniques.* Richmond, VA: Willmann-Bell, Inc.

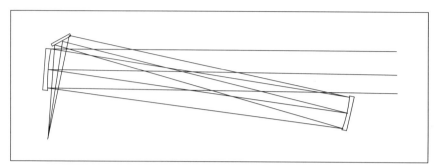

Fig. 1.2.8 *Leonard's Solano.*

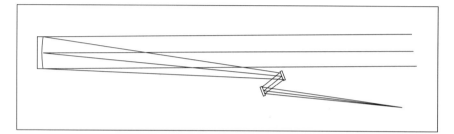

Fig. 1.2.9 *Design 1: 150 mm f/13.75 front-view Tri-Schiefspiegler.*

Although the use of double-curvature surfaces allows one to conceive well-corrected two-mirror telescopes, the design of three mirror instruments using only spherical surfaces continues to be of interest. The Tri-Schiefspieglers discussed above clearly illustrate that with three mirrors one can correct astigmatism and coma at the center of the field. However, a question that arises is: can linear astigmatism also be corrected? Please recall that this aberration is the one that most often limits the useful field of view, once on-axis astigmatism and coma are corrected.

The answer to that question is affirmative; Figure 1.2.9 illustrates such a three-mirror configuration, and Table 1.2.1 gives its specifications. This design solution was found by solving algebraic equations to account for astigmatism and coma at the field center and for linear astigmatism. A computer optimization was necessary to fine-tune the design that is limited by more subtle aberrations. A 150 mm aperture working at $f/13.75$ provides a diffraction-limited image over a $\frac{1}{2}°$ field of view. A problem with this configuration is that the image position is inconvenient, as in the Herschelian telescope. Thus, even though the aberration correction is successful, the practicality of the design is not high unless a fourth, plane mirror is used to reflect the image to a more favorable position.

1.2.3   Future Designs

It may be of interest to have a glimpse of future telescope concepts, such as that illustrated in Figure 1.2.10, for which the specifications are given in Table 1.2.2.

Table 1.2.1		**Table 1.2.2**	
Design 1: Tilted paraboloid corrected with spherical mirrors (150 mm aperture, $f/13.75$).		**Design 2: Tri-Schiefspiegler with double-curvature tertiary (500 mm aperture, $f/12.5$).**	
Primary	paraboloid	Primary	ellipsoid
Radius	3000 mm concave	Radius	10353.26 mm concave
Conic constant	$K = -1$	Conic constant	$K = -0.55$ ($K = -1.92$ aplanatic)
Tilt angle	$3°$	Tilt angle	$5°$
Spacing	1000 mm	Spacing	2438.859 mm
Secondary	spherical	Secondary	spherical
Radius	829.6276 mm convex	Radius	10353.26 mm convex
Tilt angle	$23°$	Conic constant	$K = 0$ ($K = -17$ aplanatic)
Spacing	100 mm	Tilt angle	$15.5°$
		Spacing	2438.859 mm
Tertiary	spherical	Tertiary	double-curvature, concave
Radius	1115.498 mm concave	Radius	8238.783 mm (plane of symmetry) and 8967.392 mm
Tilt angle	$25°$	Tilt angle	$10°$
Spacing	712.96 mm	Spacing	1888.527 mm
Image plane tilt $6°$		Image plane tilt $0.0°$	
Image anamorphism 2.3%		Image anamorphism 3.1%	

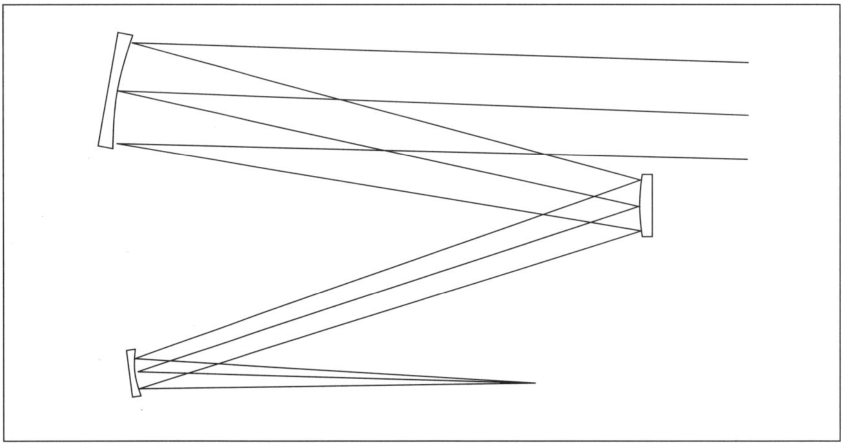

Fig. 1.2.10 *Design 2: 500 mm f/12.5 Tri-Schiefspiegler.*

This design uses an elliptical primary, a spherical secondary and a double curvature tertiary. It has been corrected for spherical aberration, coma, astigmatism, linear astigmatism, and image plane tilt. The aperture is 500 mm and the focal ratio

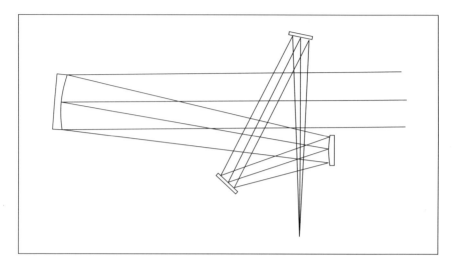

Fig. 1.2.11 *Folded version of the 500 mm Tri-Schiefspiegler.*

is 12.5. The performance of this design is limited by linear coma that can be corrected by hyperbolizing the primary and secondary mirrors. With this further correction, the design can compete with the Ritchey-Chrétien system and provide very good images over almost a one-degree field of view. The design goal was to obtain compactness in a large aperture; the distance between primary and secondary mirrors is 2439 mm for a 500 mm aperture. If the system is scaled down to a 250 mm aperture, a 150 mm secondary and a 100 mm tertiary mirror would be required. The design can be folded as illustrated in Figure 1.2.11, and the double-curvature tertiary can be null tested.

1.2.4 Conclusion

The design and fabrication of unobstructed reflectors is a very exciting part of telescope making, and there is a sufficient variety of unobstructed models to cover a large span of configurations and apertures. The known designs cover very well the span of small (3 to 5 inches) and medium apertures (6 to 8 inches) with great practicality and transportability. However, for larger apertures (10 inches or more), there is a need for very compact and moderately fast designs (f/8 to f/15). These designs will probably require three mirrors and a double-curvature surface like the large Tri-Schiefspiegler discussed in Section 1.2.3 on page 16.

Unobstructed reflectors are mainly intended for observing planetary details, the moon, and double stars. One of the most appealing features of these designs is that they can be made at home with very simple materials and tools, and yet produce the finest astronomical images.

Note: The effect of surface roughness can be estimated by considering the Strehl ratio $= 1 - (6.28 \, RMS/\lambda)^2$, where lambda is the wavelength of light and

RMS is the root mean square surface roughness.

The Strehl ratio gives the ratio of the Airy disk peak (high in the presence of small aberrations) to the peak of the unaberrated Airy disk. The formula is only valid for very small amounts of aberration such as introduced by surface roughness.

For a given surface roughness, a refracting surface introduces about one-fourth the wavefront aberration that of a mirror surface, and a sixth that of a Mangin mirror surface. This comes about because upon refraction or reflection the wavefront is deformed by the factor ($n' - n$) where n and n' are the indexes of refraction of the media before and after refraction or reflection. For light refraction the factor is $\frac{1}{2}$, for reflection in air 2, and for reflection in glass 3.

Since surface roughness enters as the square in the Strehl ratio formula, we see that light scattering by a refracting surface, a reflecting surface in air, and a reflecting surface in glass occurs in the proportions 1, 16, and 36.

1.3   An Innovative 17.5-Inch Binocular
By Paul B. Van Slyke

Most of what we are as sentient life-forms comes in twos—legs, arms, ears, eyes, etc. If we lose one we can still function at a reduced efficiency.

Since Hans Lippershey invented the telescope at the beginning of the seventeenth century, we have been pursuing visual astronomy with monocularity. Why have amateur and professional astronomers, over the centuries, been content to visually observe our three-dimensional universe with only two-dimensional clarity? We do not perform our daily tasks with one eye closed. So, why do we view our universe in such a manner—dogma, habit, tradition, practicality, cost factors, or all of the above?

To try to explain this ineffable binocular experience is almost impossible. I don't need to sell anyone on the concept of stereo sound over mono sound. Everyone knows the realism attained with a good stereo system. The same principle applies to stereo sight versus mono sight. The proof is in the experience—seeing is believing.

Before I built the 700-lb beast shown in Figure 1.3.1, I purchased a binocular viewing attachment to improve lunar, solar and planetary images for public observing sessions at the observatory. The device worked so nicely that I yearned to view deep-sky objects in stereo as well.

Being a fairly resourceful person and having read about Lee Cain's double-barreled creation[12], I took a look through the observatory's scrap pile to inventory the possibilities. I was to be able to scavenge enough material to at least build a sturdy mount.

There were two major problems. The first was bulk. Lee's design was simply

---

[12] Cain, Lee. "A 17½″ Transportable Dobsonian Binocular," *Telescope Making* #24. p. 28.

Fig. 1.3.1 *The author with his 17.5-inch binocular telescope.*

too bulky to be practical for our needs. So, a tubeless design, that would lay almost flat, became mandatory.

The other major problem was that I had only one 17½" (f/4.5) primary mirror. It appeared that it would take a year or two to get a second one from the manufacturer.

I was thinking of grinding my own mirror when a miracle happened. An amateur astronomer and a new resident of our area called my observatory and asked if I knew of anyone who would like to buy a 17½-inch f/4.5, homemade Dobsonian. Well, this telescope was a "tank" and didn't thrill me at all, but the optics were mint.

With two 17½-inch f/4.5 mirrors at my disposal, an obsessed fanatic was created. Like the transformation of the "Hulk," I had no control over my metamorphosis. I began to eat, sleep and dream about various aspects of design criteria. As with my previous telescope, no designs were ever transcribed.[13] All my plans were three-dimensionally displayed as a mental image. I don't know if this is a gift or a burden, but it sure saves a lot of time and paper. The actual construction proceed-

---

[13] Van Slyke, Paul B. "Gleanings for ATM's—Four Speed 17½-inch Newtonian," *Sky &Telescope.* Feb. 1986, p.199.

ed with no unexpected design changes or unanticipated problems.

Anyone with a fair amount of common sense can design a telescope (or whatever) on paper and then have some job shop build a prototype. To reverse this process is a rare gift many amateur astronomers possess out of sheer necessity and/or an empty pocketbook. The innovation, and at times pure genius, of the ATM has always fascinated me. How anyone can take a bunch of gizmos and whazits and create a celestial machine is indeed a miracle of our species. Society's discards, coupled with a group of motivated ATM's, could quite practically put a colony on Mars.

1.3.1   The Mount

All serious amateur astronomers understand the importance of a sturdy mount for studying the skies. A telescope without a steady support is only fun for a short while. After you've created all the quasi-laser light shows in the eyepiece from thumping various parts of your telescope, the fun is over. All you have left is frustration and disappointment.

With binoculars designed for astronomy, the need for rigidity is doubly important. No effort was spared in the support of my binoculars. The azimuth axis has a 1½-inch steel shaft on double-tapered race roller bearings which I did not consider sufficient to dampen all vibration to ground. In addition, a ⅝-inch thick ring, fashioned from steel plate (a rejected flywheel blank from the scrap yard) is attached to the T-shaped base, and roller skate type steel ball bearings are mounted on each side of the tubular fork creating a thrust bearing stabilizer. The amount of time for induced vibration to be transferred to ground by this design is negligible. Similar lateral support is also incorporated on the right ascension axis of the BFO's new 30-inch $f/9$ Cassegrain—but that is another story. The altitude axis has no bearings in the conventional form. It pivots on cannibalized hydraulic cylinders. I might add that hydraulic cylinders would make, in part or in whole, an inexpensive and extremely stable mounting, no matter what its configuration.

1.3.2   The Mirror Cradle and Cells

Again, I will stress the importance of structural rigidity in all assemblies associated with any binocular device—especially the primary mirror cradle. Structural cross-bracing is mandatory if the mount is to maintain optical alignment. Changes in altitude will create different stress points throughout the binocular structure due to gravitational effects. This will disrupt the parallelism of the two optical paths unless strict care is applied to all structural considerations.

The mirrors are supported in conventional 9-point flotation cells; except for the motorized collimation adjustments on one of them. This system was added after field tests revealed the necessity of this feature. There are two reasons, of which the latter was the most important: (1) The thermal characteristics of the optics and the various ferrous and non-ferrous metals caused structural changes during the evening cooling process. From time to time, due to the changing ambient temperature, tweaking would be required to maintain proper collimation; (2)

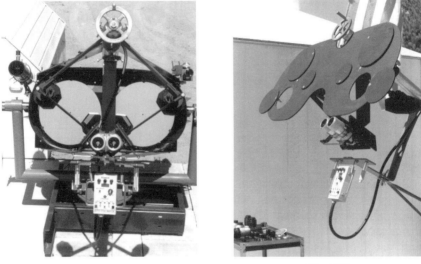

Fig. 1.3.2

Fig. 1.3.3

Some observers would see a perfectly-merged image—others would not.

We came to the conclusion that this was not the fault of the optical instrument, but the diversity of human physiology. I must conclude that each person perceives our universe literally from a different visual perspective.

To simplify the motorized collimation, only two 60 rpm 8–16 vdc reversible gear-reduced motors were used. Activating both motors together conveys the action of a redundant third motor that would adjust the primary mirror vertically.

The controls can be designed according to personal preference, and a LM317T integrated circuit (IC) can be incorporated to vary the motor speed. I recommend this IC over conventional rheostat controls as adequate current is supplied over the adjustable voltage range, thus maintaining higher motor torque at low rpm settings.

1.3.3   The Secondary-Tertiary Head Assembly

This head has coarse- and fine-focus adjustment to accommodate any type of eyepiece or combination eyepiece/expander/compressor, etc. The coarse adjustment is controlled from the brass hand-wheel at the upper center of the head (see **Figures 1.3.2** and **1.3.3**). This also shifts the head back enough to use the beam converging attachment, effectively creating a multi-mirror telescope (MMT), which will be discussed shortly. Fine focus is achieved by adjusting the 2-inch helical focusers individually at the eyepiece.

The inter-ocular distance is adjusted with the horizontal black knob on the right, just below the helical focusers. This separates the individual tertiary assemblies via pivot points just behind the center of each secondary flat.

1.3.4   A Multi-Mirror Telescope

A binocular viewer attachment (briefly mentioned earlier) can give spectacular views of solar system objects. By reversing the intended function of a typical binocular viewer, which consists of a simple pyramidal prism array with a beam splitter at its heart, we create a beam-converging system.

By utilizing this device backwards, we can take the light path of two mirrors and combine them into one image, to collectively double the light grasp of a single objective.

To my knowledge, prototype multi-mirror telescopes (MMT) designed and built by amateurs have failed to meet the final test. Admittedly, my MMT system does need perfecting; but the concept is sound, and it does work.

Of course, the concept is limited to two objectives; but after experimentation, I find I prefer binocular viewing over monocularity. This may sound odd, but until you've experienced binocular vision, don't dismiss my claim too quickly. Binocular viewing involves many subtleties that, when combined with the complexity of the human mind, go far beyond the capabilities of monocular vision.

1.3.5   Conclusion

For some inexplicable reason, binocular telescopes excel when observing dark nebulae along the Milky Way (Lee Cain also commented on this phenomenon). The three-dimensional illusion appears very strongly when viewing dark nebula.

The surrounding stars seem to be positioned at discernible distances from the nebulous clouds, and these clouds seem to be glowing around the edges—maybe from reflected starlight? When viewed from dark sky locations under good seeing conditions, all the little-known fuzzy blobs become Orions, Rings and Dumbells times two. For brighter deep-sky objects, the binos become a teleportation machine into the living color of deep space. The delicate filamentary structure of the peculiar galaxy M 82, or that of the Veil supernova remnant—display better clarity and depth than any photograph could ever capture. Scanning the constellations of Sagittarius and Cygnus reveals endless hunting grounds for elusive and mysterious phantasms that don't seem to be on the charts or in the books. What a fantastic comet hunter these deep-sky binos would make!

1.4   A Schupmann Medial Telescope For the Serious Observer
By James A. Daley

Ludwig Schupmann, a German architect of public buildings, invented a totally different type of refractor which is essentially free of color defects. He patented two forms: the Brachyt and the Medial in 1899. It is the Medial type which shows the most promise for the amateur builder.

The little known Schupmann is considered to be the definitive planetary telescope design. The few modern examples built by amateurs certainly confirm the fine, high-contrast imaging that the Medial provides. For the past 24 years I have worked to promote the Schupmann by building several successful instruments and encouraging those of my amateur friends with ray trace capabilities to further develop the concept. This effort has resulted in a fully practical design for the advanced amateur lens maker.

What is so special about the Medial, you ask? It is common knowledge that a color free refractor would theoretically provide the best possible planetary images. This is because the refractor is an unobstructed and sealed optical system and can be designed to be coma free and almost fully corrected for spherical aberration (only slight zonal retouching is necessary). The slight remaining astigmatism over small fields is usually harmless.

To design a refractor that is totally free of color has thus far proven impossible. However, by using special glasses, only recently available, we can approach a color-free image.[14] The Schupmann telescope is the only refractor which is completely free of both longitudinal and lateral color defects. The design is also free of coma and, in a few special designs, is highly corrected for spherical aberration with all spherical surfaces. Other design variations require aspheric work which can vary between just detectable to 25 fringes. It is the all-spherical, or "Super Schupmann," which I will describe in this article.

The actual theory of operation of this design is common to all Medials. Let's briefly look at each element in turn, describing its function and how it works in concert with the others to form a color-free image.

1.4.1   The Objective

The objective (Figure 1.4.1) is a single element lens made of optical grade Bk-7 or BSC-2. This glass takes a fine low-scatter polish and is resistant to weathering effects. The lens is double convex with most of the curvature on the skyward surface (R1) and just enough on the back (R2) to make it coma free. The lens can be as thin as convenient, but not too thin, since that would make figuring a good sphere difficult—a 10:1 thickness ratio is fine. A simple positive lens like this focuses the red rays long and the violet short. The focal length of the objective in green light is equal to the final system focal length in the design detailed here.

1.4.2   The Field Mirror

The next element encountered is the field mirror. This mirror is just a bit larger than the linear field-of-view required. It is concave and polished to the best pit-free surface possible. This mirror is located at the green focus of the objective, and its purpose is to image the objective onto the third element. To do this, it has just enough power (concavity) that a light source at the objective front surface would be sharply focused on the first surface of the third element, known as the

---

[14] Gregory, John. "Gleanings for ATM's—A Quest for the Perfect Refractor," *Sky & Telescope.* June 1987, p. 662.

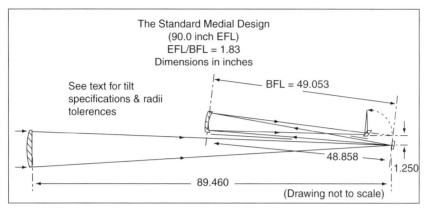

The Standard Medial Design
(90.0 inch EFL)
EFL/BFL = 1.83
Dimensions in inches

See text for tilt
specifications & radii
tolerences

BFL = 49.053

48.858

1.250

89.460

(Drawing not to scale)

	Objective	Field Mirror	Corrector
Material	Bk-7	Pyrex	Bk-7
Radius	50.997 CV	63.705 CC	13.373 CC
Radius-second surface	532.028 CV	—	26.142 CV
Thickness-center	0.625	0.375	0.450
Clear aperture	6.0	1.0	3.278

Fig. 1.4.1

corrector, or in Schupmann's words, "die Kompensator."

1.4.3   The Corrector

The corrector takes the form of a Mangin mirror (silvered back surface) and operates as a non-magnifying, coma-free, relay element. If we place a white light source at its operational focus, it will relay an image in which the red focus will be short and the violet long. If we make the corrector of the same glass as the objective and shape the curves properly, the color induced by the corrector will be precisely equal in strength (and opposite in sign) to that of the objective. If we combine the three elements, and align them properly, a color-free refracting telescope is created. Since each element is coma free, the combination of the elements is also coma free. Thus we attain a high performance instrument free of coma and longitudinal color!

Because the field mirror controls all oblique rays combining them at the corrector, there is no lateral color throughout the field. The corrector is tilted to gain access to the image surface, usually via a prism or diagonal mirror. This tilt introduces some astigmatism, which is canceled by tilting the objective slightly in a perpendicular plane. This corrects the usable planetary field for astigmatism.

By properly choosing the ratio of objective to corrector diameter, a solution for spherical aberration can be found, and we arrive at the all-spherical "Super Schupmann" design. What a beautiful solution for the color-free refractor; and just think, only 103 years have passed since Ludwig Schupmann dreamed it up!

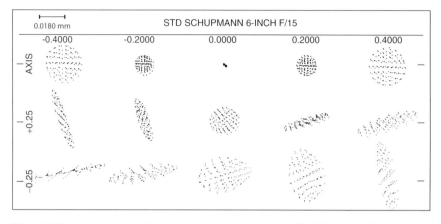

Fig. 1.4.2 *Through-focus spot diagrams for three wavelengths; 0.656, 0.589, and 0.486 microns, at three field points; center of field and ±0.25° from center in the plane of Figure 1.4.1. The diffraction disk diameter scale is shown in the upper left. Ray-trace work and diagram kindly provided by Bert Willard.*

Of course, the critical among you may argue—and rightly so—as to whether or not the Schupmann is a true refractor. Schupmann called his idea the Medial (lying in the middle), and close examination of the design shows its catadioptric nature. However, this "cat" is off in a different evolutionary corner since most of it's power is derived from refraction—the silvered Mangin back surface acting only as a retro-reflector (this surface essentially returns the rays upon themselves). Others argue that the field element optically combines the objective and corrector, forming a plane parallel plate which is color-free when used in collimated light. Thus the final image is, in effect, formed by the residual power of the silvered surface. Hmm, telescopium absurdum.

Hopefully, having dispensed with all that, lets get started on our all-spherical "Super Schupmann." I will not, in detail, instruct you how to build a Medial in this short article. It is assumed that the builder is highly skilled in the making and testing of lens and mirror systems.

I will, however, touch upon the important and interesting points of construction. I have chosen a 6-inch f/15 system in order to keep the cost reasonable and assure the best chance of a successful instrument. This system will show tremendous detail on the planets and Moon, and its pinpoint images will delight you on double stars and globulars (Figure 1.4.2). Because of the instrument's absolute achromatism over a huge spectral range, photography of the Moon and planets is unsurpassed. Figure 1.4.3 is a photograph of the author's instrument.

1.4.4   Glass Purchase

Order your optical blank (Bk-7 or BSC-2) ½-inch larger than the clear aperture specified. This will provide an adequate mounting area and also hide the inevitable slightly-turned edge. The glass must be of the best objective grade with mini-

Fig. 1.4.3 *Instrument described in this article as mounted in the author's backyard observatory in 1984. A larger version, 9-inch aperture, is currently in place and used in measuring double stars with a CCD camera. Photo by author.*

mal bubbles and a homogeneity value equal or better than ± 0.000005 for delta n and 6nm/cm birefringence. Talk to your glass supplier, and they will advise you on the precise wording of your order. At today's prices the objective and corrector will run about $250.00. Also order tools of plate glass (2 for each element). Try to get the elements and tools diamond-generated at a local optical shop.

Use your spherometer to hold the curves as close as possible. Keep wedge to less than 0.001." Make a test plate, say 3" diameter for R1, adjusting its curve by fine grinding and repeating test polishes until the radius is within $\frac{1}{32}$" of your goal. Polish R1 to test plate match, then adjust the radius of R2 in polishing to get the focal length correct. This test is in autocollimation using your flat and an Hg monochromatic tester. To measure the BFL use a stick cut to the specified value. Try to get your radii and focal points accurate to one part in 1000. This is not as difficult as it sounds, and 1 part in 5000 is possible with some additional effort. The design is forgiving of small radius errors; you simply adjust the element spacing to tune out residual color.

The field mirror is made oversize, by about $\frac{3}{8}$-inch. Give it the best polish you can, keeping the figure a reasonably good sphere. The figure of the field mirror does not normally affect the telescope aberrations. However, a zone exceeding the radius tolerance would sensibly defocus the pupil image, leading to trace lateral color for stars occupying the zoned position. Have the field mirror aluminized and overcoated as soon as it is finished since you need good reflectivity in the final testing and figuring phase.

The corrector is a fun job. Grind and test-polish the concave side until you hit the radius. Finish the convex back to spherometer accuracy and give it a quick polish. Set up your monochromatic tester at the approximate design relay distance and adjust the knife-edge to null the center zone of the corrector. Adjust the back radius to bring the null distance to the design specification. Use a wooden rod cut to this distance between the knife-edge and front surface as you did on the objective. Finish polishing the front surface to a good sphere while holding the radius right on. Finish polishing the back surface keeping the zone matched to your corrector's focal distance rod. You will see a typical donut shadow over the whole corrector.

Using lots of cold pressing and 10-minute spells of careful polishing, you will obtain a spherical back surface which will be indicated by a smooth conic section shadow. When you autocollimate the objective, you will see the same shadow—with the sign reversed and the strength doubled. Remember, the corrector was tested directly, but the objective is measured in auto-collimation.

Set the optics up in temporary wooden cells on a strong bench. Get the spacing and angles correctly set. Auto-collimate the whole system in white light with a bright source. It helps to remove the slit and align the system with the light of the bare bulb. Using an eyepiece at the final focus, just to the side of the bulb and shielded from it, tweak the flat to return an image to the eyepiece. The image will be loaded with lateral color. Adjust the field mirror tilt to zero out the color. This adjustment optically centers the objective onto the corrector, effectively removing wedge or prism effects. This is why the Schupmann is not specified as tightly for wedge as a standard achromat. The image will show a little astigmatism (stop down the bulb to a square slit for this test). Tilt the objective a bit one way or the other until the astigmatism clears up.

Now let's get to work. Replace the slit and observe the image (still using the eyepiece). Remove the remaining trace of lateral color. Examine the in- and out-of-focus images for traces of color in the extra-focal blur. Adjust the spacing of the corrector to field mirror to remove any remaining trace. To do this, reduce the spacing if the red focus is short, and increase it if the red is long. Remove the eyepiece and cut off the image with the knife-edge. You will probably see some weak zonal error.

Remember, we only tested one surface directly for sphericity. If you polished the surface carefully and slowly, the zone, if any, will be weak and soft. Simply pick your favorite surface and figure to a good null, and the job is done.

While your corrector is being silvered, build a tube for the optics, reproducing the tilts and spacings of your test setup. Install the optics and bolt the tube to a solid equatorial mount. Point the Schupmann at a star near the zenith and adjust your field mirror to zero out lateral color. I assume that you will align the system by centering the elements with targets against the daytime sky before your first night test. After a small adjustment of the field mirror, the star image should be perfectly round without a trace of color. Since you have learned the adjustments during the test phase, you can make any necessary final settings. A well-made Medial will remain well-aligned for months at a time. My own 7-inch stayed perfectly aligned for a year; the only nightly adjustment being the lateral color correction, which becomes as instinctive as focusing.

A short article like this one cannot provide a complete description of tube construction and adjustment devices needed for a first-class Medial. I feel that the advanced amateur is inventive and will enjoy the challenge of tube and cell design as much as I do. I prefer thin plywood-on-frame construction and aluminum cells. You must provide a tip/tilt adjustment for the field mirror and corrector. The objective cell can be permanently shimmed for zeroing out the astigmatism. The field mirror adjustment must be handy to the eyepiece.

One interesting feature of the Schupmann is its ability to cancel atmospheric dispersion, which colors planet and star images when viewing below 30° altitude. Just tweak the field mirror to introduce an opposing lateral color spectrum, which, to a very good approximation, removes the atmospheric color. Because the Schupmann is coronagraphic by nature, you can use a Lyot stop at the corrector to perform some very interesting experiments. I have a tiny flip-in, flat black disk on a thin wire which comes in near contact with the field mirror. When I view the star Sirius, the bright image can be occulted by the disk. As the eye is no longer dazzled, the star field pops out revealing stars around Sirius perhaps never seen before. The moons of Jupiter show beautifully using this same trick. A special field mirror perforated to let the sun's image through will let you see solar prominences with a simple #92 red filter at the eyepiece. You need a rock solid mount for this one, though.

It is great to have a telescope with tremendous resolution that is easily kept in perfect alignment. The possibilities of the Schupmann seem limitless, so why don't you get started on your "Super Schupmann!"

The author is especially grateful to Bert Willard for raytracing this version of the "standard medial" and providing the spot diagram.

Ed. Note: It would take much more space than we have available to do justice to the material Mr. Daley was kind enough to submit. Those interested might request his book, *Amateur Construction of Schupmann Medial Telescopes.*

1.5 ATM Journal Staff

Compiled by Debra Cook

1.5.1 Introducing the *ATMJ* Staff

We frequently receive letters in which the writer mentions having met a staff member or two at this or that star party or telescope making convention, but indicates that he knows little or nothing about the others. Considering the amount of time and talent expended by these individuals on behalf of telescope makers everywhere, I felt that this would be a good place to tell you a bit about those who are working to keep the highest traditions of *Telescope Making* magazine alive through the pages of *ATM Journal*

I think that it is especially good for ATMA members to realize that while some of these individuals have gone on to earn advanced degrees in optics or other types of engineering, their backgrounds have a common thread—love for astronomy and telescopes.

No matter how high the level of respect these people have achieved in the world of professional optics, they have not forgotten their roots—that first glimpse through the eyepiece—and those who went before them, who took the time to share their knowledge.

1.5.2   Richard A. Buchroeder

Richard, the oldest of three children, was born to Richard W. and Genevieve Buchroeder on August 9, 1941 in St. Louis, Missouri. Shortly after the outbreak of World War II the family moved to Los Angeles, where he attended public schools. One of his most vivid memories is of going on a field trip to the Griffith Observatory. He feels that that opportunity probably led him into amateur astronomy and eventually into optics as a profession.

Just after starting junior high school, the family, now including a brother, Bill and sister Peggy, moved to Ojai, California. Unlike Los Angeles and its bright lights, Ojai featured dark skies, and it was there that he bought his first telescope, a 5" reflector from ESCO Products. This convinced him to try his hand at mirror making. Shortly thereafter he ground and polished an 8" $f/15$. According to Dr. Buchroeder, "It was meant to be an 8" $f/6$, but I was so eager to use it I quit grinding as soon as it was smooth! I subsequently reground the mirror, but was first rewarded with beautiful views at the longer focal length."

Optics and astronomy were still just a hobby at this time, After graduation from high school, he went to the University of California at Berkeley to study mechanical engineering. While there, he joined the engineering co-op program and worked several times for the Naval Ordnance Test Station at China Lake. That's where he really became involved with optics. First, he was privileged to participate in the earliest Moonwatch Programs, timing satellites with Carroll Evans and becoming acquainted with Art Leonard. Second, he finagled a job with a group that actually DID optics! His boss, Bob Lawrence, inspired him to pursue optics as a career.

Upon graduating from Berkeley with a Bachelor of Science in Mechanical Engineering, he drove to Rochester, New York and worked as an apprentice lens designer at Bausch & Lomb. Simultaneously, he enrolled in the optics program at the University of Rochester and acquired a Master of Science degree, mainly through night classes.

From there he was offered a position as a Research Associate at the Optical Sciences Center of the University of Arizona. This was an unparalleled learning

Fig. 1.5.1 *Richard A. Buchroeder.*

experience, allowing him the opportunity to rub shoulders with the giants of optics and astronomy. While there, he earned a Ph.D. in Optical Sciences.

He subsequently worked for Hughes Aircraft Company's Missile Systems Group and then again for the University of Arizona before becoming a free-lance lens design consultant, which he has done for more than a decade now.

Dick states that his principal hobby is "optics," especially testing and viewing with older telescopes and binoculars. He also enjoys 35 mm SLR photography, Realist Stereoscopic photography and Widelux Panoramic photography. He has 9 optical patents to his credit, with 2 more pending and has authored many articles in professional and amateur publications on such subjects as telescopes, eyepieces, catadioptric objectives, lunar laser telescope systems and so forth. He has also reviewed many optical books.

Dick has 3 children, (Elizabeth, Susan and Michael) and a grandson, (Alexander).

1.5.3   William J. Cook

William, the only child of William James and Mabel Cook, was born on July 28, 1951 in West Memphis, Arkansas.

His interest in optics started in his 7th year at a one-room schoolhouse well off the beaten path between Glasgow and Tompkinsville, Kentucky. There, when class was not in session for him (which was often because there was one teacher for 6 grades), he would slip outside to the creek and make magnifying glasses out of water-filled bottles which had been discarded along the banks.

Fig. 1.5.2 *William J. Cook.*

In 1960, while living near Dover, Tennessee, he found a grocery bag filled with "dead binoculars and spotting scopes." With this bag of junk and his father's tools readily available, a life-long interest in optics was born.

In 1970, during his first week at college (Arkansas State University), where he worked on a double major (Broadcasting and Journalism), he met a student who was upset because he had just dropped the binocular his parents had given him as a high school graduation present. Upon seeing the dilemma, Bill offered his assistance. The student told him that he would give him $5.00 if he could fix the binocular. Ten minutes later, Bill left the student's room with $5.00 in cash and the thought that optical instrument repair might be a great secondary skill to possess.

Shortly after leaving college, he sought and obtained a seat at the Navy's "Opticalman" Class A school at The Great Lakes Naval Training Center in Illinois. There he furthered his optical education by being trained to work on a wide variety of instruments.

Upon leaving the Navy in May of 1978, he took a position as the senior instrument technician for a surveying instrument manufacturing company in Atlanta. Here, he was free to sharpen his skills with the lathe, milling machine and other machine tools.

Then one day he opened the door to the optics shop. There, in a room which had not been used for 10 years, he found 12 lens grinding machines which could handle glass up to 18 inches in diameter and enough Carborundum to bathe in— lunch time would never be the same again!

In 1981, he moved to the West Coast and took a position as a senior optical "mechanic" with Puget Sound Naval Shipyard, and there specialized in the calibration of sophisticated optical instruments, testing devices and fixtures.

Starting at the top of the pay scale, and with no advancement or opportunity to be "original" being possible, he determined to find a job with growth potential and one which would not be driven by invariable routine or politics. Upon leaving the shipyard, he returned to college (this time for fun) and in 1985 earned a degree in History from the University of Washington.

Today he is the manager of the Precision Instruments & Optics Division of Captain's Nautical Supplies in Seattle, Washington where he has the opportunity to work on some of the finest old and rare binoculars and telescopes in the world. Recognized as an authority on binoculars, he has authored or contributed to more than a dozen articles and brochures on binoculars, sextants and telescopes in the past two years.

Without the mathematical background needed to be anything more than a "lens designer wanna-be," he tries hard to be content with his reputation as a craftsman.

Bill considers it an honor to have friends like those who act as Associate Editors for *ATM Journal* and eagerly gives them credit for refining his understanding of optical design techniques. He also readily admits that without their unwavering support, he would never have had the nerve to take on the responsibility of forming the ATMA or producing *ATM Journal*.

Bill has a wife (Debbie), two sons (Bill and Sean), and a 3-year-old daughter (Noelle).

1.5.4   John Gregory

John was born in 1927 in Cleveland, Ohio, and had the good fortune of growing up just three miles from the Warner and Swasey Observatory which he "discovered" at the age of ten. Under the influence of the late Dr. J.J. Nassau, he ground a 6-inch mirror from ¾-inch plate glass. It was "hopelessly hyperbolic, still much better than naked eye viewing," he recalls. Later, while at Case Institute of Technology, with the inspiration of the 24-inch f/3.5 Case Schmidt, he attempted a 5-inch f/3 Schmidt which Unk Ingalls published in the November, 1949 *Scientific American*.

In 1950, after graduating with a Bachelor of Science degree in mechanical engineering from Case in Cleveland, John was a project engineer for 8 years for the Perkin-Elmer Corporation. During this time he worked on the development and testing of telescopes, aerial-camera lenses and tracking cameras and had the good fortune of knowing Dr. James G. Baker.

Seeing Baker's Super Schmidt meteor cameras and working on the Baker-Nunn tracking cameras was frosting on the cake, as Gregory saw it. It was during this time that he investigated Cassegrain-Maksutovs for the amateur, leading to the March 1957 *Sky & Telescope* article disclosing the designs.

From 1958 to 1974 he was a senior optical engineer at Barnes Engineering

Fig. 1.5.3 *John Gregory.*

Company, where he designed optical systems and components—many for the infrared, including the lunar orbiting IR scanning radiometer flown on the Apollo 17 command module; the horizon-sensor optics used in AGENA spacecraft; horizon, sun, and planet simulators for testing instruments used in spacecraft and germanium immersion lenses that speed thermistor bolometer response.

The UV lens used by GEMINI Astronauts was his modification of the Barnes UV lens. He also made major contributions to development of the photo-electric auto-collimator used in the POLARIS and POSEIDON submarines, a 10-inch $f/1$ widefield thermal imaging device, and other projects.

He recalls with pleasure the founding of Stamford Observatory at the then new North Stamford Connecticut Museum campus, by the local amateur group. He designed the Observatory's 22-inch photovisual Maksutov telescope. Dedicated in June 1965, he helped conduct the community effort for its fabrication and installation.

He is the author of several published articles on telescope optical systems and is the designer and fabricator of several 8.3-inch and 10.8-inch Cassegrain-Maksutov telescopes.

At the 1974 Stellafane Convention, he learned of a job opening that proved irresistible. He exchanged his 41° latitude for the 31° latitude where he would be chief engineer at the University of Texas McDonald Observatory. While there, he continued his investigation of apochromatic refractors, using a CDC 6000, later resulting in the article in *Sky & Telescope* of June, 1987.

Since 1978 he has owned and operated Gregory Optics, a consulting firm specializing in optical design, spherical and aspherical fabrication and testing, and optical glass sales. Like many designers, he is currently thrilled with the capabilities of the 486-33 PC combined with the ZEMAX optical design and optimization program.

John is a very busy and fascinating man, with many interests. A licensed pilot (his experimental aircraft project fills the garage), he also enjoys photography, astronomy, telescope making, gemstone faceting and music—two grand pianos and an organ fill the living room. Lately, he has combined hobbies in an annual summer cross-country flight to the Oshkosh Fly-In and Stellafane Convention, with visits in Oklahoma, Ohio, New York, Vermont and Connecticut, then on to North Carolina where two sons and two grandchildren live.

1.5.5 Diane Lucas

Diane was born in Buffalo, New York on January 31, 1931. Her interest in astronomy began during her last couple of years at Elmira College, where she earned a Bachelor of Science degree in chemistry. One day she passed a newsstand and saw the Golden Nature Astronomy Guide; it looked interesting to her so she bought it. From there she subscribed to *Sky & Telescope* magazine and thus began her long-time love of astronomy and telescope making.

She made her first telescope, a 6-inch $f/8$ back in the 50's right after graduation from college and while working for General Electric in Schenectady. It was made while attending a telescope making class with the Schenectady Astronomy Club, and she still remembers an article written about the class in a local newspaper in which a picture of her taken through the bottom of the mirror appeared. She says it was a most flattering (!!??) likeness!

By the late 1950's, with a Master's Degree in chemistry from St. Bonaventure University, a new job at NACA (predecessor of NASA) and a new husband, Jim, she became hooked on the idea of making a Maksutov after seeing John Gregory's article about them in *Sky & Telescope*. John's article said no one should try making a Mak without some experience on short-focus mirrors, so she re-ground, polished and figured the 6-inch mirror (from that first telescope) to an $f/4$ curve. Husband Jim agreed to help with the mounting, after first purchasing a gas welding outfit, a drill press and a lathe!

By 1960 she had her Mak, a 6-inch $f/23$, two daughters (Sandy and Karen), a house, three 8-inch flats made to test the Mak by auto collimation, and last but not least, several thousand variable star observations. Several of these items were shared with Jim, "still my favorite friend and husband," according to Diane.

She "retired" from working for pay at NASA to work full-time raising her daughters. Diane said, "In order to stay semi-sane, I made several dozen optical things over the years until I went back to work for pay in 1973." This included a bunch of Newtonians up to 12½", a 4-inch $f/15$ refractor, several flats, a 16-inch Cassegrain, a 4-inch Huyghenian eyepiece for the Stamford, Connecticut telescope, as well as playing around with optical design. At first her designing was done with lots of paper and log tables. Later she acquired a used mechanical Friden calculator that could multiply, add, subtract and divide 10 digit numbers.

Fig. 1.5.4 *Diane Lucas.*

Both were difficult, even painful ways to design. She had a very good correspondence with Dick Buchroeder during that time, and he helped her immensely to know what was important and what wasn't. In Diane's own words, "I owe a lot to Dick."

When Diane went back to work full time, her boss happened to go on vacation the week after she started and didn't leave much for her to do during the two weeks he was gone. She found a programmable Tektronix desktop calculator that did many of the things that the first HP hand calculators did, and since it was unused, she wrote a ray-tracing program. This was very enjoyable compared to the paper and log tables, and the Frieden calculator.

When the first computer kits came out in the mid 1970's, she bought one from Imsai and built a home computer. At first she had to enter programs using switches on the front panel, but soon progressed to more sophisticated methods and wrote another ray-tracing program. She says that for her the best way to learn how to program or learn a new programming language, is to simply write a ray-tracing program!

Before retiring in 1991, she made one more telescope, a 10-inch *f*/7.3, which she mounted in a rotating observatory in the back yard. Diane says that 10-inch was "the only mirror she ever made in which there were no side steps or 'do that grade of grit over' or 'let's grind out this ridiculous figure I have just spent 50 hours creating'." She says she has discovered that it is really very hard to completely ruin a piece of glass.

She looks forward to making another telescope, playing with computers and observing, radio astronomy, amateur radio and image-processing astronomical stuff. Her main area of interest is now radio astronomy on the Sun and Jupiter.

1.5.6   Harrie G. J. Rutten

Harrie was born on June 25, 1950 in the hamlet of Arcen in the Netherlands. Geographically it is an interesting place: it is on the most eastern curve of the Meuse

Fig. 1.5.5 *Harrie G. J. Rutten.*

River. After finishing secondary school, he studied mechanical engineering and precision technology. Following work as a mechanical engineer, he became head of the instrumentation department of a worldwide copy machine manufacturer with responsibilities that included the mechanical, optical, electronic and acoustic design of measuring and research equipment. Currently he is working on special problems in precision engineering and optics. In 1984 he joined the board of the Dutch Society of Precision Technicians. Since 1986 he has been editor of the journal, *Micron*.

He feels that his interest in astronomy was inherited from his grandfather. His earliest astronomical observation was of a partial solar eclipse in 1961. For Christmas that year he received a small telescope. Two years later he built his first telescope from army surplus optics. In 1969 he became a member of the Dutch Society for Meteorology and Astronomy. Later he joined the board of the local chapter of this organization, Section Venlo, and in 1975 he was elected chairman, a position he still holds.

In 1969, he met Jean Delsing, a man who had built one of the largest telescopes in the Netherlands at that time: a 10-inch Newtonian. Building telescopes with Jean was fascinating but even more so when he discovered that Jean had a beautiful daughter named Elly. In 1972 Harrie and Elly were married; they have one son, Maurice.

Jean and Harrie had very similar interests and built several telescopes together during the ensuing years: a 180 mm Rumak, a 130 mm apochromatic refractor, and a 305 mm Schmidt-Cassegrain. They also built a fully automatic grinding and polishing machine (Zeiss-type) for optics up to 450 mm and an optical bench with a Foucault tester capable of reading deviations to 0.001 mm. Unfortunately, Jean passed away in 1986.

In 1978, after years of reading "Gleanings for ATM's" in *Sky & Telescope*, Harrie began to study optics because he wanted to experiment with new designs in an effort to improve off-axis performance. During that year he bought his first computer and wrote ray-tracing and optical design programs. In 1980 his first article was published in the Dutch magazine, *Zenit*. Since then, numerous articles, in collaboration with Martin van Venrooij, have been published in many European magazines and in *Telescope Making*. Some years later Richard Berry persuaded Harrie and Martin to write a book on telescope optics. Entitled *Telescope Optics, Evaluation and Design*, it was published in 1988 and included software for optical design and ray-tracing. In 1989 Martin and Harrie were presented with the "Dr. J. van der Bildt" award, the highest honor in the Netherlands for amateur astronomers.

1.5.7   Martin van Venrooij

Martin was born in Rotterdam, the Netherlands on June 19, 1934. He worked for 34 years as a chemical and mechanical engineer on the design of chemical plants and the development of new chemical processes. In 1960 a large amount of natural gas was discovered in the Netherlands. During the last 15 years of his career, he worked as an adviser in the natural gas industry, his expertise being production, purification, compression and transportation of the gas. Today some 95% of all the homes in the Netherlands are connected to natural gas for heating and cooking. In 1993 Martin retired after having had, in his words, "a very interesting professional life which brought me in contact with hundreds of people with a high level of expertise and education."

His most memorable experience was as a boy of 14, looking for the first time at the craters on the Moon with a self-made telescope containing a simple spectacle glass. In 1970 he purchased a 4-inch refractor. In 1973 he became one of the first people in his country to own an 8-inch Celestron. Being a new instrument in the Netherlands, for most amateurs it was a real sensation. He did a lot of demonstrations in amateur astronomy societies and also wrote articles about the instrument. In 1975, he received a letter from Harrie Rutten, who was chairman of an amateur society in Venlo, asking him to give a demonstration of the C8 for their society. They met each other and began discussing a problem with off-axis aberrations on the C8 and came to the conclusion that for a thorough investigation of the image aberrations of a telescope, a ray-tracing program was of crucial importance. In 1978 Harrie Rutten wrote his first computer program. After that, many articles followed in magazines in Holland, France, Germany and the United

States. In researching their publications, Martin and Harrie were both in the favorable position of having access to scientific literature on optical design and a direct access to the American Patent Office as well as European patent publications on optical systems. In 1988 their book *Telescope Optics, Evaluation and Design*, was published, which brought Martin a high degree of satisfaction.

Martin and his wife, Ans, are the parents of two children, Marie Pauline and Adrian. He is enjoying his retirement very much and is very involved with not only his responsibilities as an associate editor of *ATM Journal*, but with piano lessons, which he has been taking for almost 2 years. He has also, at the request of the Historical Society of Geleen, recently written a book about the liberation of that town by the Americans from German occupation on September 18, 1944.

Issue 2

2.1 A 17.5-Inch Wood and Aluminum Split-Ring Telescope
By David E. Moerke

The desire to see more stars and my love of telescope making are the reasons why this scope was built. The optical system is an uncomplicated Newtonian, and the mount is an exceptionally stable split-ring with no motor power. See Figure 2.1.1.

2.1.1 Lower Tube, Key to Mount Size

Because minimizing the size of the mount was a top priority of mine and as lower tube design is especially important in this respect, I decided to build the lower tube first.

The location of the tube balance point, or more exactly, the distance from the very bottom of the tube to the pivot point ($14\frac{1}{2}$ inches), as well as the diameter ($17\frac{1}{2}$ inches) and focal length (79 inches) of the primary mirror, have great impact on the overall mount size. The top of the octagonal-shaped lower tube projects above the balance point about 4 inches to provide support to the 11-inch diameter by $1\frac{1}{2}$-inch thick declination bearings. These plywood bearing disks have a Formica side and a 1-inch wide by 0.060-inch aluminum wrapper band. The two sides closest to square and parallel with one another were used to hold the declination bearings. The bearings were located by matching center holes in the bearings to the pivot holes in the tube, and attached with $\frac{1}{4}$-inch by $2\frac{3}{4}$-inch flat-head wood screws. The bearings ride on ten Teflon pads mounted on the cradle. See Figure 2.1.2.

To avoid tube play and wobble when the cradle is swung from extreme east or west, the fit between the Teflon pads and the Formica sides of the declination thrust bearings was kept tight. This was accomplished by shimming or trimming the pads.

This provides stable support to the tube in any attitude. To reduce the tube weight, six large rectangular holes were cut and closed out on the inside with Formica. The rounded edges of these recesses make good places to grip the instrument during transport. The lower tube is constructed of $\frac{3}{4}$-inch plywood throughout. The 8 sides are double thickness. The top is made of two overlapping layers, contoured as shown in Figures 2.1.2 and 2.1.3. The lower end has 4 layers

Fig. 2.1.1 *David Moreke's compact Porter split-ring mount for his 17½-inch f/4.5 is portable and features Teflon RA brakes for windy evenings under the stars.* Photo by Mike Revera.

tapered to match the removable bottom.

The primary mirror, from Coulter Optical, and its cell and support make up the bottom. The cell support consists of a plywood inner shell and an aluminum outer one. The shell's outer contour is a transition from octagonal to spherical. This allows the weight of the structure, cell and mirror, to sit lower into the ring, than otherwise would be possible if a flat bottom were employed. The outer shell is made by overlapping contoured 0.080-inch thick plates, sandwiched between two ⅛-inch thick aluminum disks at the center. All are held together with double flush aluminum rivets.

Fitting into the outer shell is the inner wooden shell on which is mounted a ¼-inch thick 8-lb steel plate. Four holes in this plate are threaded for ¼-28 collimation bolts. These bolts push up against a ⅛-inch thick round aluminum plate mounted to the bottom of the cell. Four point collimation is more difficult than three point, but gives better support to the cell. A clearance hole for the center cell flotation or tension screw is also drilled in the steel and aluminum plates.

The mirror cell is built around an old 16-inch pulley. The area between the spokes was filled with plywood, and a plywood disk was mounted to the top. The aluminum 18-point flotation system is fitted on the disk. Part of the travel of the three rocker arms is into cutouts in the wood below. In the belt groove of the pulley are six rubber bushings. These hold the cell against the inner wooden shell and allow enough movement for collimation.

Fig. 2.1.2 *(Left) The declination axis and polar axis intersect, requiring two upper-ring counter-weights. (Right) The yoke ring inner contour is shaped to permit minimum tube bottom clearance. This allows the shortest possible north-south base length.*

Four fold-out hand grips and four brackets for attaching up to 10 pounds of counterweight are fitted to the bottom. The weight was not needed as the tube proved to be well balanced without it. The brackets were left on, however, as they provide good support for the bottom when separated from the tube.

2.1.2 Tubular Truss and Rotation Ring

A 12-inch high Formica and aluminum dew and light shield, attached to the lower tube, projects up inside the truss tubes. All surfaces facing the light path, including the inside half of the truss tubes, are painted flat black. The black anodized aluminum tubes have a 1-inch outside diameter with a 0.055-inch wall thickness. Rather than flatten the tube ends, I chose to use inserts made of aluminum and wooden dowels. These work as well as machined inserts and are much less difficult to make. The tubes are connected to aluminum plates on the lower tube and rotation ring. The top plates are slightly curved to match the rotation ring. The lower plates are also curved to fit wooden pads, which adds strength to them. Clearance holes, for 8-32 screws, were drilled in the plates. The screws attach to nut-plates riveted to the inserts. This method requires only a screwdriver for assembly.

The rotation ring, which forms the top of the truss, has eleven thin Teflon pads glued to its Formica top. Attached to the bottom of the ring, and projecting up around the outside, are four two-direction roller assemblies and two brakes. The upper tube rests on the pads and is held in position while the entire assembly is allowed to rotate by the rollers. This design weighs only 2 pounds which assists in keeping the declination axis close to the bottom tube end.

2.1.3   Rotating Upper Tube

The upper tube consists of two ¾-inch plywood rings separated by five ½-inch and two ⅝-inch wooden dowels. The aluminum focuser bracket is located on the larger dowels. To prevent twist during rotation, six 5⁄16-inch diagonal dowels were installed. The vertical dowels are press fit and glued into ⅜-inch deep socket holes in the rings. The diagonals are flush fit and glued. The rings are 1-inch wide between the dowel locations, where they bulge out to 1½-inch. The 1-inch area is rounded to provide hand-holds for rotation. The bottom ring and Formica top and bottom form bearing surfaces. The heavy-duty spider and diagonal holder, from Kenneth Novak, are attached to the upper ring. A thin 7-inch Formica and aluminum extension is fitted above the ring to provide better dew and light shielding. The outside of the area between the rings is wrapped in thin Formica.

2.1.4   Down-Sized Diagonal

To improve image quality, a smaller than usual 3.1-inch Coulter diagonal is used. Because the diagonal cuts the light cone at a 45° angle, the position of the part of the diagonal farthest from the focuser is the closest to the primary and thus must cover a slightly larger light cone than the part near the focuser. To allow for this, the diagonal was offset from the center (away from the focuser) about ½-inch.

Some thought was given to using a low-profile focuser, but I opted for the convenience of one I am used to, a 1¼-inch rack and pinion type from University Optics. To counteract the 3-inch racked-in focuser height, the focuser was moved to where the bottom of the eyepiece tube, which is painted black, infringes slightly on the light path to the primary. This is not a problem, as none of my eyepieces focus that far in. By minimizing the distance from the eyepiece to the diagonal, the light from the primary is intercepted by the diagonal as far out as possible. In this way, I can keep the size of the diagonal mirror to a minimum, thus increasing over-all image contrast.

2.1.5   Split-Ring

The split-ring is 36¼-inches in diameter, including the ⅛-inch thick by 1½-inch wide aluminum wrapper band. It is made of doubled ¾-inch plywood for strength. The fit between the ring and tube bottom is kept close, so no more is cut from the inside of the ring than absolutely necessary. To insure this close clearance, a cardboard template was cut to fit the tube bottom. The balance point (declination axis) was marked. This was then duplicated on a piece of ¼-inch plywood with small wooden blocks attached to the balance point on both sides. A hole for a small nail was drilled into the blocks and the tube at the balance points. Nails were inserted, and the template was swung back and forth over the tube and trimmed to fit. See Figure 2.1.3. The template contour was then transferred to the ring, and the final

Fig. 2.1.3 *Lower tube with the bottom end latched on. The plywood template pivots on nails through the declination axis to demarcate the closest contour for the inside shape of the split-ring and yoke.*

fitting of the ring was done with the tube mounted in the cradle.

Special care was taken to make the outside of the ring as round as possible, and the fit of the wrapper tight. The result was good, but could have been better. Fortunately, this is not as critical on a "visual only" telescope as on one used for photography. Two Teflon-shoed brakes for the declination bearings are fitted on the prongs of the ring, as are two pointers for the declination circle on the tube.

2.1.6   Cradle

The ring, yoke and bottom brace make up the cradle. The yoke and brace are doubled ¾-inch plywood. The width of the brace is determined by the bottom width of the ring where they connect. The inside contour of the yoke was traced from the ring. Both the ring and yoke are mutually supported where they connect and by the brace. The ring has two additional supports—the rollers. The yoke has only the one south bearing. This leaves a long unsupported distance between this bearing and its mating point with the ring. For this reason, the outside of the yoke was made wider in some areas than the ring. The narrowest point on the yoke (where it clears the RA circle) is 3 inches. A very strong attachment of the components of the cradle is extremely important. This is provided by the use of 2 x 2 x ¼-inch aluminum angles, 0.080-inch thick aluminum plates and ⅜-inch bolts and nuts. To balance the cradle, two 5-pound steel weights were added to the ring tips of the cradle to counterbalance the weight of the lower ring and the brace.

Loading and removing the tube from the cradle is no problem for two people. To make it easy for one person, a wooden cart (shaped to fit the ring and base) is used. The tube simply slides south into the DEC Teflon bearings along two inclined rails.

Fig. 2.1.4 *An illustration of Mr. Moerke's success in minimizing the ring diameter. Fancy metalwork highlights his craftsmanship. Others might choose foam contouring with fiberglass cloth and epoxy for forming the lower end shape. Either way, this mount is very close to the minimum size for a tube of this diameter.*

2.1.7   The Base

A small (1/10 scale) cardboard model of the cradle and first preliminary base design was made. The rather tall order for this little model was to help in working out the base shape and dimensions necessary to allow clearance for the swing of the yoke and brace. It also facilitated checks on ground clearance. The details were worked out with full-scale drawings of the cradle and base.

The base was built in two sections—north and south ends. These were clamped in position before final assembly and the cradle was attached. The cradle was rocked back and forth to check clearances and to check alignment of the south bearing, ring and rollers.

Two fixed casters made of hard rubber provided support points on the north end, and one swivel caster provided support for the south end. This adds height to the mount and allows room to use a wooden lever to lift the base when making large latitude or level adjustments (jacks can be turned by hand or with the help of a hex wrench).

Fig. 2.1.5 *Telescope base and RA setting circle. Note the south RA thrust bearing with the caster wheel removed.*

2.1.8   North and South Bearings for RA

Hard rubber casters serve as the two rollers for the north bearing. The casters are spaced about 70° apart, being 21½-inches between centers. I first mounted steel casters, but these tended to allow the ring to move too quickly. The original plan was to use the Teflon jacks, mounted just inboard of the casters, to provide a Dobsonian feel to the cradle motion in most of the sky. The steel casters were to be used in the polar area, where there is little leverage available from the long tube. Unfortunately, the jacks proved to be too stiff, requiring too much pressure to move the cradle. Plan B, rubber casters, was much better, supplying smoothness very close to Dobsonian performance in RA. The jacks have been relegated to taking the weight off the casters for storage.

The south RA bearing is a modified heavy-duty 3-inch swivel caster. It has proven to be very functional at my latitude of 34°. Located around the bearing is a wooden RA setting circle, which rotates on bearings similar to the upper tube. See Figure 2.1.5.

2.1.9   Setting Circles

As most of my observing is done in light-polluted skies, the setting circles have proven to be very valuable. They have dramatically shortened object location time. No finder is mounted. Almost all sweeps and object locations are done with my favorite eyepiece—a 20 mm Erfle from Meade. It gives a 4.4 mm exit pupil and allows use of a 1¼-inch focuser which provides more than ample field diameter.

2.1.10 Weight Data

The approximate weight of the complete mount is 203 pounds. Like most split-rings, this mount can be broken down into components for transport. The weight in pounds of each as follows: tube–120, cradle base–32, blocks–5, mirror cover–3. Tube weight breakdown is: rotating upper tube–12 pounds, rotation rings–2, truss tubes–8, lower dew shield–2, lower tube–36, and the bottom (including the mirror)–60.

This scope has proven to be well-balanced and stable in all areas of the sky. It is the product of learning what others have done before, and adapting or modifying an idea, and adding a few of my own. The hope here is that a few ideas have been presented that can be of value to other ATM's.

Although not easy to construct—at least for me—I believe that if you take the time and make the effort required to design and build one, you will find that the Porter split-ring is really a wonderful telescope mount.

2.1.11 References

1. Harris, Luis A. "A Split-Ring Newtonian for 9° Latitude," *TM #43*. Mount is rigged for photography and has rotating tube. A comparison of this beautiful 12½-inch *f*/6 scope with the one it is based on by Joe Pearson (lat. 35°) well illustrates the affect of latitude on the mount proportions, load distribution between the north and south bearings and access to the southern horizon.

2. Moerke, David E. "A Split-Ring Mounting with that Dobsonian Feel," *TM #39*. My first split-ring that still gets plenty of use. It is a 12½-inch *f*/6, (lat. 34°). Rotating upper tube mainly visual.

3. Walker, Gary. "Dobsonian Performance from an Equatorial Mount," *TM #31*. Two pages of photos and captions, no story but well worth a look. This is an exceptionally well thought out 14-inch scope. It has adjustable tube length for photography and intricate bracing at the bottom. An ultra-light design.

4. Pearson, Joe. "An Equatorial Split-Ring," *TM #27*. Excellent portable design. This 10-inch *f*5.6 scope is also in *Sky and Telescope*, March 1986. Rotating tube, powered for photography.

5. Clark, Tom. "Supertune Your Telescope," *TM #39*. Good on down-sized diagonals plus 18-point flotation.

6. Ingalls, A. *Amateur Telescope Making*, Book One (4th ed). Russell Porter's patent drawing for the split-ring is on page 349; also see pages 27, 133 and 348 for more split-ring information. Eighteen-point flotation covered on page 230.

7. Chandler, Dave. "Flotation Cell Design," *TM#26*. I used this formula for the 18-point spacing.

8. Peters, William T. "Downsizing Diagonals," (letter page 10), *TM #22*.

Full of information, very interesting.

9. Cox, Robert E. "Better Mirror Cells," *TM #6*. Many types covered. My cell is partly adapted from the one illustrated on page 37, upper left.

10. Kreige, Dave. "Low-Tech R&D and Off-the-Shelf Parts for Super-Big Telescopes," *TM#44*. A bonanza of information. The idea of using silicone wax was found here. Kreige's low-tech 18-point flotation and cell, which is very compact, could easily be adapted to the split-ring.

11. Daley, Richard F. and Sally Daley. *Build Your Own Customized Telescope*. TAB books, paperbound. Very good on woodworking. Features a polar disk mounted fork type, from which the split-ring evolved.

2.2   Industrial Fabrication of Small Lenses

By Dennis Lithgow

Industrial fabrication of spherical lenses does not differ much, in its most basic form, from the steps used by amateur telescope makers. First, the curve of the element is roughed out (a step referred to as "generation") followed by grinding, and finally polishing. These steps will normally take place in sequence for first one side of the element and then the other, although there is no hard and fast rule here. Once both sides have been polished, the lens will be centered, edged, and coated.

Several things will influence the exact method to be chosen in the fabrication process. One has to first look at the number of elements to be manufactured in order to decide if the cost of special tooling to manufacture them is warranted. Also to be considered are the physical size, design parameters (like radius of curvature of both surfaces), and tolerances.

One of the best examples of the type of special tooling used is the spot block. (Figure 2.2.1) If the lenses to be worked are small enough and their radius of curvature is not too short, several of them may be mounted together with a common center of curvature. This has the effect of lowering the cost of an individual lens as the labor cost for working a spot block full of parts is not significantly higher than for working a single part mounted by itself. So instead of spending 2 hours polishing one lens, we can spread that two hours out over 17 lenses for a significant savings. Of course, the cost of such specialized tooling (over $2,000 for a typical tool) has to be balanced against the labor savings to be incurred.

2.2.1   Generation

In an industrial setting, each of the fabrication steps is conducted at a separate machine. Thus, we will see a machine used exclusively for the generation of curved surfaces. (Figure 2.2.2) If parts are to be run individually (i.e., without the use of a spot block), then they will have to be held individually. One of two methods is normally used here. The more common of the two is for the parts to be held in the generator by means of a vacuum chuck. The vacuum chuck is nothing more than

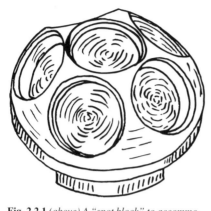

Fig. 2.2.1 *(above) A "spot block" to accommo-date 6 lenses.*

Fig. 2.2.2 *(right) Automatic curve generator courtesy of Wilhelm Bothner GmbH, Germany.*

a cylindrical tool having threads to fit the generator on one end and a radiused edge, usually of Teflon, on the other end. The lens is then placed on the chuck, with no vacuum applied, and placed in motion. Then, a "centering stick" is brought into contact with the rotating lens and with a slight amount of pressure on the edge, the lens centers itself. Once centered, vacuum is applied and generation may begin.

The second method employs a holding fixture called a runner. The runner is similar to the vacuum chuck in that it too is a cylindrical tool threading the generator on one end and a radiused edge on the other end. Normally it is made entirely of aluminum. In mounting the lens, the runner is placed on a hot plate and allowed to warm up. Having reached the proper temperature, a wax stick is brought into contact with the radiused edge and a good layer of wax laid down. A common alternative method is to first spray the back side of the lens with hot wax, allow it to cool, and then seat the lens against the warmed runner. Specific formulations of blocking wax are used so as to minimize problems with the lenses changing shape, or "springing," when they are de-blocked. The lens is centered on the runner in a manner similar to that for the vacuum chuck. Once mounted in this manner, the lens will remain attached to the runner through the generation, grinding and polishing of one side of the lens. Once the first side has been completed, it is de-blocked either by re-heating the runner to melt the wax bond or, in some cases,

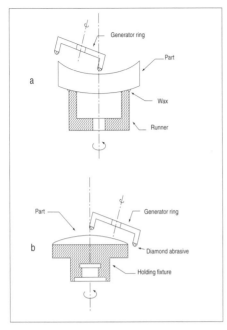

Fig. 2.2.4 *Setup for curve generation courtesy of Wilhelm Bothner GmbH, Germany.*

Fig. 2.2.3 *a. Curve generator for concave surfaces (a), and convex surfaces (b).*

by shocking it free from the runner by mild impact. That choice has largely to do with the lens geometry and material.

In the generation process itself, the proper curvature is generated on the lens by the use of a counter-rotating diamond tool set at a precisely determined angle, with both the block and cutting tool being powered. (Figures 2.2.3 and 2.2.8) When completed, the shape will be very close to the final desired curve, but a significant amount of sub-surface damage will be present, along with severe scratches which will need to be ground out. Industry will almost never use grinding compound at the generation stage due to the long cutting time it imposes. Also, even though the price of a diamond generator ring is several hundred dollars, it can generate tens of thousands of lenses, making it much more cost effective than grinding in the initial curve with loose abrasives.

2.2.2 Grinding

The grinding operation can take one of two different forms depending on the number of lenses to be produced. If volume is sufficiently high to warrant its use, a diamond pellet grinder will be built, and the part will be ground in a machine called a pellet grinder. Otherwise, grinding on a cast iron tool with loose abrasives is the method of choice, and a tub grinder is the machine used. (Figure 2.2.5) During this operation, the lower (convex) part (the tool or the lens) is powered, and the upper (concave) item (part or tool) is moved back and forth by means of an arm on the

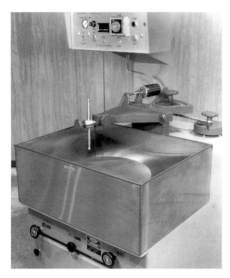

Fig. 2.2.6 *Spherometer courtesy of R. Howard Strasbaugh, Inc., Huntington Beach, CA.*

Fig. 2.2.5 *Conventional polishing machine courtesy of R. Howard Strasbaugh, Inc., Huntington Beach, CA.*

machine. In the case of the pellet grinder, this arm is power operated.

Occasionally during grinding, the curve of the part will need to be checked to see if and how far it has deviated from the desired curve. This curve will be tested using a spherometer which is nothing more than a hollow cylindrical "bell" of known diameter with a very sensitive dial indicator, or "Mahr" gauge, mounted on its central axis. (Figure 2.2.6) This instrument is used in conjunction with a test fixture to test the radius of the part. A test fixture is a measurement standard made of glass that has a curvature on one side corresponding to the curvature of the lens to be fabricated. By setting the "bell gauge" to zero with it resting on the test fixture and then placing it on the lens to be tested, the difference in height (sagitta or "sag") can be measured. From this one can calculate the true radius of the part, but normally that is not necessary. The stroke of the upper arm is simply changed to force the lens curve back into compliance.

2.2.3 Polishing

As in grinding, there are two methods available for polishing, dependent upon the volume of production to be run and the lens geometry. If the volume is low, the techniques familiar to the amateur telescope maker are used. A polishing lap will be made using optical pitch that has been applied over a substrate of metal or glass after it has been cut to an appropriate curve. A polishing machine is used with the same setup as was used at the loose abrasive grinding step, except that here polishing compound, (usually a rare earth oxide), is substituted for the grinding compound. (Figure 2.2.7) After a sufficient amount of time has passed, the parts are removed from the machine, and the lens surfaces are cleaned. A test plate is then

Fig. 2.2.7 *Manual loose abrasive grinding and polishing machine courtesy of Wilhelm Bothner GmbH, Germany.*

Fig. 2.2.8 *Diamond grinding wheels and generator rings courtesy of Abrasive Technology, Inc., Westerville, OH.*

placed in contact with the lens and viewed under a monochromatic test lamp. The test plate is a radius and figure standard made of glass having a known curvature on one side that is the same as that of the part being tested, except that it is a curve of the opposite sense so that the two will mate, This test shows how well the part fits the test plate curve, and consists of two values. The first is the spherical power, which measures the variation in radius from the test plate to the lens. The second looks at irregularity in the lens. The "test" is read by looking at the interference bands visible through the rear surface of the test plate.

For lenses that meet lower figure requirements, high speed polishing equipment can be used. In high speed polishing, the pitch polisher is replaced with one made of polyurethane or Polytron. The speeds and pressures applied by the machine to the lens are much greater with this process, and the polishing times drop drastically as a result. The reduction in labor costs of this method, as compared to pitch polishing, can easily exceed 3:1. The geometry of the machinery generally limits the lenses that can be run to those with a radius range of 0.5" to about 8.0". Because the polishing material is now hard and cannot flow to match the curve of the part, the substrate of the polisher must be formed with a great deal of precision. This is one of the major reasons that the tooling for this type of machine is much more expensive than that for conventional machines. Because of the pressures involved, the process will not work for parts of high aspect ratio (diameter/center thickness).

After a lens has been generated, ground, and polished on both sides, it is centered and edged on an edging machine. Edging and generation are currently the only lens fabrication operations that make use of numerical control (NC) machinery (although NC polishers are under development). If an NC machine is used, any bevels to be applied to the lens can be ground in; whereas, if a manual machine is used, this operation is normally done by hand later. Before the lens is centered, it is coated with a protective material. In centering the lens it will be mounted with wax to a mandrel, similar in shape to the runner from earlier operations. The mandrel will have had its lens-mounting surface trued to its axis of rotation. An indicator is then set-up on the other surface near the outside diameter and monitored as the mandrel is heated slightly, and the part is adjusted to run true by the indicator. At this point the part is edged by spinning it and slowly feeding it against a rotating diamond grinding wheel (Figure 2.2.8) to bring it down to the proper diameter. It is then de-blocked, cleaned and sent to coating.

2.2.4   About the Author

Dennis Lithgow has worked for Texas Instruments as a process engineer in a precision optics fabrication facility in support of the Defense Systems & Electronics Group. He is currently in Little Rock, Arkansas with SIU.

2.3   Front-Loaded Collimation

By Henry T. Van Gemert

If you transport your Newtonian telescope some distance to observe, chances are that you have become quite adept at collimating your optics once you arrive at your "spot". But if you are like me, you still have a problem remembering which screw moves the primary in which direction. It seems that no matter how many times I align my telescope, I keep forgetting. Therefore, I devised a way to collimate while looking through the eyepiece, eliminating the need to remember where my center spot and Cheshire ring were before the last tweak.

I had seen some examples of this, notably Steve Watkins' full-tube design (*S&T*, 1/91, p. 31) and Steve Collett's Dobsonian mirror-box arrangement (*TM* #45, p. 25).

In designing my mirror cell, I had a number of requirements:

1. Ease of fabrication (I am not a good carpenter and have no machining skills).

2. Ease of use (this was intended to replace a tiresome procedure, not substitute another one).

3. Rigidity (to preserve the collimation that I so easily had accomplished).

4. "Tool free" operation (I didn't need another "gadget" filling up my eyepiece box).

Unfortunately, I only satisfied the first three requirements, though I am sure that someone will finish off the list on their own cell. I found that the tools I need-

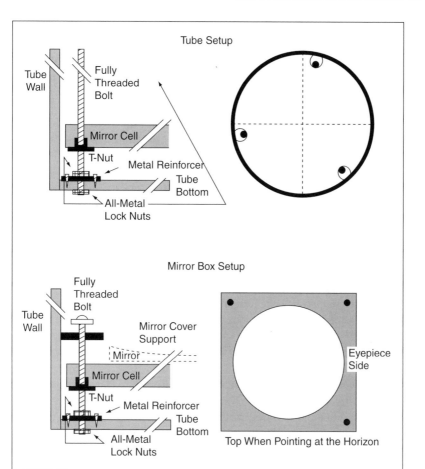

Fig. 2.3.1 *Tube Setup and Mirror Box Setup. Diagrams courtesy of Ed Stewart.*

ed simply replaced the screwdriver and socket driver I was already using.

The heart of my front-loading cell is a trio of threaded inserts or T-nuts arranged around the perimeter of the mirror cell (see Figure 2.3.1). In a round, tube arrangement, these are almost always going to be at 120° intervals. For a boxy Dobsonian style, they should be placed at the bottom two corners of the mirror box and at one of the top corners (this works a lot better if you put the top nut on the eyepiece side of your cell), giving you an orthogonal arrangement.

In the tube style, threaded rods (¼-20 work fine up to 17½-inch mirrors) are run through the nuts, an all-metal lock nut, the base of the telescope tube, and another all-metal lock nut. The hole in the base of the tube should be covered with a thin (aluminum or steel plate, say ⅛-inch), and the hole drilled through this to the exact size of the rod. This reinforcement maintains rigidity and keeps the collimation adjustment points from wandering.

The top of end each threaded rod passes through a restraining bracket a couple of inches from the top of the telescope tube (scrap pieces of ½" copper pipe about 2" long, painted black, work great—though they require a piece of foam stuffing to keep from rattling) and are topped with another all-metal lock nut about an inch from the end of the tube.

Adjusting the rods is a job for a ¼-inch nut driver, but be sure to epoxy the socket on so that it never ends up meeting your mirror on an intimate basis. If the rods are arranged 120° apart, you still have to figure out which nut moves the alignment which way, but you watch it "live."

In the Dobsonian arrangement, it is even slicker. The T-nuts or inserts are used the same way; but instead of a threaded rod, you use a fully threaded ¼-20 hex bolt that is ½" longer than the distance between the base and the mirror cover. Put holes in this cover to fit your bolts and do all the other construction steps the same for the base of the box (take care to reinforce the holes with a metal plate). Attach a ⁷⁄₁₆-inch socket to the end of a suitable rod (again, a piece of ½-inch copper pipe works great) with a length equal to the distance from your hand, while standing at the eyepiece, to the bolt, plus about 6" for good measure.

This adjustment goes like lightning! By using the two outside bolts and leaving the center bolt alone, the motions of the mirror tilt are at right angles and very easy to control and predict. You can usually bring a center-dotted primary into collimation with a Cheshire eyepiece in less than 10 seconds—even faster with two such control rods.

If the Dobsonian's arrangement allows your mirror cell to sag, it will negate this easy collimation procedure so often that it will not be a blessing at all. However, with the metal plates reinforcing the base holes, there is no need to have the fourth corner also threaded to allow a push-pull locking ability. The collimation holds well during the observing session. Also, the complete mirror cell is enclosed in the telescope and runs no risk of falling out if you unscrew the rods or screws too far.

2.4   Four-Collimation-Point Mirror Cell
By Ed Stewart

When I decided to use a rotating tube design for my 12½-inch Dobsonian (see *Telescope Making* #45, pages 10-16), I decided that it did not lend itself to a commercial mirror cell. Besides, there were references in several issues of *Telescope Making* that said there were no good ones available. Also, it did not lend itself to the usual sling found in the traditional Dobsonian. Since it was obvious that the cell would have to support the mirror in all dimensions as a single unit, I decided to try something different in constraining the mirror's side-to-side movement. Also, I opted for a square layout of four collimation points similar to that mentioned in *Telescope Making* #6, and used push/pull bolts to make the collimation adjustments.

2.4.1   Side-to-Side Mirror Alignment

Since I had a surplus of Formica laminate on hand from construction of bearings for my scope, I decide to make use of it. Two disks of ½-inch Baltic birch were cut on a router table to ⅛-inch larger diameter than the mirror and glued together with very exact alignment of their edges. Work inside the cell, such as the flotation plates and ventilation holes, was completed first. Then a strip of laminate was cut to the length of the disk's circumference and to a width sufficient to span the full width of the combined mirror cell base disk and the mirror disk to within ¼-inch of its surface—about 3 inches in my case.

To insure that the laminate would wrap around the disk accurately, I spaced above the mirror cell a scrap plywood disk (of the same diameter) on a bolt through their centers (the hole used on the router table) in the position of the upper part of the mirror. The laminate was then glued and screwed around the mirror cell disk forming a bowl-like container for the mirror. A 3 x ⅛-inch curved piece of aluminum was screwed in place over the laminate joint to strengthen it. Additionally, a line of ⅛-inch holes, for inserting room temperature, vulcanization silicone rubber adhesive (RTV), had previously been drilled at one-inch intervals on the laminate, and spaced across its width so that they were centered along the mirror's thickness (see Figure 2.4.1).

I rubbed silicone grease on the outside surface of a 5 mil plastic strip, and then wrapped it around the mirror to create a small gap for the next step. After centering the mirror in the laminate circle, I squeezed RTV through the line of ⅛-inch holes to form a pad of RTV between the laminate and the mirror. Even though the silicone grease prevented the RTV from sticking to the 5 mil plastic, the mirror was very slow to dump out of the cell. Once out, I removed the plastic strip and put the mirror back into the cell by lowering the cell onto the upside down mirror. No mirror clips were necessary as the mirror came out in a face down position very slowly. This laminate/RTV method has worked beautifully during eighteen months of use. Hint: for heavier mirrors, two or more layers of laminate could be glued around the cell for greater strength.

2.4.2   Four-Point Collimation Adjustment

The usual triangular arrangement of collimation points does not allow for adjustments at a right angle to each other, as on an "$x - y$" chart. An arrangement of four points in a square does provide for right angle adjustment, and the advantages are more than would at first seem apparent—six axes of adjustment instead of three using the triangular arrangement (see Figure 2.4.2). By moving in or out with any side-by-side pair of adjustment points the cell can be caused to pivot on the axis running through the other two for a total of four axes. (It would seem that the parallel axes are redundant, but should one axis run out of inward adjustment, the other parallel axis can be used.) With diagonally opposite pairs, one is moved in and the other out, resulting in a pivot on the axis running through the other two diagonally opposite points.

Fig. 2.4.1 *Mirror cell is shown with Forstner bit used to drill the holes for the tubing and to countersink the T-nuts. The line of holes for the RTV can be seen circling the Formica.*

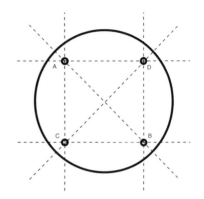

Fig. 2.4.2 *Mirror with four collimation points arranged in a square.*

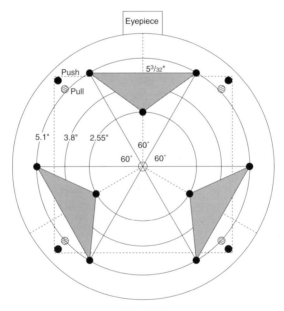

Fig. 2.4.3

   In practice, a given collimation adjustment bolt is pulled (turned counter-clockwise) to move the mirror spot image toward the direction desired. Turning a collimation bolt the width of a screwdriver slot (about $3/32$ inch) will move the mirror spot about $1/10$ of its dimension as seen in the Cheshire eyepiece. The results are very predictable and long lasting. After transporting the scope in a single axle trailer dozens of times over eighteen months, the collimation has not changed even in the autocollimator eyepiece. Probably an equal share of credit goes to the rigid-

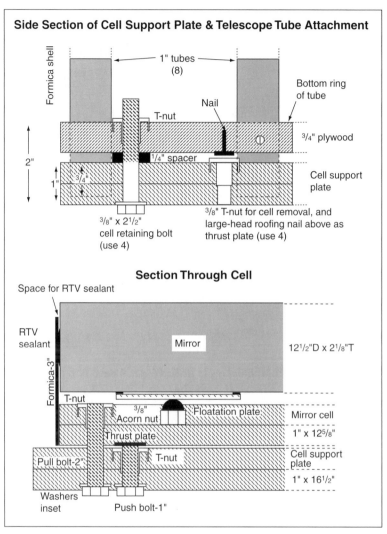

Fig. 2.4.4 *Axial View of Four-Collimation-Point Mirror Cell*

ity and strength of the Baltic birch plywood. The original article suggested using three points, but in a right-angle arrangement. That bothered me as half of the mirror's weight would be unsupported.

2.4.3 Making the Cell Removable

Since the rotating tube design does not break down like a truss tube one, I believe the heavy mirror and cell unit should be removed for transport to avoid over-stressing the cell. As it turned out, it would have been too heavy for me to lift into the trailer by myself. Using a drill jig for the tube's rings (see *Telescope Mak-*

Fig. 2.4.6 *The black plastic mirror spot is glued with RTV under the intersection of 3 diagonal threads. The rectangles beside the stub holes are the ¼-inch spacers shown in the top part of Figure 2.4.4.*

Fig. 2.4.5 *Author shown with the "Gatlin Gun" rotating tube Dobsonian at a state park one mile from the site of the annual Texas Star Party. Entire tube assembly, from eyepiece to mirror cell, rotates as a unit.*

ing #45), blind (flat bottom) holes were drilled ¾ inch deep into the 1-inch plate of the cell support with a Forstner wood bit (see Figure 2.4.1). When the cell is mounted onto the skeleton tube, the eight stubs of 1-inch tubing set firmly at the bottom of these cell support plate holes; and, at the same time, the support plate presses against ¼-inch spacers attached to the last ring in the skeleton tube.

The use of the drill jig makes the fit onto the tubing stubs so exact that something more than arm strength is needed to remove the cell. This was solved by using retaining/removal bolts around the outer edge of the cell support plate in line with the surface of the last ring in the skeleton tube. T-nuts were placed close to the tubing stubs to apply the force near the points of resistance. A full set of eight pairs of bolts were installed, but in practice only four were needed to hold the cell on and three to remove it. The bolts used to hold the cell on are removed at the end of an observing session and transferred to the removal nuts, to save duplication. Construction details are given in Figures 2.4.2, 2.4.3 and 2.4.4.

2.4.4   Mirror Spotting

The instructions that came with the Tectron 3 eyepiece collimation set said to put a 5/16-inch square black spot in the center of the primary mirror. I recently read a suggestion that a circular hole reinforcement (used on paper in a 3-ring notebook) painted black worked better as it would allow you to see the secondary's spot inside the opening rather than superimposed on top. To find the center, I stretched three lengths of black thread across diameters of the mirror 120° apart, securing their ends with masking tape on the side of the cell (see Figure 2.4.6). Then I measured and adjusted their intersection repeatedly until confident that the center was accurately indicated. Next, I used tweezers to place a square of black plastic into position at the center of the mirror surface, and secured it with a small drop of RTV. I used a similar approach on the secondary, except the spot was ⅛-inch square, and the threads were elevated slightly from the diagonal's surface with pieces of toothpicks laying on its outer edge.

2.4.5   Conclusion

One of the joys of using this scope is that I seldom have to make collimation adjustments. But when they are necessary, the right-angle or parallel movement of the adjustment bolts makes the process straightforward for one person to accomplish. Any comments or questions are welcome. Email: stargazer@skymtn.com

2.5   Sub-Aperture Maksutov Correctors
A Comparison of Some 8-inch *f*/10 Designs and an *f*/7 Photo-Visual Telescope
By Ralph W. Field

In August 1981, *Sky and Telescope* published my telescope design which featured a sub-aperture Maksutov corrector. Although Maksutov stated that for the complete correction of coma, the meniscus should be increased in thickness; I found that the astigmatism and transverse color then became too large. As my design was intended partly for photography, it was necessary to make the corrector of a thickness which kept both the astigmatism and transverse color reasonably low.

In *TM* #29, David Shafer presented an *f*/18 version with a very thick corrector (advised by Maksutov); and later in *TM* #38, he offered some *f*/12 and *f*/15 designs. Unfortunately, such instruments require extra surfaces and spaces which result in larger obstructions with no improvement in transverse color. I have computed some *f*/10.5 designs of 8-inch aperture to make valid comparisons of design variations. Because of the fairly high obstruction ratio, these are not ideal for planet observing.

Of the six examples shown, numbers 1 through 5 are sub-aperture correctors for *f*/10.5 Cassegrainian systems. The sixth operates at *f*/9. All dimensions are in inches. Glass used: BSC2. $N_c = 1.51462$; $N_e = 1.51899$; $N_F = 1.52264$; $N_g = 1.5269$. Surfaces are numbered in the order in which the rays meet them. LA′, ChrA′, LZA′ and spherochromatism are indicated under Ray heights. Primary mirror is 8-inch *f*/2.5 for all designs. Hole in primary mirror is 1.25" diameter for designs 1 to 4.

2.5.1   Design No. 1

This is similar to the design suggested in *TM* #29, except for a ¼-inch space be-

No. 1: Thick Maksutov					
Minimum clear corrector diameter 2.4; back focal length 19.571; EFL 83.866; aberration scale 0.001" = 1 mm.					
	Surface	**Radius**	**Distance**		**Ray heights**
	1	−40.0000	16.000	F e C	4.000
	2	5.3986	−14.130		2.980
	3	5.3986	−1.238		1.325
	4	−13.5000	−0.250		

tween the secondary and corrector which slightly reduces the necessary thickness for coma correction. However, at nearly 1¼ inches thick, it is really only appropriate for an 8- to 10-inch telescope due to the cost and difficulty in acquiring a blank of the necessary thickness. Transverse color in this design is just over 0.001" at the edge of a 1-inch diameter circle, corresponding to a half-angle of 0.34°. At this angle coma is 0.00023" and astigmatism is 0.013".

2.5.2   Design No. 2

This solution to the thick corrector problem was suggested by Shafer in *TM* #29 and *TM* #38. The different spacing and thicknesses in my design are partly due to the greater relative aperture. Like David's version, it also has similar radii on the corrector surfaces. Coma, transverse color and astigmatism are virtually the same as in design No. 1.

The obvious disadvantage is the two extra surfaces to produce; although they do not require special tools, they must be finished to a good figure.

2.5.3   Design No. 3

To partly overcome the difficulty of obtaining thick glass, a thinner corrector may be spaced as shown. Because the meniscus is now closer to the primary, its diameter must be increased to catch the marginal rays from the primary. Although the volume of the glass is only slightly less than that of the first design, it should be more readily available.

Obviously an infinite number of thickness/spacing combinations are possible. With the thickness specified here, coma is above the Raleigh limit of 0.0002" but OSC' is less than 0.001". Astigmatism is 0.009" which is just below the tolerance of 0.01".

No. 2: Split and Spaced Maksutov						
Minimum clear corrector diameter 2.6; back focal length 19.6; EFL 84.327; aberration scale 0.001" = 1 mm.						
	Surface	**Radius**	**Distance**			**Ray heights**
	1	−40.0000	16.000	F e C		4.000
	2	5.5269	−13.900			3.000
	3	flat	−0.500			1.400
	4	flat	−0.075			
	5	−5.5269	0.600			
	6	−13.5000	−0.500			

No. 3: Spaced Medium Maksutov						
Minimum clear corrector diameter 2.85; back focal length 18.763; EFL 83.814; aberration scale 0.001" = 1 mm.						
	Surface	**Radius**	**Distance**			**Ray heights**
	1	−40.0000	15.350	F e C		4.000
	2	−3.6916	−12.900			3.000
	3	−4.3480	−0.400			1.400
	4	75.0000	−1.400			

No. 4: Field Visual					
Minimum clear corrector diameter 2.75; back focal length 18.654; EFL 84.856; aberration scale 0.001" = 1 mm.					
	Surface	Radius	Distance		Ray heights
	1	−40.0000	15.350	C e F g	4.000
	2	−3.6916	−12.900		3.000
	3	−4.3480	−0.400		1.400
	4	75.0000	−1.400		
	5	−13.5000	−0.450		

2.5.4   Design No. 4: Field Visual

To overcome the problem of transverse color, I designed an f/10.5 sub-aperture corrector system which is not based on the Maksutov principle and avoids excessively thick or split elements.

The improvement is achieved by making the secondary a back-silvered lens (sometimes called a Mangin mirror) with a positive meniscus in front. The space between the elements is necessary in order to reduce the zonal spherical aberration (of the overcorrected type) to within 1/10 of the Raleigh limit.

This distance also helps to keep the thickness of the lens elements to a minimum while achieving excellent correction for coma and transverse color. Unfortunately, astigmatism is now higher, but preferable to color fringes.

2.5.5   Design No. 5: Field Photo-Visual

For those who wish to use a 35-mm reflex camera for astrophotography, this design is corrected for astigmatism, but has a slightly curved field of approximately 30.5-inch radius. However, due to the tolerance for focal range, images will be excellent across the full 35-mm format.

The design may be scaled up, but for increased coverage it would be preferable to spring the film slightly inwards to conform to the focal surface.

2.5.6   Design No. 6: Field Photo-Visual, *f*/7

None of the preceding designs has been constructed, but this f/7 telescope was

No. 5 Field Photovisual EFL 84.39					
Minimum clear corrector diameters 2.94 and 1.75; hole in primary mirror = 1.75; back focal length 1.904; aberration scale 0.001" = 1mm.					
Curvature of field	**Surface**	**Radius**	**Distance**		**Ray heights**
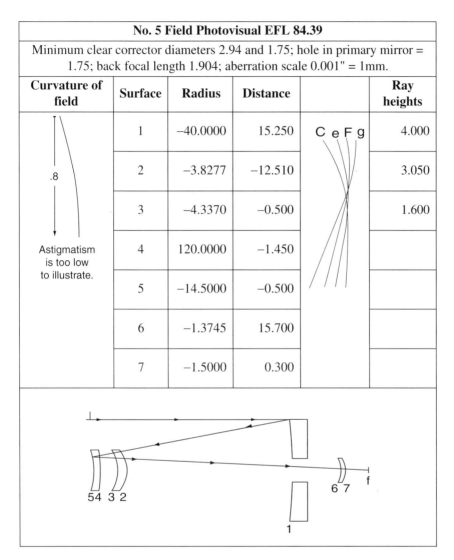	1	−40.0000	15.250	C e F g	4.000
.8	2	−3.8277	−12.510		3.050
	3	−4.3370	−0.500		1.600
Astigmatism is too low to illustrate.	4	120.0000	−1.450		
	5	−14.5000	−0.500		
	6	−1.3745	15.700		
	7	−1.5000	0.300		

made about four years ago. Unfortunately, my location is quite unsuitable for any kind of astronomy, and the only tests have been on the moon and Jupiter with its satellites. However, these results left nothing to be desired.

2.5.7 General Remarks on the Specifications

For those with computers who wish to check these designs, the following information is relevant:

The first figure in the distance column refers to an 8-inch diameter diaphragm position and may be ignored for the first 4 designs. The distance given is that from the primary to the back of the secondary in each case. The only effect of

No. 6: Field Photovisual, $f/7$ EFL 56.85					
Minimum clear corrector diameters 3.50 and 2.00; hole in primary mirror = 2.00; back focal length 0.7; aberration scale 0.001" = 3 mm.					
Astigmatism and Curvature of Field	Surface	Radius	Distance		Ray heights
s t .8	1	−40.0000	00.000	g Fe C	4.000
	2	−3.8576	11.160		3.116
	3	−3.6550	−0.580		1.850
	4	−40.0000	−0.500		
	5	−26.0000	−0.450		
	6	−1.8092	17.500		
	7	−2.0000	0.300		

the diaphragm is a very slight reduction in coma and astigmatism.

However, for Design No. 5 the diaphragm (or stop) should be used because of the correction of astigmatism, and its distance should be approximately 15.25". At an angle of 1.16°, the linear coverage is 1.7 inches—the diagonal of the 35-mm format.

Design No. 6 was corrected for astigmatism without a stop, which means that the main tube should have a minimum diameter of 8½ inches at the sky end for an 8-inch primary.

Figures given under the heading of "ray heights" are those used for the correction of chromatic and spherical aberration. As there are no paraxial rays, the

Fig. 2.5.1 *Ralph Field's 8" f/7 photovisual telescope.*

lowest height given is ½ the diameter of the corrector/secondary obstruction. This height, with the marginal height, is used to effect correction for spherical aberration. The second, or middle height, is used for the chromatic correction and represents a point which gives equal areas above and below it.

All surfaces are spherical or flat in all designs and secondary color is virtually zero. Bk7 glass may be substituted for the specified BSC2 without alteration of any parameters.

Ed Note: Mr. Field has earned congratulations for some fine designs; all the more so because he performed his calculations on an HP 71B, running at 4.5 seconds per ray, through a 10 surface system!

The photovisual design at $f/10$ is diffraction-limited to the edge of a 1.2° field of view. Couple this with its 26-inch concave radius focal surface, and it is clearly a fine design.

Centering and axial positioning in telescopes with sub-aperture correctors is critical and deserves an in-depth study of mechanical design.

If starting from scratch, the amateur designer would do better to use as slow a primary as feasible, since aberrations drop rapidly with increased primary focal length, thus requiring less powerful corrector lenses. For example, compare the design in *TM* #29 (scaled to 8 inches) with the Field #1 design using an $f/2.5$ primary. The $f/1.9$ primary requires so much lens power that the final image is color

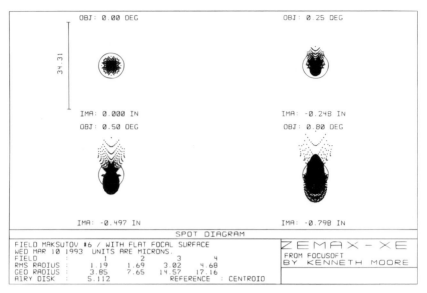

Fig. 2.5.2 *Sub-Aperture Correctors for f/10.5 Cassegrainian Systems.*

corrected over only about $\frac{1}{10}$ of the visual spectrum. The *f*/2.5 primary leads to a full range of color correction. Also, the diffraction-limited field of view is extended from 0.08° to 0.2° as the coma is cut in half. The extra tube length is a small price to pay for centering and image tolerance relaxation.

2.6   Easy Finder Alignment

By Rich Coombs

For anyone who has struggled with the three small adjustment screws on their finderscope—tighten, loosen, tighten, loosen—this tip should provide some welcome relief. The solution to easy finder alignment comes from a device called a spring plunger which replaces the bottom alignment screw. The device is basically a setscrew whose center is almost completely drilled out. A short plunger is inserted that protrudes from one end, followed by a spring, and a plug to hold the spring in place. The device serves to constantly press the finder tube against the other two adjustment screws. Adjustment is then as simple as turning one or both of the remaining adjustment screws. Once you have tried this arrangement I doubt you will ever be content with the old three-screw method.

The spring plunger has rather short travel, but this is not a problem since the plunger can be screwed in or out until it is close to the correct position. Since there are many spring forces to choose from, here is an example to use as a guide. My finder weighs 1.3 lbs, and I use two spring plungers, one at the front and one at the rear. I have found that Vlier's part number M-57 with a 1¼-lb initial force and a 4-lb final force is adequate, but I would recommend going a bit heavier on the

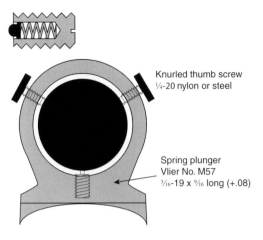

Knurled thumb screw
$1/4$-20 nylon or steel

Spring plunger
Vlier No. M57
$5/16$-19 x $9/16$ long (+.08)

Fig. 2.6.1

forces for this particular weight finder. A shallow groove around the finder tube would aid the plunger in keeping it in place. A catalog of various spring plungers is available from Vlier Enerpac, 2333 Valley Street, Burbank, CA 91505. Their phone is (818) 8431 – 922. The plungers are usually available from bearing supply stores. Individual units cost about $5, and you will find it well worth the money to improve the user-friendliness of your scope.

2.7 The Perfect Telescope Is...

The perfect telescope is a 10-inch *f*/6 Newtonian with a Dobsonian mount. The reason for this is the fact that in this one instrument, the factors of cost, aperture, portability, and optical performance come together in a way not shared by other designs.

2.7.1 Cost

It is well established that the Newtonian telescope provides maximum aperture for the dollars invested. This can be further optimized by using it in a Dobsonian configuration. My telescope was made to the specifications outlined by Richard Berry in his book *Build Your Own Telescope*. The paper tube cost a whopping $11.53, and the plywood components were picked up, pre-cut, from a lumber company for about $45.00. Considering that this instrument could be made for less than $500.00 (even with a full-thickness mirror), it cost only about 25% as much as a Schmidt-Cassegrain of the same aperture.

2.7.2 Aperture and Portability

Being one of those people who feels that the best telescope is the one that gets used the most, I have never fallen victim to aperture fever. I have seen the wonders of the universe through some very large telescopes. However, there is something about being able to set up in 30 seconds and view with both feet planted firmly on the ground, that is very comforting to me.

But why 10 inches? For one thing, a 10-inch instrument is the largest instrument that I would consider portable. The 10" $f/6$ fits easily into my van without having to remove any of the seats; whereas, a 12½-inch instrument would require special arrangements, even when chopped to $f/4.5$.

2.7.3   Aperture

We have all heard that money can't buy happiness. It is often true that aperture can't buy "good seeing". As an instrument increases in aperture, it must look through a larger column of air. If the atmosphere is unstable, instruments of larger aperture have a harder time reaching their performance potential. Of course, those who live in desert regions or on a mountaintop might take exception to this. However, for most of us it holds true.

For those who still think that potential magnification increases with aperture…well, you are only partly correct. Since our atmosphere rarely allows us to improve image quality by using magnifications greater than 600x, and since 60 power per inch of aperture is the limit for most observations, it stands to reason that a 10-inch instrument will allow one to see virtually all that can be seen …right??

2.7.4   Optical Performance

Since the Newtonian telescope has no "correcting" lens, such as in Schmidt and Maksutov designs, (and since mirrors reflect all wavelengths of light equally) the only chromatic aberrations to be encountered are those created by the eyepiece. In a low power, short focus instrument, they are seldom a bother.

What about coma? 'Thought you would never ask! Coma is a bothersome, but often overrated, aberration. It is true that short focus instruments which utilize paraboloidal mirrors suffer from coma. However, coma is by far more troublesome to the serious astrophotographer than the casual observer. Quite often, we see spot diagrams illustrating the coma present in the image produced by an $f/4$ or $f/4.5$ mirror, and we wonder how such instruments could be of any value at all. The fact is that for visual applications, coma is only about ⅓ as large as might be represented by computer generated spot diagrams. In order for spot diagrams to correctly illustrate the effects of coma (usually found interacting with astigmatism), a large number of rays need to be traced, and the observer must consider the intensity of the light as well as its linear measurement.

Finally, the Newtonian telescope, with its eyepiece on the side of the tube, needs no special baffling system. Stray light can't come barreling down onto the focal plane as can happen in Cassegrains, Schmidt-Cassegrains or Maksutovs.

Issue 3

3.1 The Stevick-Paul *Off-Axis* Reflecting Telescope
By David Stevick

The technical advantages of unobstructed aperture telescopes have long been appreciated, and the current popularity of apochromatic refractors is testimony to their efficacy. Unobstructed reflective and catadioptric telescopes are known[1], but have attracted only a small following. This is attributable to their comparatively large size, technical difficulties of fabrication, often slow speed, and peculiar aberrations that limit their useful fields of view. However, progress has been made in overcoming these obstacles, and the present design provides a wide-field telescope of virtual optical perfection, consisting of just a parabolic primary and two spherical corrector mirrors.

I wanted to build a faster, better, more compact, all-reflective, tilted-component telescope; wherein simple, rotationally-symmetrical components (in my case, mirrors) are aligned center to center along a single reflected axis. Although I had no prior experience in optical design, I wrote a computer program to optimize such a telescope and began my search.

Working by trial and error, I tried to refine a version of the Buchroeder Tri-Schiefspiegler, creating an instrument in which the secondary and tertiary mirrors would have equal radii. As the computed image improved, it tended to fall in front of the primary mirror. At first, I balked at using a fourth mirror to throw the image clear of the instrument, but changed my mind when I accidentally made a discovery that seemed too good to be true!

I found that if I placed the convex secondary at the focus of the primary, producing collimated light between it and the tertiary mirror, coma and astigmatism, both on-axis and in the field of view, vanished simultaneously when the separation equaled the radius of curvature of the corrector mirrors! Of course, spherical aberration also vanishes, leaving us with only image tilt, field curvature and distortion.

Exact spot diagrams show extraordinary image quality over any imaginable visual field, but image tilt sets a limit on how fast the telescope can be and still work well with an eyepiece. Making the image accessible required the use of a fourth, flat folding mirror.

---

[1] See References 1 through 4 on page 78.

Enthusiastically, I constructed a prototype and finding that it worked, I wrote up my discovery and sent it to workers in the field. They tempered my enthusiasm by kindly explaining that my discovery was not entirely unknown in the lens design community.

Maurice Paul[2] essentially derived the same design, by different means, as a corrector for large observatory telescopes. The Stevick-Paul is, to a first approximation, equivalent to an off-axis section of the Paul telescope, but modified by adjusting parameters slightly to optimize higher-order image quality around the principal ray which has now become the effective axis of our telescope.

James G. Baker[3], responding to a telephone inquiry, reported that he, likewise unaware of Paul's 1935 paper, had similarly re-derived the Paul solution and reported on various aspheric improvements in a 1969 issue of the *Proceedings of the IEEE*. Dr. Baker also mentioned that Robert E. Fischer had studied properties of the Baker and Paul designs for a master's thesis presented at University of Rochester.

As explained by R. A. Buchroeder, the Paul, and therefore the Stevick-Paul, can be understood most easily by regarding it as the joining of a Mersenne telescope which is free of spherical aberration, coma and astigmatism, but fails to focus, with the spherical mirror of a Schmidt camera, which focuses but has spherical aberration. When combined, we have a telescope that focuses, is free from coma and astigmatism, but has spherical aberration. Therefore, by changing the secondary mirror from a paraboloid to a sphere to eliminate the spherical aberration of the tertiary, we produce the Paul telescope.

Since the two spherical mirrors produce perfection only at the third order level, an error remains that can be balanced by a slight spacing or radius change. Image tilt is a manifestation of the principal ray being inclined to the parent focal curve; if the design made that curve concentric to the exit pupil, there would be no tilt in the Stevick-Paul. Distortion is normally very low in the Paul-type of design, but perceived distortion is due to interpreting the image on a plane normal to the principal ray rather than against a plane perpendicular to the Paul parent axis.

Similar designs for a four-mirror Schiefspiegler family investigated by Michael Brunn[4] were brought to my attention, but curiously Brunn's patent did not include the ideal design that constitutes the Stevick-Paul telescope, nor does he cite Paul as prior art.

3.1.1   How to Make a Stevick-Paul Telescope

The Stevick-Paul is a form of Tri-Schiefspiegler. Figure 3.1.1 illustrates the configuration. It begins with a paraboloidal primary mirror of the same *f*/ratio as that desired in the final instrument. The secondary and tertiary mirrors are spherical and of mating curve. The radius of this curve is 0.40 times the radius of curvature of the primary mirror.

---

[2] See Reference 5 on page 78.
[3] See Reference 6 on page 78
[4] See Reference 7 on page 78

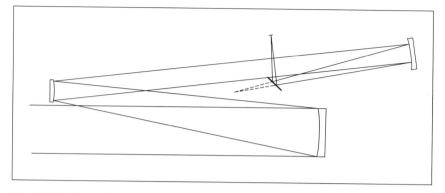

Fig. 3.1.1 *The layout of the Stevick-Paul telescope; consisting of a paraboloidal mirror of the same focal ratio as that desired in the completed instrument, two spherical mirrors (with mating curves), and an eliptical flat to bring the focus outside the instrument.*

The convex secondary mirror is placed just to the side of the light entering the telescope, leaving a little room for a light baffle. It is positioned afocally, so as to send parallel light on to the tertiary mirror.

The concave tertiary is positioned exactly twice as far to the side of the entering beam as was the convex secondary, and its own radius of curvature distant from the secondary. Because the tertiary mirror receives parallel light from the secondary, it forms an image at its focus, just as it would with starlight.

The focal plane lies within the system of mirrors, but is accessible to the eye with the inclusion of a flat diagonal. This fourth mirror does not obstruct light and corrects the reverted image that plagues telescopes having an odd number of reflect ions. It is angled to throw the image out of the telescope at 90° to the secondary-tertiary beam.

Computer modeling revealed that diffraction-limited performance is possible down to about f/5.6, but that focal plane tilt increases with decreasing f-ratios. Therefore, the tilt of the focal plane constrains the design to more moderate f-ratios.

I have chosen to restrict my designs to f/10 or longer. This way, image anamorphism is kept under 2% and focal plane tilt need not exceed 10°. While this system delivers an excellent image, eyepieces won't tolerate being tilted in such a steep light cone. Any tilting of the eyepiece will leave the top and bottom of the field out of focus. It is for this reason I recommend more moderate f/ratios be considered.

The prototype Stevick-Paul was a 10-inch f/6.2 design. This size and focal ratio was chosen because I already had the paraboloidal primary mirror on hand, and I wanted to test the system quickly. Thus, only two spherical mirrors had to be made.

I employ a value for the R.O.C. (radius of curvature) of the spherical mirrors that is 0.40 times that of the primary R.O.C. This 'radius ratio' has been found to

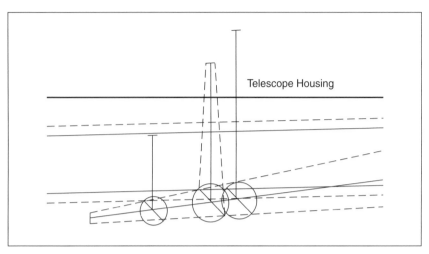

Fig. 3.1.2 *The designer may position the diagonal mirror so as to make the desired compromise be-tween a small amount of vignetting and focal plane accessibility. The circles are to illustrate the rela-tive sizes of the diagonal mirror for three possible positions near the focal plane.*

be optimum, but other families of telescopes are possible with different radius ra-tios. Telescopes with radius ratios less than 0.40 are more compact, (at least down to 0.33 where increasing the primary to secondary spacing lengthens the instru-ment again), they have a smaller primary tilt and require smaller spherical mirrors. The trade-off here is that they suffer from having a more steeply tilted focal plane. Telescopes with radius ratios greater than 0.40 are longer. They have a larger pri-mary tilt and larger diameters for the spherical mirrors. They do, however, enjoy a less steeply tilted focal plane. By a stroke of good fortune, the eyepiece falls quite close to the balance point in an instrument designed with the preferred ratio of 0.4. It is in a convenient position for viewing, and its arc of travel is short.

You may have noticed that there are elements of the Schmidt camera in-volved in this telescope. The secondary mirror acts like a corrector for the tertiary suggesting that the aperture stop should be located there. This, however, is not the case. Leaving the system stop at the primary preserves maximum aperture.

Because of the Schmidt-like construction, the tertiary mirror must be larger than the secondary by twice the clear field. Required minimum diameters of the spherical mirrors are determined with these equations:

$$\text{Secondary diam.} = A \times D + (1 - A) \times d \qquad \textbf{(3.1.1)}$$

where: A = radius ratio.

$$\text{Tertiary diam.} = A \times D + (3 - A) \times d; \qquad \textbf{(3.1.2)}$$

where D = primary diameter and d = diam. of clear field, and A = radius ratio.

Mirror tilts have a fixed relationship among themselves. The primary tilt de-termines all the others. In a Stevick-Paul the paraboloidal primary mirror must have a tilt a bit larger than that required for freedom from obstruction, to allow for an effective light baffle. The equations are as follows:

$$\text{Primary tilt: } T_1 = \frac{C}{N}; \qquad (3.1.3)$$

where C is selected from the following table, and N = primary f/ratio.

Mirror diam.	C
4.25"	37.08
6"	36.24
8"	35.67
10"	35.29
12.5"	34.97
16"	34.67

$$\text{Secondary tilt: } T_2 = -k \times T_1 \qquad (3.1.4)$$

where $k = \dfrac{1 + A}{2 \times A}$ and A = radius ratio

$$\text{Tertiary tilt: } T_3 = T_1 + T_2 \qquad (3.1.5)$$

$$\text{Focal plane tilt is approximated as } -4 \times T_3. \qquad (3.1.6)$$

The last three equations are valid for all possible values of the radius ratio. The numbers for the primary tilt are empirically derived and are valid for a ratio of 0.40 only. Positive values represent a clockwise tilt; negative values are for counter-clockwise tilts. In a Stevick-Paul, the primary mirror and focal plane have positive tilts while tilts for the spherical mirrors are negative. The positive tilt of the focal plane reverses after reflection from the flat so that in a finished instrument, this tilt is counter-clockwise also. The spaces between the surfaces can be found simply as:

$$\text{Primary-secondary e.f.l.: } e_1 = F_1 + F_2 \qquad (3.1.7)$$

where F_1 = primary f.l. and F_2 = secondary f.l.

$$\text{Secondary-tertiary e.f.l: } e_2 = R_3 \qquad (3.1.8)$$

where R_3 = tertiary radius of curvature.

$$\text{Tertiary-focal plane: } e_3 = F_3 \qquad (3.1.9)$$

where F_3 = tertiary f.l. Note: F_2 is negative.

When this telescope was first diagrammed, it was obvious that two problems had to be solved: getting the focal plane free, and baffling against stray light—all while minimizing tilt angles. Let's look first at the problem of freeing up the focal plane.

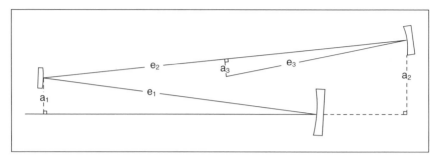

Fig. 3.1.3 *Design variables for the Stevick-Paul Telescope.*

Assigning the minimum tilts necessary to prevent obstruction of incoming light by the secondary mirror, here is what we find. The diagonal mirror is positioned in the cone of light converging to the focus (Figure 3.1.2). If it is placed to catch all the light for a 1-inch diameter field, and at the same time, kept from infringing on any of the field rays between the secondary and tertiary mirrors (upper position), not enough residual cone will exist to get the focal plane free. The only solution to this dilemma seems to be an increase in the tilts of all the mirrors. This would lengthen the available cone of light but would increase the focal plane tilt dramatically.

The solution I reached nestles the diagonal mirror up against the central beam of light passing between the secondary and tertiary mirrors (middle position). Therefore, the diagonal intrudes into the light at the lower edge of the field, but is out of the way at the center and the upper part of the field. The diagonal mirror is then slid along the central beam toward the tertiary mirror (lowest position), until it catches only enough of the rays forming the top edge of the field to produce a ½ magnitude light loss due to vignetting. These two compromises, acceptable intrusion at the lower part of the field and acceptable vignetting of the top of the field, have kept tilt angles down while providing a sufficient amount of light cone to allow the insertion of a focuser.

To allow for vignetting and sufficient free cone, I design for the following clear field diameters:

Mirror diam.	Clear field
4.25"	0.28"
6.00"	0.45"
8.00"	0.62"
10.00" and up	1.00"

In this design, free cone is more important than residual cone. The length of cone which can be fielded by the diagonal mirror is the residual cone. I define free cone as that part of the residual cone which exists on the observer's side of the parallel secondary-tertiary beam. I look for at least 4 inches. A low-profile focuser would be required if less free cone were available.

The second problem, that of the light baffle, was solved in the following

Fig. 3.1.4 *Cutaway of the Stevick-Paul Telescope.*

manner. The baffle surrounds the converging cone of light from the primary mirror, fitting into the notches made where the several light paths cross. Again, trying to keep the baffle out of the road of all field rays results in high tilt angles. Following the philosophy of permitting a small loss in illumination at the edge of the field, the baffle was sized to just pass the central beam intact. This means that because the baffle is so far from the focus, it blocks less than 10% of the light at the field edge. Vignetting arising from the placement of the diagonal mirror is far greater than that from the baffle.

With the baffle thus sized, the required primary tilt is not much larger than the theoretical minimum. By ray-tracing the light that just makes it through the baffle, I look to see that it approaches no closer than 1-inch to the field center. This means a field 2 inches in diameter is free of stray light. Even 2-inch O.D. eyepieces would thus be protected from washout. The table of primary tilts reflects this design philosophy.

3.1.2   Summary

Let me emphasize that you do not need a computer to design a Stevick-Paul. Don't worry about getting tilt angles right to three decimal places. The equations are here only for other designers to check my work. The tables for primary tilt angles and clear field are all you need. Work only with spacings and axis distances. The full-scale drawing must be the last word on the design anyway, as it is the pattern against which the finished telescope is assembled.

This article conveys the knowledge gained from a year and a half of intense labor. I have believed from the very beginning that there is something inherently basic about a Stevick-Paul. A telescope with this optical perfection and geometric simplicity must surely be an expression of some fundamental laws. I like to think that I discovered it, rather than designed it. If you are serious about building one of these telescopes and have questions, please write. My profound thanks to Don Caron who has meticulously read the many versions of this paper.

> **Editors Note:** This article is dedicated to those ATM's who feel that a person must have years of experience (and untold knowledge), to truly make a contribution to our hobby, and who have looked at articles on the more advanced aspects of telescope making as being for some supposed elitist group.
>
> Mr. Stevick has had no formal training in optics. However, his keen interest in telescopes has spurred him on to acquire the knowledge needed to go forth with a hammer, saw and a few calculations to create what appears to be (at least on paper and in the computer) a world class telescope. The Associate Editors of *ATMJ* are all experts in optics and people who, because they have "seen it all before," rarely get excited about anything "new." However, Diane, John and Dick were all excited about this one. Dr. Buchroeder had mentioned to me that he thought this article would go well without providing the customary spot diagrams. I questioned his reasoning, and the following was his reply:
>
> "...I see no point in using any of his spot diagrams; they take up space and prove nothing—The spot diagrams for his particular design are so good that they reveal nothing. The words in the text are sufficient."
>
> As of this date, the instrument has yet to be finished. However, *ATM Journal* will be following this design closely, and we will report our findings when the instrument draws "first light."

3.1.3 References

1. Buchroeder, R. A. "A New Three-Mirror Off-Axis Amateur Telescope," *Sky & Telescope*, Vol. 38, No. 6, December 1969, pp. 418–423.

2. Buchroeder, R. A. "Technical Report Number 68," May 1971, Optical Sciences Center, University of Arizona.

3. Cox, R. E. "Telescopes with Unobstructed Light Paths", *Sky & Telescope*, Vol. 43, No. 2, February 1972, pp. 117–120.

4. Kutter, A. "A New Three-Mirror Unobstructed Reflector," *Sky & Telescope*, Vol. 49, No. 1, January 1975, pp. 46–49.

5. Paul, M. "Systemes correcteurs pour reflecteurs astronomiques," *Revue d'optique*, Vol. 14, No. 5, May 1935, pp. 169-202.

6. Baker, J. G. "On Improving the Effectiveness of Large Telescopes," *IEEE Transactions*, Vol. AES-5, No. 2, March 1969, pp. 261–272.

7. Brunn, M. "Unobstructed All-Reflecting Telescopes of the Schiefspiegler Type," U. S. Patent 5,142,417, Aug. 25, 1992.

8. Schmidt, R. A. "Constructing a 10" Kutter Schiefspiegler," *Telescope Making* #4, Summer 1979, pp. 8–13.

9. Cox, R. E. "Better Mirror Cells," *Telescope Making* #6, Winter 1979/80, pp. 32–39.

10. Cox, R. E. "The Kutter Tri-Schiefspiegler," *Telescope Making* #16, Summer 1982, p. 10.

11. Woods, A. L. "How to Build a Tube for the Tri-Schiefspiegler," *Telescope Making* #16, Summer 1982, pp. 11–17.

12. Woods, A. L. "Collimating the Kutter Tri-Schiefspiegler," *Telescope Making* #16, Summer 1982, pp. 18–21.

13. Wessling, R. J. "Building a 12.5" Buchroeder Tri-Schiefspiegler," *Telescope Making* #28, Fall 1986, pp. 32–43.

14. Johnston, S. W. "Construction of a Second 12.5" Tri-Schiefspiegler," *Telescope Making* #28, Fall 1986, pp. 44–51.

15. Sasián, J. M. "Variations on the Schiefspiegler," *Telescope Making* #43, Winter 1990/91, pp. 18–25.

16. Sasián, J. M. "The World of Unobstructed Reflecting Telescopes," *ATM Journal*, Fall 1992, pp. 10-15.

3.2 A New Equatorial Platform Design
By Donald W. Davies, PhD.

I will describe what I believe to be a new design for an equatorial table for Dobsonian telescopes. I needed a table for the 17.5-inch instrument that I am building, but the center of mass of my system seemed too high for the standard equatorial platform designs I had seen, and required rather complex (by my standards) curved bearing surfaces. I decided to see if a table using only plane surfaces and "hardware store parts" could be designed and made so that the system was neutrally stable (wouldn't tip over or rotate back to vertical if not held—i.e., rotate about the center of mass). To my surprise, a design was found that met all of my requirements.

The heart of the system is four non-swiveling casters running on four pieces of aluminum "angle iron." By proper choice of the spacing and angles of these four planes, the mount moves in such a way as to rotate around an axis parallel to that of the earth.

3.2.1 How it Works

The first key to finding a solution was to allow the telescope to "translate" (i.e., shift position) as it rotates. If the motion of the telescope mount is such that the center of mass of the system translates horizontally as the rotation occurs, then the system will be neutrally stable.

The second step was to find a geometry that provided a horizontal translation of the center of mass as the table rotated. Figure 3.2.1 shows the design geometry that does this. If a platform is constructed so as to have two points in contact with the outside of an isosceles triangle, the point on the platform corresponding to the apex of the triangle, when the platform is horizontal as shown, will translate exactly horizontally! This is not at all obvious at first glance, and was determined

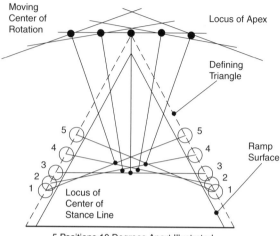

Moving
Center of
Rotation

Locus of Apex

Defining
Triangle

Ramp
Surface

5
5
4
4
3
3
2
2
1
1

Locus of
Center of
Stance Line

5 Positions 10 Degrees Apart Illustrated

Fig. 3.2.1 *Basic geometry of the north and south triangles. The equatorial axis translates horizontally as it rotates. The "stance" is the distance between the centers of the casters when the table is horizontal.*

empirically at first from fairly elaborate computer modeling which calculated the motion of various points on the platform. Once it was apparent what was occurring, an analytic proof was developed.

Since the rotation of the platform has to occur around an axis pointed at the north celestial pole and is not horizontal, the triangle has to be tipped (by an amount equal to the observer's latitude) relative to the vertical. There actually need to be two triangles, one at the north end of the mount and one at the south. Since the polar axis is not horizontal, these triangles must have different heights. Figure 3.2.2 shows a side view of the system, looking to the west.

There is one additional constraint that could have foiled the whole plan. Once the north-south separation of the two triangles is fixed, the heights of the two triangles are not adjustable. The "stance" (the distance in an east-west direction between the caster centers) of the two triangles is adjustable. The constraint that must be met is that as the mount rotates, the horizontal translation at the north and south ends has to match. If it does not match, the polar axis will twist. This problem could have had no solution, or the solution could have involved unreasonably large stances or near-vertical planes. Fortunately, this is not the case. Figure 3.2.3(a) displays the north triangle angle and stance for my telescope. I decided that angles up to 60° on the north triangle could be tolerated (the south one is always shallower in the northern hemisphere). That results in a south stance of 95cm, a north stance of 82cm and a south angle of 41°. Other configurations with lower centers of mass have solutions that involve shallower angles. Figure 3.2.3(b) shows the relationship between N–S distance and the height of the center of mass above the base of the triangle (the line through the caster centers). For this example the angles of the north and south triangles are fixed at 60° and 45°, respectively.

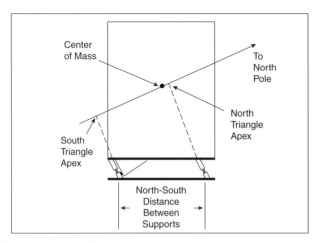

Fig. 3.2.2 *Side view of mount looking west with the bottom section of the telescope resting on it. The line between the apexes of the two triangles passes through the center of mass of the combination of the telescope and the moving part of the equatorial table.*

The appendix gives the analytic formulas to calculate the angles of the plane surfaces and the listing of a BASIC program to perform the calculation.

3.2.2   Test Mount

Since I really did need an equatorial mount, and I had an uneasy feeling about some hidden "gotcha" in the design, I built a working model from ¾-inch particle board and 2 x 4's. The two triangle sections consist of 1.5-inch wide extruded aluminum angle iron (L-shaped cross section) attached to pieces of wood to hold them at the correct angle. Casters with 2-inch diameter plastic wheels run along these surfaces. Since these surfaces are tipped to the south, the casters would roll off the tracks if not prevented. To counter this thrust, there is a Teflon pad riding on the other leg of the "L" on the south triangle. This also removes the side load from the casters. Teflon was used in place of another caster because the point of contact describes a shallow arc instead of a straight line.

The mount was "motorized" by running a jackscrew between the upper and lower plates (the surfaces with the casters and the triangle bases, respectively). Both ends of the jackscrew are pivoted since the angle between the screw and the two surfaces changes as the mount tips. A geared-down 6-volt DC motor was used to turn the screw. (Similar to Edmunds part #Y31, 827).

When I placed it on the sidewalk in front of my house and put my old 6-inch Newtonian on it, I was able to keep Saturn at 120x in view for about 20 minutes by simply pulsing the DC motor when it drifted to the edge of the field of view. The mount worked!

I finally received a 17½" mirror for which I had waited 18 months, and, after constructing my telescope, I built a new equatorial platform for it. The aluminum tracks are now held at the correct angle by casting them into a pyramid of

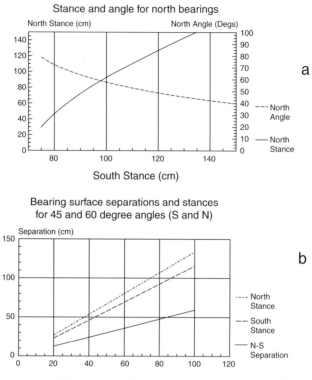

Fig. 3.2.3 *Relationships between spacings and sizes of the north and south triangles as calculated for a latitude of 34°. (a) gives the size and angle for the north triangle if the south one is fixed. (b) shows the north-south separation of the two triangles if the angles are fixed at 45° south and 60° north.*

"Quick-Fix All-Purpose Patching Compound"—Downman's FixAll would probably work just as well. This allowed a lighter and lower-profile mount. Figure 3.2.4 shows the two halves of the mount before the jackscrew was installed.

The new mount provides about 30 minutes of descent tracking with the jackscrew driven by a computer-controlled stepper motor. I have used it in conjunction with a home-made spectrometer; and with a simple image stabilizer near the focus, I can walk away from the telescope for several minutes, and the system will stay locked onto a star.

I would like to thank Andy Saulietis for a constructive review of the paper and for allowing me to use his much clearer version of Figure 3.2.1.

3.2.3 Appendix

Equations and computer program
Definitions:

    a   angle of the triangle (90° is vertical)

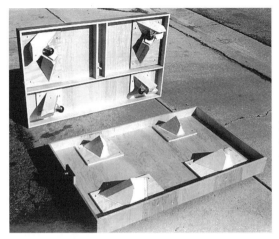

Fig. 3.2.4 *Two halves of the mount before installation of the jackscrew.*

Fig. 3.2.5 *Completed mount with the 17½" telescope in place.*

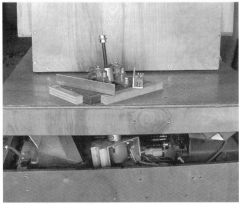

Fig. 3.2.6 *Closeup of jackscrew on west side of telescope.*

h   height of triangle (base to vertex)

s   stance (base of triangle, caster center-to-center)

y   north-south distance between triangles

xcm east-west position of the center of mass

zcM vertical distance of the center of mass from base of triangle (not height of triangle)

t   theta, the rotation angle around the polar axis

e   epsilon, the latitude of the observing site

An appended subscript "s" or "n" to the above definitions refers to the north or south triangles.

The relationship between the north-south distance between triangles, the latitude and the height of the center-of-mass is easy to determine from Figure 3.2.2 on page 81:

$$h_s = zcm_n - y * \frac{tangent\ (e)}{2}$$

$$h_n = zcm_s - y * \frac{tangent\ (e)}{2}$$

The constraint that the translation be the same at the two ends of the polar axis amounts to setting the derivative,

$$A\frac{d(xcm)}{d(t)} = \frac{h * \cos(t)}{\sin^2(a)} = \frac{h}{\sin^2(a)}$$

equal for the north and south ends at theta=0.

After a few lines of messy algebra, the equation for the stance of the north triangle in terms of the other variables is

$$S_n^2 = S_n^2 * \frac{h_n}{h_s} + 4*h_n*(h_s - h_n).$$

Following is a listing of a simple BASIC program that will calculate the stance of the north triangle for other geometries, or give the north-south separation for fixed angles for the north and south triangles.

```
pi = 3.14159
GOTO ycalc
     This section calculates north stance if other variables
          are given
     zcm = 68.6     height of center of mass above bearings
     y = 55.9       N-S distance between supports
     ss = 95        stance of south bearings
     epsdegrees = 34latitude in degrees
     eps = epsdegrees * pi / 180
yloop:
```

```
    INPUT "zcm,y,south stance"; zcm, y, ss
         calculate N and S slant distances
    hs = (zcm - y * TAN(eps)/2)* COS(eps)
    hn = (zcm + y * TAN(eps)/2)* COS(eps)
         calculate north end stance
    snsq = ss * ss*((zcm + y * TAN(eps)/2)/(zcm - y*TAN(eps)/2))
    snsq = snsq - 4 *(zcm + y * TAN(eps)/2) * y * TAN(eps)
         *((COS (eps))^2)
    sn = SQR(snsq)
angles of north and south triangles
    tanalphan = 2 * hn/sn
    tanalphas = 2 * hs/ss
    alphan = ATN(tanalphan)
    alphas = ATN(tanalphas)
    alphandegrees = alphan * 180/pi
    alphasdegrees = alphas * 180/pi
print out results
    PRINT zcm, y, ss
    PRINT hs; hn, ss; sn, alphasdegrees; alphandegrees
    PRINT "angle difference="; alphandegrees - alphasdegrees
    LPRINT zcm, y, ss
    LPRINT hs; hn, ss; sn, alphasdegrees; alphandegrees
    GOTO yloop
ycalc:
    This section calculates N-S separation of planes vs
         center-of-mass height if angles are given
    eps = 34 * pi/180
    alphas = 45 * pi/180 : alphan=60 * pi/180
    LPRINT "angle difference="; alphandegrees - alphasdegrees
    LPRINT
    LPRINT "zcm vs y-spacing for 45° and 60° wedges"
    sn = SQR(snsq)
         angles of north and south triangles
    TANALPHAN = 2 * hn/sn
    TANALPHAS = 2 * hs/ss
    alphan = ATN(tanalphan)
    alphas = ATN(tanalphas)
    alphandegrees = alphan * 180/pi
    alphasdegrees = alphas * 180/pi
         print out results
    PRINT zcm, y, s
    PRINT hs; hn, ss; sn, alphasdegrees; alphandegrees
    PRINT "angle difference="; alphandegrees - alphasdegrees
    LPRINT zcm, y, s
    LPRINT hs; hn, ss; sn, alphasdegrees; alphandegrees
    LPRINT "angle difference="; alphandegrees - alphasdegrees
    GOTO yloop
END
```

3.3   The 1993 Texas Star Party and Riverside Conference

By Dean Ketelsen

I consider myself among the fortunate in astronomy. Living in Southern Arizona, on any given night of the year I can drive for a half hour and be under some of the best skies anywhere. Also, come the month of May Tucson is midway between the Texas Star Party and the Riverside Telescope Maker's Conference, favoring the California event by about 30 miles. Why drive 500 miles, you ask, when you have great skies on your doorstep? As almost everyone will tell you, a big draw is the reunion with old friends, the making of new ones and the camaraderie of observing with them through the latest behemoths on the telescope field. There are also the famous and not so famous that the organizers bring in to share their work. As for me, there are the new ideas in telescope making that have hatched into working models of innovation. As an optics professional, I am nearly always amazed at some of the ideas that find their way into telescope projects.

As TSP approached, a month of perfect weather was interrupted by an eruption of clouds and moisture over the southwest and northern Mexico. Great, I thought. Get it out of the sky, and it will be clear all week for TSP. Bob Goff, another optician from Tucson, soliciting for Hextek (makers of lightweight mirrors) on this trip, accompanied me at dawn Sunday as we headed east. At the halfway point, near El Paso, it was obvious we would be playing tag with clouds as the Rio Grande marked the cloud boundary—the Mexico side remaining clear.

We arrived at the Prude Ranch, a few miles from McDonald Observatory, in time to catch dinner. We then rushed to set up camp while being distracted by the occasional acquaintance stopping by to say "hello." Our sleeping quarters in place, there was time for a quick tour of the telescope field before dark. There were some impressive scopes there, capped by a 33-inch and a 30-inch, among a sea of SCTs and big Dobs. It stayed cloudy, and for a travel day, I was actually relieved. With a slight drizzle, I turned in early, guilt-free.

Monday dawned clear, some witnesses saying observing had started at 3 a.m. I experienced a pang of guilt, but well-rested and with blue-sky overhead, there was always tonight. With no activities scheduled until Wednesday, the first few days were all my own. Without having a week off for a year, I planned to relax with a capital "R." My bicycle even made the trip for some leisurely rides through the area. Local roads are great with wide roadside lanes. I encourage more riders to pack their bikes along with the scopes for this event.

Thus, one settles into life at TSP. It goes something like this: Sleep late, rise and shower in time for lunch, visit with friends, stop by the vendors area, afternoon talks, dinner, visit some more and catch the evening program, review the telescope field before dark, observe, night lunch with Chip Prude's great burritos, observe some more, and fall into bed at the start of twilight.

On cloudy nights there is a slight schedule change—observing is shelved in favor of "laser wars!" Started two years ago, to have something to do on cloudy nights, the Ft. Bend and Houston groups dress up in mylar suits and welder's goggles and chase each other around the telescope field with high intensity lights. Barbara Wilson was armed with the dreaded "Stealth Nagler", a 2-ft. long array of flash tubes and batteries, conveniently fitting into a 2-inch focuser to attach to her killer Dob and attack; see Figure 3.3.1.

My personal project at this TSP was to try out a new lens for all-sky photography. I had rented a 35-mm fisheye lens for the Pentax 67 camera, to evaluate its performance. If exposed on 4 x 5 film, it would provide a 3.3-inch image circle— recording the entire sky. This has been tried before with a 16mm fisheye on 120 film, and the longer-focus fisheye needed evaluation. So, on every clear night, I would set up my CamTrak and take a series of exposures at various *f*/stops to check the images. Since you can't guide a CamTrak, I had lots of time to survey scopes on the field.

The statistics on telescopes were startling. On one evening stroll through the upper football field, I counted over 20 scopes over 17.5". There was literally a sea of telescopes. One of the organizers said there were over 350 instruments present, and about two observers for every instrument. But, the surprising thing was that there was very little crossover observing. Even on the 33-inch telescope, the longest line ahead of me was 3 people!

Though TSP is truly an observationally oriented event, I was somewhat disappointed in that the large majority of the instruments were of commercial manufacture. There is nothing wrong with that or with those who use them, but you get a lot of innovation when people have to use the materials and tools at hand to solve construction problems. A case in point is the Crayford focuser. Fifteen years ago everyone had a rack and pinion focus unit. Now, the majority use a derivation of this very straightforward idea.

Mike Benz and Dan Bakken made the trip down from Spokane, Washington with "the Beast", the 33-inch *f*/6 telescope documented in the September, 1992 *Sky and Telescope* (Figure 3.3.2). They gave an afternoon presentation on its manufacture. The long-focus Newtonian provided some great views—from the top of its 16-ft. stepladder, if not through the telescope! All kidding aside, being able to simultaneously view 20 galaxies in the Hercules cluster demonstrated that there is no substitute for aperture. Mike and Dan did all the optical work themselves, and pushing a 20-inch diameter lap is no easy feat.

Steve Watkins, of Houston, brought a 10-inch *f*/8 entirely of his own making; see Figure 3.3.3. The optics were exquisite and, when we had great viewing on Tuesday night, it provided absolutely the best view of Jupiter I experienced. Steve incorporated a number of new ideas throughout the mechanics. He used a tangent arm for tracking. Now, almost everyone knows that a tangent arm will not track uniformly over its full range. However, Steve has a small computer-controlled stepper motor, to allow for accurate tracking. Collimation is a breeze, as he has geared the adjustment via chains to the outside of his tube where it is carried up near the DEC axis with shafts that are readily reached. For transportation, the

Fig. 3.3.1 *Barbara Wilson and friends during "Laser Wars." With a Stealth-Nagler riding high over the eye-piece, this puppy can wipe out 8 to 10 enemies per nano-second.*

Fig. 3.3.2 *Mike Benz of Spokane, stands before his (and Dan Bakken's) 33-inch f/6 "the Beast."*

Fig. 3.3.3 *Steve Watkins with his 10-inch f/8.*

Fig. 3.3.4 *Steve Watkins' telescope folded up for loading into his trailer.*

north RA bearings fold down, the struts are removed, and the entire scope collapses into a wheelbarrow-type unit, easily rolled to his trailer or garage.

    Ed Szczepanski built a 10-inch *f*/4.5 binocular that gets around a problem which often afflicts similar arrangements; see Figures 3.3.5 and 3.3.6. When changing interpupillary distance (IPD), the collimation usually is disturbed, and since the operator cannot see with the new IPD, the view suffers. Ed mounted the tubes far enough apart that the converging beams come toward each other. The IPD can be changed by adjusting one or both of the Crayford focusers. The colli-

Fig. 3.3.6 *Illustrating the IPD adjustments for Ed Szczepanski's binocular.*

Fig. 3.3.5 *Ed Szczepanski (white t-shirt) with his 10-inch f/5 binocular.*

mation is unaffected, and the new user merely re-focuses using the helical focuser coming off the tertiary mirror.

The 1993 TSP week ended with a total of three memorable nights of observing. We witnessed a spectacular storm, during which it rained 2 inches in 15 minutes. More pleasantly entertaining were an imaginative presentation by Tony and Daphne Hallas, and an amazing glimpse of the future of astronomical imaging by Don Parker. Over 800 amateurs had pre-registered, and 750 showed up making this the largest Texas Star Party ever! Next year's event occurs during the week of the May '94 annular solar eclipse, for which side trips to the center line are being planned.

3.3.1   On to Riverside

Four days later my wife Vicki and I, along with four friends from Tucson, were heading west into the sunset for an encounter with the Riverside Telescope Maker's Conference (RTMC). After an overnight stay in Indio, we pulled into Camp Oakes at 12:30, just in time for the annual vendor feeding frenzy. By the time we parked the van, slapped on some sunscreen, and headed up to the telescope field, it was obvious that no one was following the "no sales until after 1 pm" rule. Yet, there were still many bargains to be had in the Celestron and Meade "seconds" offerings.

By the time we arrived a good many telescopes had already been set up. As at TSP, the Dobsonians ruled. And, of course, a new aperture record was reached for RTMC: a 40-inch constructed from scratch by the Valley of the Moon Astronomical Society. A stroll down telescope alley brought every type of optical aid into view—binoculars (large and small), refractors (from the new apochromats to the 116 year old Clark), and a wide variety of reflectors and mountings.

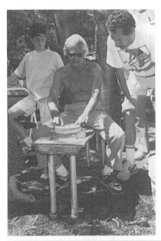

Fig. 3.3.7 *(Left) John Dobson gives instructions on mirror making.*

Fig. 3.3.8 *(Right) Gerry Logan's 12-inch f/15 Maksutov.*

Fig. 3.3.9 *(Left) Jim Hannum with his 24-inch f/4 Newtonian.*

Fig. 3.3.10 *(Right) Randy Steiner with his electronically driven 8-inch f/4.5 Newtonian.*

Fig. 3.3.12 *Dean Ketelsen prepares to look down the bore of Steven Overholt's 30-inch f/3.75.*

Fig. 3.3.11 *Andy Meyer stands beside his 3.5-inch f/33 Herschelian.*

There was nothing I couldn't live without from the vendors, so we returned to set up camp. By twilight, and a couple more trips through the field and telescope alley, a mental checklist was made of which scopes should be visited later that night. Rich Livitsky had his 20-inch binocular there again. It had been a highlight of last year's RTMC, where viewing had been curtailed by clouds. Gerry Logan brought an excellent 6-inch Schupmann the last couple of years, but this time had a 12-inch *f*/15 Maksutov. I had never observed through a Mak that large and looked forward to it.

Finally, with darkness upon us, the observing started in earnest. Unlike at TSP, the line for each telescope was very long. With a quarter Moon high in the sky, it was obvious that serious deep-sky observing would not start anytime soon, and it was easy to tell that the scopes were directed at the Moon, Jupiter, or M13. I immediately came upon my personal pet peeve this year—magnification that's too low. I was examining Jupiter through a 24-inch instrument and as I racked the focus in and out a bit to check the images, I found I had to move my eye around to see the entire image. Backing my head up a bit, there was at least a 9 or 10mm exit pupil! It was not an isolated incident. Dobson defiantly stated, in a talk on the weekend, that one needs at least 12x per inch of aperture for maximum detail, but there were lots of 20- to 30-inch scopes using less than 100 power. True, the seeing was average at best, but there were many times when I wished for much higher power, but did not want to ask with 20 people behind me.

Gerry Logan's 12-inch Maksutov provided some wonderful images; see Figure 3.3.8. To help the system come to thermal equilibrium, he installed small fans to pull air through holes cut in the top and bottom of the normally sealed tube. He claimed that 30 minutes of that treatment ends the problem usually affecting large Maksutovs. The tube assembly was of wooden composite construction—¼-inch thick wooden strips fiberglassed with Kevlar inside and out, and a leatherette exterior finish. All cells used push-pull adjustments for stable collimation.

The mechanical marvel of the conference was the 24-inch *f*/4 telescope and mount made by Jim Hannum which was based on the Schaeffer design; see Figure 3.3.9. Jim did all the machining on the equatorial mounting, as well as the foam core fiberglass tube. The scope uses commercial optics, but those are about the only parts he did not make. Weighing in at 660 lbs., it takes three people to assemble.

Randy Steiner won an award for "first telescope project," with an instrument he designed for the handicapped and for kids and school groups; see Figure 3.3.10. Confined to a wheelchair himself, eyepiece height was very important, as well as ease of assembly and breakdown. The mount was designed around a Coulter 8-inch *f*/4.5. He has about $400 invested in the mount and drive electronics which utilize push-button controls.

One of the more unusual telescopes was that of Andy Meyer (Figure 3.3.11). The 3.5-inch primary was the central core left over from the Tucson club's 30-inch project. Polished on all surfaces and sold as a fund raising item, Andy got it and vowed to build a Herschelian telescope around the mirror. It retains the focal length of 30 inches, and at $f/33$, is tilted a few degrees to the incoming beam to provide an unobstructed view. A diagonal is utilized for ease of viewing, and a tube is used to block background light. Views of the Moon and Jupiter were very sharp and of high contrast.

Steven Overholt's 30-inch $f/3.75$ was a distinctive telescope (Figure 3.3.12). It seemed to be almost all tube. Steve stretched the limit in providing ideas for large ultra-lightweight instruments. He admitted that the key to the success of his 30-inch is in minimizing the weight of the top end. The top-rotating component of his scope weighed only 7 lbs with a thin secondary and small baffles to reduce stray light. By minimizing the telescope's cross-section to the wind, steadier images result. Most of the scope is constructed of $\frac{1}{8}$-inch ash doorskin glued to closed-cell Styrofoam.

John Dobson provided an impressive demonstration on Saturday. Sponsored by the Pomona Valley Astronomy Association, he created a 10-inch telescope from a precurved blank and pile of wood; see Figure 3.3.7. Starting at 11 a.m., with a crowd around him (volunteers were put to work grinding and polishing), the telescope was finished shortly after dinner. Even with an uncoated mirror, it provided good views of the moon before the evening program, for which it was the prize for a PVAA raffle.

Richard Combs provided a similar display this year. As an outgrowth of meetings held during the last couple of RTMCs, Richard showed how to grind and polish your own primary in a weekend-long demonstration. Using a Ronchi tester, those interested were instructed in its use and interpretation of the bands. Installed in a finished telescope on Sunday night, observations of the Moon and, when seeing permitted, diffraction rings around bright stars were observable inside and outside of focus.

In one of the more interesting talks of recent years, Peter Ceravolo presented plans for the construction of an interferometer. Using a laser beamsplitter, GRIN (gradient index) lens and reference lens, the entire tester can be built for $500, or less if the amateur already has some parts or can make his or her own reference element. Effectively of the Fizeau type, the wavefront reflecting back from the reference element is compared to the return beam from the surface under test. The resulting interference pattern can be viewed or photographed coming off the beamsplitter. For those unfamiliar with fringe analysis, Doug George has written an analysis program for Peter to aid in interpretation. The combination of the two provides a powerful tool available to all optical manufacturers for much less than the tens of thousands of dollars for other systems. He is making a booklet of plans and helpful hints available for $20 from his new company, Ceravolo Optical Systems, P.O. Box 1492, Ogdenburg, NY 13669.

RTMC has always been the Mecca of telescope makers, and this year was no exception. The viewing was limited somewhat by the waxing moon, but not by the weather. The crowds seemed smaller than usual, but the food was better. There were 38 entries in the judging and 20 prizes were awarded—a good indication that a lot of innovation was evident. In Sunday's round table discussion on the future

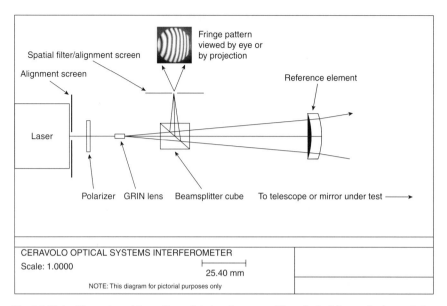

Fringe pattern viewed by eye or by projection

Spatial filter/alignment screen

Alignment screen

Reference element

Laser

Polarizer   GRIN lens   Beamsplitter cube   To telescope or mirror under test ──▶

CERAVOLO OPTICAL SYSTEMS INTERFEROMETER
Scale: 1.0000

25.40 mm

NOTE: This diagram for pictorial purposes only

Fig. 3.3.13 *An illustration of Peter Ceravolo's interferometer. Plans for building and using such devices may be obtained by sending $20 to: Ceravolo Optical Systems, P.O.Box 1492, Ogdenburg, NY 13669.*

of telescope making, there seemed to be no clear consensus on direction. Some thought size would continue to grow unabated; others thought size would be limited by coating facilities. Some thought mounting technology and surface accuracy would improve driven by the constraints of CCD technology. Well, we will see the future when we get there—see you there next year!

> **Ed. Note:** Perhaps the most important event at Riverside this year was the announcement of Peter Ceravolo's setup for performing interferometric testing. For years, I have labored over the thought of trying to build an optical testing device which would be more reliable than my old Foucault tester. However, each instrument I have considered has been too time consuming to build, expensive or required machine tools that I did not have at my disposal.
>
> If Peter's testing arrangement and software performs as expected, it will give amateur telescope makers a tool for testing optics more accurately than ever before.

3.4   The Eye and the Use of Telescope Optics, Part I
By Dr. Richard A. Buchroeder

This abridged and edited interview asks an opthamologist questions about the eye pertaining to the design and use of telescopic optics. Dr. Thall, M.D., F.O.O. already a distinguished practitioner, is presently taking graduate courses in "Real Optics," and telescope-making lessons from a professional telescope maker.

Q. Please give us a brief professional bio.

A. I got my undergraduate degree in physics at Penn State, went to the University of Pennsylvania for my MD, and did my residency in opthamology at Case Western. I was invited to join the staff of the International Eye Institute. Later, I was in private practice until I got the idea to participate in the graduate program at the Optical Sciences Center, U of A, where Prof. R.V. Shack is my program advisor.

Q. Mr. Max Bray, 80 years old, says he has just gotten intraocular implants, and sees high distortion resembling the waist of an hour glass.

A. I've never heard a complaint like that. I wish all my patients were optikers, because they can quantify their problems better than others. This problem should be investigated; it shouldn't happen. For some high myopes, it is sometimes possible to remove and leave the lens out entirely and still achieve good focus, but they complain of distortion.

C. But Mr. Bray does have lens inserts. He reports problems now in interpreting Ronchi ruling patterns.

Q. Do all healthy eyes have same angular resolution, assuming they all have the same number of detector elements?

A. This depends on what you consider to be a "normal" eye. Do you consider a person who just has a refractive error to be a normal eye? Five years ago that would be called a healthy eye, but now in an era of refractive surgery, it is no longer the case. This person might be a radial keratonomy candidate. But, if you accept that as a normal eye, the answer is "not quite." Overall, the answer is "yes".

Q. Does everybody have about have the same number of photoreceptors?

A. Yes.

Q. Do men and women have any known optical differences?

A. No, not that I know of. But men find large pupils attractive in women (blue-eyed people tend to have larger pupils) so the drug Bella Donna was once used to dilate pupils of women for obvious reasons. There are racial differences. For example, oriental people have smaller eyes, but only by a few millimeters.

Q. Is there any reason to think that brown-eyed people have any better vision than blue-eyed people?

A. Actually, that's irrelevant because everybody has a very dense layer of pigment behind the iris. That's the layer that blocks the light. Otherwise you'd have a stray light problem, even brown-eyed persons. If you look close to the pupil margin, even with blue eyes, you'll see a serrated brown area that is the edge of that blocking layer. If that's not present, you have a problem. It can be depleted by diabetes. There is also a form of glaucoma that can cause the layer to rub off. Albinos simply don't have the pigment to begin with, and are consequently very sensitive to light. There are artificial iris techniques to replace it.

Q. What are the several types of color blindness?

A. Color blindness is a bad term. It isn't really color blindness; it's altered color perception. "Color Blind" persons still perceive color, their world is not B&W. What happens is that we have these three different receptors that each have a different color sensitivity and together we get a weighted average that simulates a color perception. We don't specify colors by their spectral wavelengths because you perceive a lot more colors than there are wavelengths (like brown). But there are only three different receptors. If you have only two, you still have mixture that leads to a different sense, not the same color sense that other people have. You can still distinguish colors. The color deficiency is based on which receptors are missing.

Q. But are they missing?

A. Well, total density is the same, and they have the same angular resolution. If you look at the cones, they are all similar but they have pigments in them that absorb differently. We have relatively few blue sensitive cones, and about equal numbers of green and red. In a color-defective person, anatomically those cones are still there, but they function differently. Usually the red receptor is off.

In protonopia, which is deficiency in red, we have good sensitivity in blue and green. Sonic protonopes can't even see red. It is ten times more common than other types, and is linked to chromosomes, affecting 7% of men, but 0.7% of women. Then there is deuteronopia (green missing) and tritonopia. If you are missing one, the other two are almost always still present. Very rarely two of the three can be missing, but it's questionable that such people see in monochrome. They are able to pick out many colors, just like anyone else, but they get confused on certain colors and under certain conditions.

Q. Do color blind people know they are color blind, or is it something somebody points out to them?

A. In advanced societies, they know. Then there is anomal. To be color anomalous, the 'filters' are tuned to a different wavelength. They may not know they have a problem, because they color match just fine and don't get confused.

Q. Do color blind people have the same light sensitivity?

A. I think it's pretty much the same,

Q. What can you say about how we see fine detail with large refractors that have enormous secondary color? Is there not some sort of spectral filtering going on in your eye, because monochromatic MTF would indicate zilch resolution?

A. Well, there certainly is. For one thing, the eye has enormous primary color, but you don't perceive it. I know what you are saying, but I have no idea how to explain it.

Q. What about a color blind person viewing with a refractor?

A. I'm not sure. But let's suppose you were missing your blue receptor; that might be very useful. Each receptor sees every color from the blue to the red, but it is most sensitive to a particular color.

Q. I notice you wear eyeglasses, Dr. Thall. Would you comment on the relative efficacy of eyeglasses versus contact lenses?

A. I'm going to do a little demonstration: here I am, my glasses are on. Now, here I am with my glasses off, and it takes about half a second to make the change. Ever try to put contact lenses on? Takes about five minutes. I grew a beard to save those five minutes each morning!

Q. But when you have the time and patience, then what?

A. Many people wear contacts mainly for vanity reasons. They have many disadvantages, yet 20% of people in the dating age group do wear contacts. The first thing we look for is, "is this person motivated to wear contact lenses?" If they just want to try them on a lark, they probably won't be wearing them two or three years down the road. Rigid lenses can give very good correction, better than glasses (especially if you have large refractive errors, like my six diopters of spherical power) but are more uncomfortable. Soft lenses tradeoff visual acuity for comfort, compared to hard lenses. You lose contrast too with soft lenses. You can read 20/20, but it isn't as 'sharp' or contrasty. In near-sighted people, the range of accommodation is reduced. With eyeglasses, you can have a larger range of accommodation.

As you get older, fitting you for contacts can push you over into presbyopia, which means you have almost no range of accommodation left, and patients hate you for that. If you now wear bifocals, you will not be happy with contacts.

Q. But let's suppose you only want to use them for telescope viewing, where you only need one state of accommodation, would you be better off with contacts?

A. Absolutely not. First, if you only have refractive error, it makes no difference anyway. If you have astigmatism, you won't be too happy with soft contacts, so you'll probably have to get hard contacts. There are medical problems that occur with hard contacts, and it isn't worth it.

Q. Deep-sky people have asked, "How can we enhance our visual sensitivity, as by eating carrots, for instance?"

A. My advice is that you should drink, and drink heavily before you observe.

Q. So there's really no sense to 'going into training' if you are a deep-sky observer?

A. No, not at all. It's an old wive's tale.

Q. How long does it take to reach sensibly full dark adaption?

A. Thirty minutes. The retina is an interesting detector. The range in luminance the eye has to deal with in the course of 24 hours is about 12 orders of magnitude. Still, within this huge dynamic range, we can see slight differences in brightness. How do you do that? The retina at any given point in time actually has a limited dynamic range, but it constantly resets the threshold for that range, providing a 'moving window'. So what you're doing when you're dark adapting, you're moving the window. After 30 minutes, you're down about as far as it's going to go.

Q. Has there been any change in the measured spectral sensitivity of the human eye in recent years?

A. It hasn't been measured since 1955. And that's one of the biggest jokes in the world. It's fun to listen to the myths that get perpetuated. One myth is that somebody measured hundreds of subjects to get the photopic sensitivity curve, but actually it was done for just six people. The "normal" observer is an average of six people!

These psychophysical studies are usually done on about a dozen people if you're lucky, and in the last twenty years at NIH they've almost all been done on paid Mormon volunteers.

Q. Is that because Mormons are regarded as healthy people?

A. No, it has nothing to do with that. It's because there's a big Mormon church in Bethesda, and two of the things Mormons have to do are service to the community and to contribute money to the church.

Q. Some people use narrow band interference filters to see spectral lines such as Calcium K in the near UV and in the near infrared regions that are usually considered invisible to the human eye. Is there any danger in doing this?

A. Well, you might want to be careful that your overall filter system is properly designed; otherwise, you could have a spectral leak. I'm particularly worried about leaks in the near infrared. Be sure the cutoff filters are really good.

Q. One effect of aging is to reduce transparency of the eyelens. Anything else?

A. My experience is that the cornea can lose transparency, but by an insignificant amount compared to your lens; but the aqueous and the retina are OK unless you've had disease.

Q. The eyes have a 5° blind spot. Is one reason for amateurs favoring binocular viewers related to this?

A. Only if you have a scotoma (other blind spots). But to cover the normal ones, I don't think so. One reason for using a binocular attachment is to reduce your squinting, which can adversely affect vision. People are used to coordinating their eyes, so for those who are not accustomed to using monocular vision, there may indeed be an advantage. I've been looking through monoculars for a long time, and after awhile you get used to it so that it isn't a big deal.

Q. My own experience with giant binoculars strongly suggests that I can more easily see fainter objects this way than with a monocular telescope. Is there something going on in the mind that integrates the light?

A. Yes, I believe there is some integration at that level, but have no evidence to support it. That's a branch of the medical literature that I'm not familiar with.

C. There might be some amateur astronomers who would be willing to perform a study on that, but it's important to avoid subjectivity.

A. Yes.

Q. When using low-power, large exit pupils on dark starfields, my perception of the field is not that it is completely dark, but consists of little cells that are in constant motion. Can you comment?

A. They're white blood cells running through the capillaries of your retina, and you are getting a shadow cast on the photoreceptors. You can also see it by looking at a blue sky. They'll come close to the center of your vision, but never cross it, because there are no capillaries there.

(To be continued in Section 4.6 on page 133.)

> **Ed. Note:** All too often, amateur astronomers dwell only on the properties of their particular telescope, ignoring the fact that the eye and the atmosphere are just as important to good "seeing" as the instrument. In the next few issues of *ATM Journal*, we will be offering information to help give you a better understanding of some of the less frequently considered detractors of optical performance.

3.5 A Photographic Tele-Compressor For Newtonians
By Ed Jones

One of the problems with the Newtonian telescope is that it suffers from strong coma, especially with the faster f/ratios that are so common today. There have been a number of lens systems designed for correcting coma, such as the Ross or Wynne correctors, and there are now a number of correctors which are available commercially. Many commercial units eliminate coma, but like their Ross and Wynne counterparts, do not significantly change the f/ratio of the primary. Photographically, there is a big advantage to having a faster f/ratio. Considering this, I decided to explore the feasibility of decreasing the f/ratio of a "standard" f/5 Newtonian while maintaining good correction for coma.

Utilizing the Zemax lens design program, I came up with a tele-compressor that reduces the f/ratio of a Newtonian from f/5 to f/3.3, using only three elements (see Figure 3.5.1). It will cover a 35-mm frame with only a 20% drop in illumination in the very corners of the film and with spots on the order of 30 microns. This increase in speed will reduce exposure times by a factor of 2.2 as well as give a 50% wider field. CCD users will also find the tele-compressor useful since it is f/3 and is optimized farther into the red where CCDs are most sensitive. The length of the lens system is under 2 inches, which shouldn't require any extra length outside the tube of most Newtonians. However, since the photographic tele-compressor is fairly close to the image plane, it would interfere with the flip mirror of single lens reflex cameras. Also, since the field coverage is much wider, there will

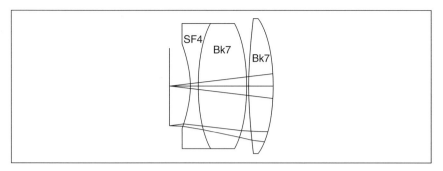

Fig. 3.5.1 *A cross section of Ed Jones' tele-compressor.*

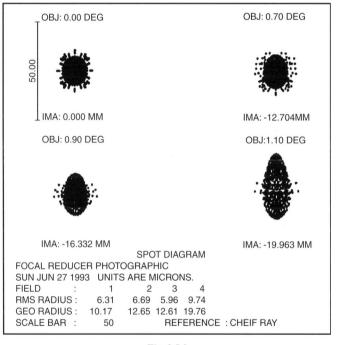

OBJ: 0.00 DEG

OBJ: 0.70 DEG

50.00

IMA: 0.000 MM

IMA: -12.704MM

OBJ: 0.90 DEG

OBJ:1.10 DEG

IMA: -16.332 MM

IMA: -19.963 MM

SPOT DIAGRAM
FOCAL REDUCER PHOTOGRAPHIC
SUN JUN 27 1993   UNITS ARE MICRONS.

FIELD	:	1	2	3	4
RMS RADIUS :		6.31	6.69	5.96	9.74
GEO RADIUS :		10.17	12.65	12.61	19.76
SCALE BAR :		50		REFERENCE : CHEIF RAY	

Fig. 3.5.2

be some additional vignetting with a 2-inch focuser. The CCD system, however, will fit into a 2-inch focuser.

3.5.1   Design

The system consists of only three lenses: a cemented doublet and an additional element. The two positive lenses are Bk7 and the negative lens SF4. I was able to make four lens curvatures the same to simplify tooling without changing the performance, and all lenses have good edge thicknesses.

Since the size of the diagonal affects illumination, I chose a diagonal that

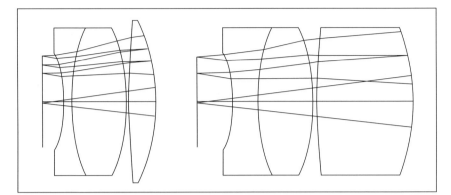

Fig. 3.5.3 *Lens "A" for Photography.*   **Fig. 3.5.4** *Lens "B" for CCD Work.*

Lens "A" Prescription

Surf	Type	Radius	Thickness	Glass	Diameter
OBJ	STANDARD	Infinity	Infinity		0
STO	STANDARD	-3048	-1211.977	MIRROR	318
2	STANDARD	Infinity	-254		86
3	STANDARD	-86.91	-12	Bk7	70
4	STANDARD	289.55	-1		70
5	STANDARD	-86.91	-24.6	Bk7	64
6	STANDARD	86.91	-3.7	SF4	64
7	STANDARD	-59.91	-10		44
8	STANDARD	Infinity	0		40
IMA	STANDARD	Infinity	0		40

Lens "B" Prescription

Surf	Type	Radius	Thickness	Glass	Diameter
OBJ	STANDARD	Infinity	Infinity		
STO	STANDARD	-3048	-1197.896	MIRROR	318.1159
2	STANDARD	Infinity	-254		78
3	STANDARD	-71.13	-26.4	Bk7	44
4	STANDARD	209.85	-1		44
5	STANDARD	-71.13	15.9	Bk7	44
6	STANDARD	71.13	-6.7	SF4	44
7	STANDARD	-45.11	-10		29.6
8	STANDARD	Infinity	0		26.40316
IMA	STANDARD	Infinity	0		26.40316

gives 80% illumination in the corners of the frame, and then optimized the system using this size diagonal. The result was a fairly uniform image quality over the field. Both the photographic and the CCD correctors use a 3.5-inch diagonal.

3.5.2   Performance

The tele-compressor allows a ¼-wave error on-axis. However, since it is intended for photography and not visual use, this is not a problem. This is also typical of other types of correctors. The spot diagrams and other performance are shown in Figure 3.5.2.

3.5.3   Construction

I have not yet started making this tele-compressor, but plan to do so. I was working on a Wynne corrector to be used on a Cassegrain primary. However, while this unit will cover a wider field, my new corrector will fill the field provided for in the 35-mm film format and will facilitate switching the instrument back to visual use.

   With only three elements this system is fairly simple; perhaps a commercial firm will be interested in producing it for those who do not have the facilities to make it themselves.

3.5.4   Proposed Tele-compressors for Fast Newtonians

Figure 3.5.3 shows a photographic design while Figure 3.5.4 shows a design to be used in conjunction with a CCD setup. The lenses were designed to enhance coma correction and change $f/5$ instruments to $f/3.3$.

3.6   Buried Gold in that Old SCT

By Jeff Beish

My first "large aperture" telescope was a popular 8-inch Schmidt-Cassegrain (SCT), a big step from an old "Christmas Special" 60 mm refractor and homebuilt 3½-inch reflector I had used. While the SCT provided many hours of enjoyment, its performance fell short of many of my friends' 6-inch and 8-inch Newtonians. My 6-inch $f/4$ Richest Field even outperformed the SCT. We live and learn.

   After many years of observing and experimenting with various types of optical systems, I learned to appreciate the subtle differences in telescope designs. Yes, aperture plays an important role in selecting a telescope; however, one should not forget that image contrast is also a very important ingredient in telescope performance.

   In a reflecting telescope, the secondary mirror is a controlling factor in image contrast because it obstructs the optical path, scattering light throughout the image field. This obstruction causes light from the center spot of the Airy disk to be scattered among the outer rings of the image. This is the real villain causing the loss

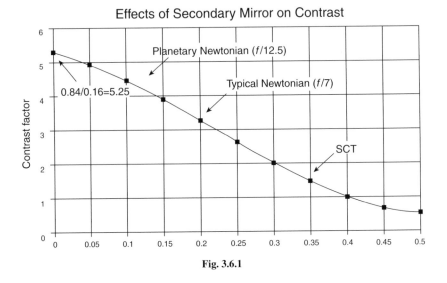

Fig. 3.6.1

in contrast in most reflecting telescopes (see Figure 3.6.1). This fact is often forgotten by commercial telescope makers and some homebuilders alike. Image contrast is lost by increasing the secondary obstruction—the larger the secondary, the less the contrast—period!

As a telescope builder and tinkerer, my thoughts were on ways to increase the SCT's performance, and in doing so, I found some buried gold in an old Dynamax-8. The story begins a few decades ago when, while camping and observing in the Everglades, I dropped my telescope! With a loud thud the optics were, shall we say, realigned! The following is what I subsequently learned about the Schmidt-Cassegrain telescope and how to increase its performance. Also discussed are tips on how to maintain your SCT.

3.6.1 Reduced Central Obstruction

In the process of realigning the optical components of the D8, the corrector plate had to be removed and replaced several times. To save time, the secondary baffle was intentionally left out, and the secondary holder was loosened to make adjustments in centering. For the coarse optical centering and alignment, a bright star and/or a Cheshire eyepiece type alignment tool was used. While using Saturn's rings to fine tune the optics, I noticed something different about the images—they appeared sharper and higher in contrast than before, even though the scope was not in perfect collimation! I could see Cassini's Division quite clearly, and more rings appeared in the intra- and extra-focal star images. Also, the background field appeared darker.

The various components and distances within the telescope were measured, and it was determined that the 2.75-inch diameter secondary baffle was a major contributor to the central obstruction (Figure 3.6.2). The 2-inch secondary was

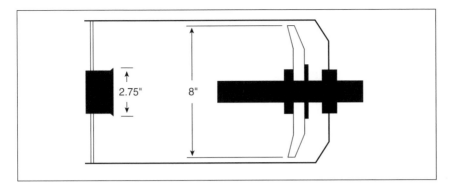

Fig. 3.6.2 *General design of the Dynamax-8 Schmidt-Cassegrain Telescope—optical components and secondary baffle. Resulting obstruction, 34%.*

found to be attached to an aluminum housing that was secured to the corrector plate with a ¼-inch overlap. Reducing the housing diameter from 2.5 inches to 2.25 inches left enough aluminum to safely secure the secondary and holder to the corrector plate. The corrector has a 2-inch hole for this purpose (a 1.5-inch or 1.75-inch hole would have been more than adequate). The increase in image quality was immediate. Looking around inside the D8, no logical reason was apparent for such a large baffle. Even my baffling calculations revealed no reason for it (no pun intended)!

While some direct light from the Moon did leak by the secondary into the image with the secondary baffle removed, it was easily eliminated by placing a glare stop near the end of this tube and lining the inside of the primary baffle tube with flocking paper. Besides, those shiny SCT baffles often cause flares on photographs. While the image size was reduced a bit, it did not cause any apparent loss in image brightness, even on photographs. The inside diameter of a typical SCT baffle is around 1.25 inches—it seems some manufacturers falsely claim that images will fully illuminate a 35-mm frame. However, how can it be larger than the I.D. of the rear end of the baffle? Maybe they design these scopes for daytime use. Can you see someone carrying around a 14-inch telephoto?!

The overall effect of my modification was to increase the contrast efficiency of the D8 by reducing the central obstruction from 35% to 28%. This produced a 41% increase in the contrast efficiency by increasing the Contrast Factor (CF) from 1.68 to 2.37 (where 0 is the lowest and 5.25 maximum). A 5.25 CF is usually found in unobstructed telescopes as indicated in Table 1[5]. Varying Contrast Factors are illustrated in Figure 3.6.1 and can be calculated with the following equation:

$$CF = 5.25 - 5.13\left(\frac{S}{D}\right) - 34.17\left(\frac{S}{D}\right)^2 + 51.1\left(\frac{S}{D}\right)^3$$

[5] Johnson, Lyle T. "Improving Image Contrast In Reflecting Telescopes," *J.A.L.P.O.*, Vol. 18, Nos. 7–8, 142–146, July-August 1964.

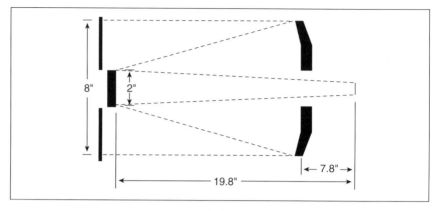

Fig. 3.6.3

where S is the secondary mirror diameter, and D is the primary mirror diameter.

Shown on Figure 3.6.1 are the CF values for the typical SCT, a typical *f*/7 Newtonian telescope (20% obstruction), and an optimized planetary Newtonian (12.5%).

If it were not for the 2.75-inch blackened area around the center hole in the primary, a 4% increase in light gathering power would have been realized as well. Since much of this blackened area had flaked off, I simply removed the remaining paint from the mirror. These results may appear to be small; however, the apparent increase in performance of the Schmidt-Cassegrain with this modification is obvious.

Last but not least, be sure to mark everything for proper reassembly. I didn't the first time around and had to learn how to do all those optical testing procedures!

3.6.2   Additional Tips for Improving Telescope Performance

1. It may come as a surprise to you, but those mosquito sprays that everyone uses in summer may be detrimental to the health of your mirror or corrector coatings. Watch out for the people who like to spray everything in sight! Other enemies to mirror coatings are; air pollution, chlorine vapor from swimming pools, and salty sea breezes.

2. Most star diagonals which come with popular SCTs, are far too small. The D8 I once owned came with a 16 mm aperture diagonal that proved to be seriously undersized. Several dealers sell a 30–32 mm aperture diagonal that works very well. I think a 2-inch is a bit too large for the usual 1-inch image diameters found in the SCT.

3. Prevent your SCT primary from shifting by keeping it tight on the primary baffle tube hub/mirror cell. Of course, do not over-tighten it. Take the focusing mechanism to a machinist for re-working; they are often too

loose for efficient focusing as they come from the factory. Much of the workmanship found in these SCTs could be improved upon, so do not be afraid to have them reworked or replaced by higher quality material and workmanship! Another way to solve this problem is to lock the primary mirror in place and install a rack and pinion focuser. I drilled and tapped #10 holes in the small plate of the D8's back housing and drove screws up against the aluminum mirror housing/cell to lock it in place. This also provides a method of slightly adjusting the collimation of the $f/2$ primary.

4. The newer SCTs use a worm gear for polar rotation. To prevent the worm housing from getting too loose, try some Loctite on each of the mounting screws that secure the worm housing components. By the way, SCTs typically are out-of-balance so do not be afraid to add weights; this will also help dampen the excessive vibrations! The finder scope is only one cause for the unbalance. However, it is sometimes wise to off-balance the R.A. axis of the mount in favor of the drive direction to lessen the backlash in drive gears.

Editor's Note: Jeff Beish is a Mars Coordinator for the Association of Lunar and Planetary Observers and an avid telescope maker. Jeff is a retired electrical engineer, formerly with the United States Naval Observatory.

3.7   Q and A For the Telescope Maker

Q: My father is in the window glass installation business. From time to time he comes up with some pretty thick pieces of scrap glass from windows which have been broken. Some of my friends with telescopes say that I could use some of these pieces to make telescope mirrors; others say that I would be foolish to use anything other than Pyrex. I really do not want to spend the money for Pyrex if I can use the glass I have on hand. What do you think?

A: It sounds like this may be your first telescope-making project. If that is the case, you would probably like to have a simple and definitive answer. I can't give you that. I will, however, give you some facts and opinions, from which you may draw your own conclusion.

If you are making your first telescope, you will probably want to stay with a smaller instrument, such as a 6- or 8-inch. If this is the case, you may be better off opting for ready-made Pyrex blanks. That is, if you consider your time to be worth anything at all. The last Pyrex blank I purchased cost less than $25.00, and if I were to value my time at minimum wage, the last 6-inch disk I cut from a 0.75-inch glass fragment cost more than $70.00.

If you think that you would like to continue in the hobby of telescope making, cut the plate glass. The experience (rewards and headaches) will do you good. If you decide to have your father cut the glass as well as provide it, you are pretty much home free. Then, the only thing to watch out for is internal stresses and stri-

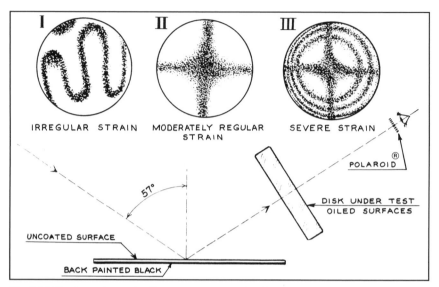

Fig. 3.7.1 *Simple test for strain. Illustration courtesy of Willmann-Bell, Inc. from Jean Texereau's* How to Make a Telescope.

ae which could cause uneven expansion and contraction during temperature changes. You may check for strains by using the simple process outlined on page 155 of *How To Make A Telescope* and shown here in Figure 3.7.1.

3.7.1   As for the Glass Itself

Plate glass is a type of crown glass and many people have made fine instruments with it. In Europe, it is often the medium of choice. Furthermore, the first corrector plate for the 48-inch Schmidt camera on Palomar was made of plate glass, and it was one of the most useful astronomical instruments ever conceived.

Plate glass is easier to work than Pyrex and takes a beautifully smooth polish. However, it reacts to thermal changes to an extent 3 times greater than Pyrex. This means that when you are in the figuring and testing modes, you will need to insure that the temperature remains fairly constant, and you will need to wait longer for the mirror to stabilize between testing sessions.

Finally, John Dobson showed us that mirrors don't require a thickness of ⅙ their diameter to provide good imagery. Nevertheless, I would recommend that you stay with the ⅛ to ⅙ ratio, if you can.

Q: Is there a rule of thumb for determining how much to bevel a telescope mirror before grinding? I want to leave as much reflecting surface as possible, but it seems that I never bevel it enough, and I wind up chipping the edge. The last time, a big chip came loose at the surface of the mirror and extended down the side.

A: Many telescope makers choose to allow a ¹⁄₁₆-inch bevel. You need to re-member that this is the desired *final* bevel! Experienced mirror makers start with bevels which are considerably larger or re-bevel a number of times during the pro-cess.

For the purpose of gaining peace of mind, you might want to re-think some old laws of mirror making. First of all, while an instrument with a chipped edge may not be aesthetically pleasing, and thus cause embarrassment to the optician, stars and planets do not know the difference. If you chip the edge of the mirror, just paint the spot black, so that it will not scatter light, and get on with the busi-ness of polishing, coating, and observing.

Of course, the best thing to do is avoid chipping the mirror in the first place. To do this, you may bevel both the mirror and tool to about ³⁄₁₆-inch and never let the bevel fall below the ¹⁄₁₆-inch goal. While it is true that a wide bevel will rob you of a little light grasp in the most important region of the mirror, you are more apt to end your project with a smile by taking the extra precaution. Weak bevels are not only likely to allow damage in grinding, but in testing and transportation as well.

Finally, with so many mirrors falling victim to the "turned edge," a healthy bevel may have the additional benefit of relieving you of this problem, especially if the mirror receives a light beveling at the end of the figuring process.

Q: I am making plans to build a Schmidt-Cassegrain telescope, and I think I have figured out how to do everything except how to cut the hole in the main mir-ror and window. Because this is to be my first telescope, I am going to keep it to a sensible 6-inch aperture. Even so, boring through 1-inch of Pyrex does not seem to be an easy thing to do. Can you tell me how this is done?

A: Instructions for perforating optics can also be found in *How to Make a Telescope,* pages 156–159. However, your problem is akin to that of the gentle-man who had such a hard time choosing a stateroom on the Titanic. Some people profess that it is easier to make a 6-inch mirror and a 12½-inch mirror than it is a 12½-inch mirror alone—because of the learning curve. If this is true, and with most individuals it certainly is, then it is probably fair to say that it would be easier to make three 12½-inch mirrors than it would to make one *good* set of Cassegrai-nian optics. On this subject, John Gregory writes, "A Schmidt-Cass is *not a first* telescope. It is a *last* telescope."

The problem arises in creating the "window." This element is not a simple piece of plano-plano glass. While one side is flat, the other has an aspheric curve which is incredibly hard to accurately figure and test.

When I was a much younger man and had decided to make my first *real* tele-scope, a classical Cassegrain, I had all the old-timers telling me why I should not and could not do it. Initially, I resisted this advice. Today, considering the premi-um I place on my time, I thank these gentlemen for raining on my parade.

You may think that we are trying to discourage you from the project. If so, let me clarify. *We are trying to discourage you from the project—at least for the time being!* After the successful completion of a few less complicated instruments, and a thorough investigation of the requirements for fabricating a Schmidt-Cassegrain, you will be ready to begin in earnest. Either way, keep us apprised of your progress.

3.8   From the Bench

By William J. Cook

"I have a pair of World War II binoculars I would like to restore. Can cemented lenses be AR coated, or do they have to be separated first? I have heard that the coating process generates high temperatures which could melt the cement, but have not found any definitive literature on the subject. I was also wondering if someone could recommend a coating house that will deal with ATMs. I realize that the cost would be in the hundreds of dollars. I am really looking for someone who would have the time and patience, and careful handling, required by the amateur who is coating only a few components which, I might add, would be irreplaceable."

This is a multi-faceted question. So, let's see if I can come up with a multi-faceted answer.

Yes, cemented lenses can be re-coated. However, just as you suspect, the lenses will have to be separated first. And, this does require the application of high heat.

In order to keep the cost of your project manageable, I would like to make a couple of recommendations. First, you might consider separating the lens yourself. Secondly, I recommend that you have the coating done by a local coating house—the same business that any number of your local optometrists use. With a little caution, patience and a lot of respect for old glass, you may obtain results equal to, or better than, the large coating houses to whom one- and two-piece jobs are a nuisance. The fellow who does coatings for my shop is quite friendly, and if I choose to stay and watch the operation, or even assist, I am welcome.

Before giving you a detailed description of what you need to do to separate your objectives, I would like to state that one of your fears concerning this process may be unfounded. While many individuals with binos from the Second World War feel that they have a one of a kind item which would be impossible to repair or replace, such is usually not the case. More than 30 of these instruments pass through my shop every year. The most common of the binoculars to still be around are those which were made under contract for the Navy. If you have an MK 28, MK 32, 39 (which is an MK 28 with a reticle), you should be able to get parts without much trouble. The older MK 21 and newer MK 45 (a submariner glass) will be somewhat more difficult. Still more rare is the Mk 41. In all the time I was doing optics for the Navy, I saw none of these, and have only seen two since I have been at Captain's.

Most binoculars made during the war years had objective lenses which were cemented with Canada balsam, a resin from the balsam fir tree. Separating lenses cemented with Canada balsam is not much of a chore—here's how:

Place the lens on a board (8″ x 8″ x ½″ will do nicely) with the crown element up. The crown lens will be the one that light strikes first as it enters the instrument. It will also be the element with the least edge thickness. Now, with an indelible marker, draw a line straight down both elements of the lens. This will aid you in re-cementing.

Place three cleats around the lower lens in a fashion similar to that for a tool from a mirror grinding setup. The cleats should be close enough to the lens to keep it from moving about, but not so snug as to apply undue pressure. At this point, you will notice that while the lens cannot move laterally, it can rock in any direction because of its contour. This can be corrected by boring a slight indentation in the center of the holder or by lightly peening the wood a few times. Finally, place a small piece of any soft material, which will not ignite at 325°F, in the center of the holder. This will give the lens something soft to rest on during the decementing process. While it is much more important to your peace of mind than it is in protecting the lens, you will work more effectively if you feel that you have total control of the operation.

Place the board and lenses in the oven and set the temperature for 250°F. When this temperature has been reached, grasp the edge of the board with channel lock pliers and use a pencil with a fresh eraser to press downward and inward on the edge of the top element.

If, after the oven has been at 250°F for 10 minutes, the lens does not begin to move when moderate pressure is applied, close the oven and raise the temperature to 300°F. Repeat the process. If you have no luck at this temperature, raise the setting to 325°F and allow the oven to maintain that level for at least 10 minutes before your next attempt to separate the lenses.

Caution #1: Avoid open windows, air conditioners and cool drafts of any other kind. Warm lenses, especially old ones, do not react well to cold air.

Caution #2: The lenses should begin to separate at a temperature between 275°F and 300°F. If you are unable to see progress after taking the lens to 325°F, the chances are that the lenses have been re-cemented, at some point, with a thermosetting plastic. If this is the case, be advised that you will still be able to separate the lenses. However, it will require temperatures high enough to cause much consternation between yourself and the person who thinks that an oven is for baking bread and that optics is for the birds. At this point, sending the lenses to a professional will appear to be a very good investment—the hourly rates for most coating houses are considerably less than those of divorce attorneys.

If the lenses do begin to separate, slide the crown element slowly to the edge of the flint. Here gravity will take over, and the top lens should fall gently over the side. If so, close the door, turn the oven off and allow time for the lenses to cool to room temperature. This will take longer than you think—give it at least an hour! Once the lenses have cooled, remove them from the board and clean them thoroughly with acetone. Keep in mind that acetone is flammable and will melt any number of petroleum-based products. It will also take your centering marks away instantly.

With your lenses separated and cleaned, you are ready to have them coated. Now you must decide whether to send them off to a major player in the coating business or to a local concern which deals primarily in eyewear. This is a decision you alone can make. A few phone calls will give you a good idea concerning cost and turn-around time. The local company will probably be less expensive and easier to work with. However, the larger coating houses speak real optics, and they will return them to you centered and ready to install in your binocular. If you opt for having the lens coated locally, I recommend that you reserve the centering for yourself.

We cannot assume that the mechanical center of the lens combination is, in fact, the optical center. In other words, one size does not fit all. However, since most people do not have a lens-centering machine at their disposal, I propose a method shown to me by my associate, Cory Suddarth.

With the lenses *absolutely clean* and aligned with the marks you made before separating them, place a large drop of Norland Adhesive #81 (available from Edmund Scientific Co., Barrington, NJ) between the elements and let the crown element settle in. Place the lens in an area where it will not be subjected to sunlight and find an empty aluminum soft drink can. Cut a ½-inch strip (cross-section) from the center of the can, place it around the lens and secure it with several strong rubber bands. At this point, the adhesive is still quite viscous, because it has had no interaction with the ultraviolet light needed to cause it to set-up, and the rubber-band-wrapped strip of soda can is acting as a centering machine.

Walk out in the sunshine and expose the lens to its rays for a few seconds; the lens will be cured and ready for a final cleaning with acetone and subsequent installation into your binocular.

I have not bothered to mention the process of how to cement lenses with Canada balsam because it would take much more space than we have and cause a sharp degradation in my vocabulary. There is something about trying to pry your fingers loose from a hot piece of glass oozing 300° tree sap that can cause one to have thought processes which might go against one's value system.

Now, if your prisms are clean and properly affixed on the prism shelf, you are ready for collimation.

From The Bench was created to offer helpful hints concerning lens and instrument care, instrument testing and calibration and elementary telescope design. However, since the question on lens coating and cementing fit in so well with our proposed format, we decided to treat it in a Q & A fashion. Furthermore, we feel it more important to answer meaningful questions from members in a timely manner than to generate any particular article and hope it is of general interest.

3.9 The Perfect Telescope Is a 20 X 80 Binocular

By Cory Suddarth

In the last *ATM Journal*, you expounded how your 10-inch *f*/6 is the perfect telescope. On several points, I would say "right on;" I have built one myself to the specifications Richard Berry published in his book, *Build Your Own Telescope*. However, I must also disagree with you in several areas.

Yes, while this scope fits inside most any van for transporting, portable it is not! I am constantly whacking either the finderscope or the focuser on the door-frame when I take it out to set up, or bring it back into the house. This "portable scope" requires muscle power and multiple trips to set up! On nights with good seeing, it is well worth all the effort. However, here in Washington State, one must utilize a "hit and run" tactic when planning one's viewing sessions. If it's not too cold, it's raining. If it's not raining, it's still too cloudy. But at times, during nature's transitions, the skies are clear and beautiful—prime time to enjoy the "perfect telescope."

Rule No. 1 states, "the best telescope is the one that gets used the most, thus bringing the most enjoyment." When applying that rule, I must put my Celestron 20 x 80 binoculars on top. Now fellow *ATM*'s, before you grab your least favorite eyepiece to throw at me, hear me out, please.

Portable: (port' able) adj- the ability to transport one's optical device, carried in one hand, while opening the sliding glass door with the other! I like to keep it simple. My binoculars are set up on an extra-tall photographic tripod with an alt-azimuth slow-motion control. This set-up is great for those twenty-minute observing sessions.

Stereo: Our brains are wired for both right and left channels. Anyone who has ever viewed anything through binoculars has quickly come to enjoy the comfort of using both eyes. Our brains perceive a full 40% increase in resolution by virtue of stereoscopic versus the mono-mode. Ever watch a beautiful sunset or a rare bird with one eye closed? Of course not! Then why settle for viewing the splendors of the heavens that way!

FOV: Let's talk field of view. Basically, we're looking through two, 3.1-inch *f*/3.75 refractors, identically matched and collimated, yielding a generously wide 3.5° field of view. Recently, when Venus' crescent phase matched that of the Moon's (separated by only 2°), the 10-inch *f*/6, with a moderately low power eyepiece (30 mm), could not come close to viewing both in the same FOV. Even the Pleiades must be viewed by starhopping, Yet, the 20 x 80's embrace the entire cluster!

Why 20 x 80's? Bino's are bino's, are bino's right? I say—"not!" When compared to the highly recommended and touted 7 x 50's, the latter simply pale by comparison. But what about the 7 mm exit pupil, you ask? By simple math the 20 x 80's only yield a 4 mm exit pupil. But, I will guarantee you that, because of sheer light-grasp, those 3.1-inch objective lenses will triple the apparent brightness of the Beehive, M42, M13, and the Andromeda galaxy—exit pupil or not! When *Astronomy* magazine (Nov. 92) ran an article on binoculars for astronomical use, they implied that the 20 x 80's were not a very good choice to use for astronomy. Really now!

We have covered what the objective end of these bino's do; now let's get into what the 20x end does for you.

Skyfog and light pollution: The 20 x 80's seemingly ignore much of these urban viewing annoyances. While an 11 x 80 gathers the same amount of light, the background viewed through the 20 x 80 is much darker! The magnification works out to your advantage. As increased magnification makes the background much darker in a telescope, the 20x eyepiece performs in a similar manner for the binocular. When viewing from areas north of Seattle, looking south in the summer months (over that familiar, menacing orange glow), my 20 x 80's "cut through" much of the pollution to reveal many of the nebulae and star clusters in Sagittarius. This is true, even when few stars are visible to the naked eye.

Price: It's a bang for the buck kind-a-thing. I went with the Bak-4 prisms and the multi-coated optics offered by Celestron; both are definite upgrades over the standard Bk-7 prisms and magnesium-fluoride coatings. This was a one hundred dollar increase over their generic cousins, but an option I thought worth going for.

Well, until things either warm up, or dry out around here, my 20 x 80's are locked and loaded in the stand-by position. And okay, no doubt, they are great spotters for the 10-inch *f*/6, for when I might want a little closer view.

Issue 4

4.1   A 17.5-inch Telescope That is Easy to Set Up
By James Stewart

I have come up with a different method for making a portable telescope. So far, all the ads I have seen for portable Dobsonian telescopes feature a very short rocker with a long mirror box attached to rotating rings. For large apertures, these units are too heavy for one person to handle. Wheels are offered for these so that the unit can be wheeled down a ramp and moved about. The upper end of the telescope is usually attached to the mirror box by a number of aluminum tubes.

I wanted a large telescope. However, I was concerned with the high cost and extreme weight associated with large aperture instruments. I was able to take care of my first concern by building the telescope myself, and the problems associated with weight and portability were solved by the methods used in construction.

There are more separate parts to my plywood-tube telescope than in most of the usual designs. The middle section is bulkier and heavier than those built of aluminum pipe, but no single piece weighs more than 40 lbs——about half that of corresponding parts of other instruments of similar aperture. I can load and unload all the pieces in my van and take them out myself, and do not need a ladder to fit the pipes (used in truss-type telescopes) into the upper section.

The component parts of the tube are very easy to build. The only power equipment needed is a good table saw. The telescope is truly Dobsonian in philosophy and suitable as a project for the amateur astronomer or astronomy club.

The major innovation in my design is the use of a removable support structure on which to assemble the sections of the telescope. (Such support structures were used by the Romans when they built their stone arches, so the idea is not new.) A removable board is placed inside the rocker; and the telescope is stacked, piece-by-piece on top of it (Figure 4.1.1).

The middle part of the tube consists of four panels (Figure 4.1.2). The two that ride in the trunnion bearings are put into place first.

Another innovation is the way these telescope sections are held together by a system of interlocking rails compressed by toggle clamps (Figure 4.1.3). The four tube panels are temporarily held together with bungee cords. The support is then removed, allowing the telescope to drop gently into the arms of the rocker box. Next, the telescope is tilted, and the section housing the secondary mirror is

Fig. 4.1.1 *Rocker box with removable board used for assembly.*

Fig. 4.1.2 *All major components ready for assembly. Tube side panels in the background.*

Fig. 4.1.3 *Placing a foot on the end of the removable board (or jack) will raise the mirror to the point where it can be pulled snugly into place by the toggle clamps.*

slid into position and clamped with two more toggle clamps (Figure 4.1.4). At this point, the telescope is ready for viewing the heavens (Figure 4.1.5).

Study the drawings and photos of the telescope to see these innovations. If you are familiar with how telescopes are built, the schematic drawings and photos are all you need. If not, you might need to do some additional reading. Richard Berry's, *Build Your Own Telescope,* (Willmann-Bell, Inc.), is the book I recommend.

I used a battery powered screwdriver and drywall screws to assemble the separate sections. They are easy to insert, hold well, and come out easily if you make a mistake. I would be happy to hear from anyone who plans to build such an instrument and who might have additional questions.

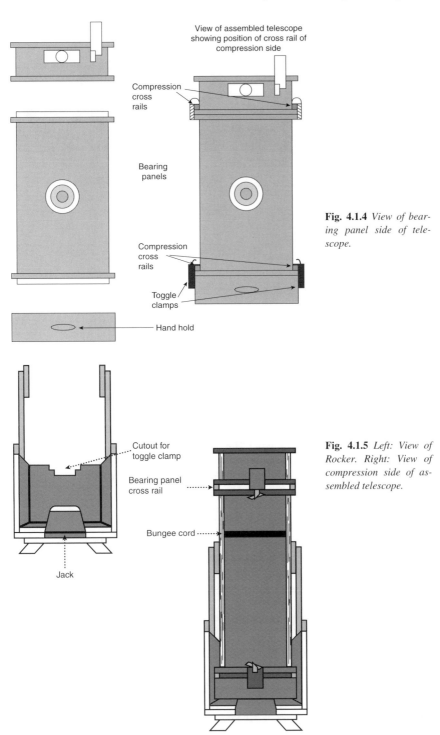

View of assembled telescope showing position of cross rail of compression side

Compression cross rails

Bearing panels

Compression cross rails

Toggle clamps

Hand hold

Fig. 4.1.4 *View of bearing panel side of telescope.*

Cutout for toggle clamp

Bearing panel cross rail

Bungee cord

Jack

Fig. 4.1.5 *Left: View of Rocker. Right: View of compression side of assembled telescope.*

Fig. 4.1.6 *Dr. James Stewart with the latest refinement on his 17.5-in telescope.*

Fig. 4.1.7 *Dr. Stewart places the top of the tube onto the telescope and secures it with two more toggle clamps.*

Fig. 4.1.8 *The clamps used for this design came from DE-STA-CO, Troy, MI.*

4.2 Improve Your Telrad

By John Shelley

When I first used my Telrad® finder I found that the reticle was rather bright, and the control had to be turned down repeatedly as my eyes dark-adapted. Also, I frequently left the switch on when I finished viewing, only to find dead batteries the next time I tried to use it. I soon determined that a momentary push-button switch was needed. After installing one in the upper front corner of the right side close to the battery, I found that I could both steer the scope tube and hold the switch on with one hand.

Later on, I found that in warmer weather the weight of the Telrad would sometimes cause my Dobsonian to plunge. Its two AA batteries seemed to be the heaviest part of the unit, and I wondered if a smaller size could be used. When I measured the current drain of the LED light source (after reducing the brightness to a suitable level), I found it to be only 3 milliamperes (0.003 amperes). This was only ¹⁄₁₀ of the normal operating current of a common LED, and it indicated that the use of a much smaller battery was possible.

During a visit to a local Radio Shack, I found a 3-volt lithium coin-battery made for digital watches. It had a capacity of 45-milliampere hours and was smaller than a dime. Calculating for 5 seconds of on-time for each use, I determined that one of these would serve for many uses. The price was $1.99, and the shelf-life is quite long!

Since the lithium cell is primarily a watch battery, Radio Shack does not provide a holder. I have not found one in any catalog, but have made a few for various

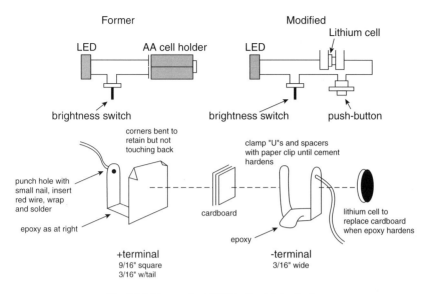

Fig. 4.2.1 *Diagram by ATMJ Staff and John Shelley.*

purposes. For this application, two "U"-shaped pieces of thin sheet-metal were glued into the bottom of a Telrad. The space between them was held to about ⅙-inch by inserting small pieces of cardboard. Bridging the bottom legs of the two "U's" with epoxy cement, a simple but effective holder was made as shown in Figure 4.2.1. Tin-can material is quite suitable for making the parts and is easily cut with small shears or scissors. Cutting sheet metal is as easy as cutting cardboard, but the consequences of losing control when the blades snap closed can make a nasty cut. Use care.

4.2.1 Radio Shack parts:

275-1547   pkg. of 4 switches, normally open, spst, momentary contact.
23-167   1 each lithium cell, 3 volt

4.3   The Houghton-Cassegrain Telescope
A Diamond in the Rough

By William J. Cook

4.3.1   Discovering The Houghton

I first learned of the Houghton telescope through an article in *Astronomy* magazine's Equipment Atlas section titled, "What's Inside Your Telescope?" The article was the fourth in a series by the late Robert E. Cox and described the layouts of several types of unusual astronomical telescopes. The Houghton was eye-catch-ing for a number of reasons; the most important being, it was a high-performance instrument that utilized all-spherical optics. Telescopes requiring the figuring and testing of paraboloidal mirrors or aspheric correctors always take longer to make than those using all-spherical surfaces and, once fabricated, do not always provide the desired results—those expected for the additional cost and effort.

At the time I knew almost nothing about lens design. I did know, however, that the more surfaces, curvatures, spacings and glass types available to the designer, the more "degrees of freedom" were available for the correcting of aberrations. The article stated that lens designer, Robert D. Sigler, was fabricating a Houghton, and that it would be completed in the near future. As it turned out, I saw nothing more on the subject until Mr. Sigler published his design in *Telescope Making #30*.

4.3.2   Exploring Various Designs

My next step was to read as much as possible on the Houghton and talk to a number of professional lens designers about the pros and cons associated with making such a telescope.

I learned that although little had been published concerning the design, it was respected by virtually every lens designer familiar with it! After a few letters and phone calls, and learning as much as possible from optical books on hand, I was certain the Houghton was the telescope for me. And, with my primary interest being planetary observing, I was sure I wanted to work up a long-focus Cassegrain design. Mr. Sigler's 10-inch Houghton-Cassegrain was diffraction limited over a 2-degree field of view, was compact, made with matching—and inexpensive— glass types, and featured very forgiving tolerances. It seemed too good to be true.

Realizing a 10-inch telescope project would tax my abilities and finances, I decided to scale the instrument down to 6 inches. When I did, I noticed the central thickness of the second element (double concave in the Sigler design), would be less than ⅕ of an inch. This caused me to wonder if a design could be created with a thicker second element. Planning to do all the grinding and polishing by hand, the thought of working with such a thin lens stirred up doubts about my chances for success since flexure in thin lenses and mirrors in grinding and polishing usually leads to astigmatism in the final product. Upon sharing my concerns with Mr. Sigler, he sent me a new design in which the central thickness of the second element had been increased to about ⅓-inch—a more reasonable dimension considering the constraints of my skill, environment and tools.

While preparing to order the glass, I was overcome by another problem. The thought of simply using the Sigler design left me with a hollow feeling. Even though his design was very good, I still wanted something I could call, at least partially, my own.

4.3.3   Enter ZEMAX

I bought a string of ray-trace programs to use on the project. With the first few programs, I found myself spending much more time in trying to learn the software than in designing lenses. It became maddening. I saw that while some of the programs were relatively powerful—especially to a first time lens designer—most were not user friendly and did not even begin to utilize the computer's capabilities.

The last design program I purchased was ZEMAX from Focus Software, Incorporated. ZEMAX had the speed and power to do everything required and was so simple to use that the manual was needed more as a "security blanket" than as an essential part of using the program. With ZEMAX in hand, I was again ready to begin designing in earnest.

4.3.4 The Sigler-Houghton

Even though Mr. Sigler is a professional lens designer, and well known for his work in investigating the performance of compound telescopes with all-spherical surfaces, I felt there was a good chance I could improve upon his design. The reason for this was that he designed a corrector lens in which the surfaces of one element could be ground and tested against the mating surfaces of the other. In doing this, he greatly simplified grinding, polishing and testing and put a great design well within the grasp of most experienced amateur telescope makers. The down side to this approach was that he had to give up some "degrees of freedom" in order to do so.

Anyone constructing a telescope according to Mr. Sigler's specifications will have a fabulous instrument. However, I not only wanted to develop an original design, but gain the experience that comes with such an undertaking. Furthermore, since the proposed telescope was to serve me for years to come, dealing with a few extra problems during construction did not seem to be unreasonable.

4.3.5 Designing the Cook-Houghton Planetary Telescope

As with every optical design project, this one had its own set of compromises. The first compromise to confront me involved understanding that spot diagrams-the most common graphic used by ATMs—do not tell the whole story when attempting to illustrate practical performance.

In discussing this with others, I was confronted with two schools of thought. One acquaintance said, "Diffraction limited is diffraction limited is diffraction limited. If all the rays fall inside the Airy disk, it doesn't matter what OPD plots indicate." Another stated that even though fine instruments have been designed with spot diagrams providing the most useful graphic analysis, most professional lens designers use OPD goals for optimization. This is true because while spot diagrams may illustrate the linear extent of an image blur, they do not reveal the intensity of the light involved. Thus the linear extent of a blur produced by a combination of coma and astigmatism might be 0.005-inch and still not pose a problem visually or photographically, if relatively few rays reach the outer portion of the blur.

With this in mind, I decided to use spot diagrams on the pre-designs and then switch to optimizing routines based on OPD calculations to make the final adjustments. Within the first few attempts, I came up with systems in which all rays fell into a tiny portion of the Airy disk, even at the edge of a 2 degree field of view. I wanted so much to be impressed with my accomplishment. However, while the system was "diffraction limited," the diffraction was far from negligible. The secondary mirror and its baffle created a 42 percent obstruction!

Before long, I realized that when ZEMAX optimized with default settings, it would struggle to sharpen the image at the expense of increasing the diameter of the secondary and lengthening the telescope. This could have been handled nicely by changing the merit function setup. However, since this was to be a learning experience, I decided to simply handle the problem manually and observe the iterations during optimization.

My first designs were based on the stop being at the corrector. Then, setting the stop at the secondary and re-optimizing, I came up with telescopes with improved OPD plots and smaller secondary obstructions—down to ~32.5% in the $f/15$. This improved contrast across a slightly vignetted field. This, of course, is inconsequential in telescopes designed for observing the planets.

It should be pointed out that in both telescopes presented here, secondary mirrors could be made even smaller by: placing them closer to the corrector, shortening back focus or tolerating a smaller fully-illuminated field.

4.3.6   Here and Now

Over the last few years, I have spent a great deal of time working with variations of the Houghton-Cassegrain telescope and have developed many diffraction limited combinations. This is not, however, a by-product of some great lens designing capability on my part, but rather the fact that the Houghton has possibly the most forgiving tolerances of any compound system.

Why then have the major manufacturers not put a Houghton into their product line. The answer seems to rest more in dollars and cents than in optical performance. It costs tens of thousands of dollars to tool up for a new product and, with the design still virtually unknown to the telescope buying public, tooling up for this particular product could be quite risky.

4.3.7   The Down Side

As with any design, the Houghton-Cassegrain telescope has its share of drawbacks. Even so, each may become acceptable with a slight shift in philosophy.

4.3.8   The Cost of the Glass

A few telescope makers have told me they would not consider making a Houghton because of the high cost of optical glass needed for the corrector. However, the cost of the glass should not be prohibitive for telescopes of small to moderate aperture.

Certainly, if one considers that Newtonians utilize no correctors at all and

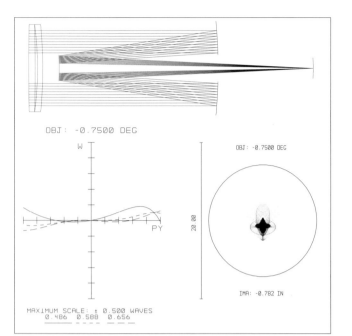

Fig. 4.3.1 *Shown her is the spot diagram and OPD (Optical Path Difference) plots for rays at the edge of a 1.5-degree field of view in the 6-inch f/10 telescope. This system was designed with Zemax configured to produce the smallest possible image blur, without regard to wavefront analysis.*

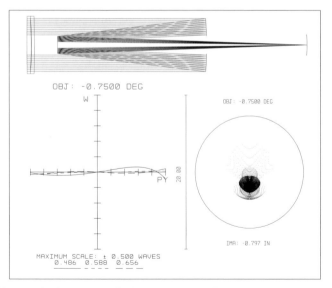

Fig. 4.3.2 *Shown in this figure is virtually the same system as shown in Figure 4.3.1, but one in which RMS wavefront calculations were used as the basis of optimization. Both graphics indicate this telescope would be a very good visual instrument. Still, in cases where only the best option will do, care should be taken in choosing optimization criteria.*

Surface	Radius	Thickness	Glass	Diameter
1	200	0.5	Bk7	6.5
2	−92.186	0.157		6.5
3	−24.735	0.4	Bk7	6.5
4	−40.725	13.456		6.5
5	−36.005	−12.174	Mirror	6.0
6	−16.572	19.5	Mirror	1.953
Image Plane 7	−15.063			1.578

Surface	Radius	Thickness	Glass	Diameter
1	196.926	0.491	Bk7	6.5
2	−62.927	0.054		6.5
3	−38.456	0.3214	Bk7	6.5
4	−196.927	20		6.5
5	−51.022	-17.272	Mirror	6
6	−22.937	29	Mirror	1.940
Image Plane 7	−20	29		2.357

Fig. 4.3.3 *Differences in the diameters of the corrector elements have no effect on the aberrations. Data in that column is only relative to the drawing of the elements. All clear apertures are exactly 6 inches. Keep in mind that the over all central obstructions will be slightly larger than the diameter of the secondary mirror alone. Digits beyond three places may be ignored.*

that most SCTs have thin correctors made of soda-lime plate glass, the cost might seem a bit high. However, the 6.5-inch Bk-7 blanks for my telescope came pre-generated for about $160.00 each and, considering the prospective performance of the telescope they are to yield, I didn't find the cost to be inordinate.

Mr. Sigler states that while planning to build his 10-inch Houghton, he considered making the corrector out of plate glass and reminded us that even the 48-inch Schmidt Camera on Palomar originally had a plate glass corrector. Still, while plate glass could be used to make the corrector, striae, bubbles and inconsistencies in index would make the use of optical glass the only real choice for those planning a truly high-performance telescope.

4.3.9 The Central Obstruction

Whether the Houghton's secondary and baffle create a 40% obstruction or only a 25% obstruction, some lunar and planetary observers will never find the design acceptable because contrast is slightly reduced in all obstructed telescopes— Schmidts, Maksutovs, Gregorians, Houghtons, and even Newtonians.

I take a somewhat different approach to the matter. Complex mathematics aside, a well-made Houghton-Cassegrain, with a clear aperture of 6 inches and a 2-inch obstruction, should exhibit contrast to that of a 4-inch apochromatic refractor. This leads us to balancing philosophies. A 6-inch Houghton-Cassegrain will not provide contrast equal to what one expects from an apochromatic refractor of the same aperture. Yet, when comparing a 6-inch Houghton to a 4-inch apochromatic refractor of the same focal length, the Houghton will offer a number of desirable advantages. These features include:

1. The contrast of the 4-inch apochromatic refractor
2. Better color correction than most apochromatic refractors
3. The ability to use a much shorter tube
4. The ability to use a much lighter mount
5. The light grasp and resolution of a 6-inch telescope

4.3.10 Testing the Corrector

While the Houghton telescope may be constructed using all-spherical surfaces, the testing of the corrector is not to be taken lightly. In my designs, three of the four surfaces are convex and cannot be tested directly without the aid of test plates or sophisticated testing setups. Some ATMs claim good test results may be obtained using only a high quality spherometer. However, anyone considering a design utilizing surfaces that cannot be ground and tested against each other might consider having plate glass tools generated along with the corrector blanks. These may then be tested with the inverse figure being applied to the surface that cannot be tested directly. The additional cost of test plates would not be prohibitive if ATMs banded together (regionally) to order corrector blanks for telescopes of the same aperture and focal length as was done with group purchases of Maksutov corrector blanks 40 years ago. Also, keep in mind the purpose of this article is not to provide specifications for the Houghtons that are the easiest to build, but rather to show that a telescope far easier to make than Schmidts and Maksutovs—and capable of out-performing them as well—is within the reach of every serious amateur astronomer.

4.3.11 Conclusion

I enjoy designing Cassegrains telescopes. However, my purpose has not been to promote one Houghton design over another. As with any family of telescopes, every instrument will have its own strengths and weaknesses—depending on how it is to be used. Instead, it has been my intent to encourage a few intrepid telescope lovers to build a Houghton for themselves. It is my belief that when amateur versions of these instruments start making their way to Stellafane, Table Mountain, Astrofest and Riverside, a new sensation will be created in the telescope market place like the sensations created by the Gregory-Maksutov or the advent of the affordable SCT. And with CCD cameras growing ever smaller, the Houghton—especially the Lurie derivative—could easily become a top-notch prime focus, wide-field astrograph.

Features of the Houghton-Cassegrain telescope include:

1. Compact design.

2. Excellent image quality over a large field of view.

3. All spherical surfaces (in some variations the two elements of the corrector may be ground and tested against each other.)

4. Ghost images diverge.

5. Corrector may be made from matching and inexpensive glass types, such as Bk-7.

6. The corrector is much easier to make than those required to make a good Maksutov or Schmidt-Cassegrain.

7. The design is very forgiving with respect to spacings and curvatures.

8. A closed tube.

9. Spiders are unnecessary since the secondary may be mounted either on the back of the corrector or on a holder passing through it.

4.3.12 Additional Reading

1. J. L. Houghton, US patent 2,350,112, May 30, 1944.

2. Robert D, Sigler, Compound Catadioptric Telescopes with All-Spherical Surfaces, *Applied Optics 17,* (1978), pp. 1519–1526.

3. Robert D. Sigler, Two Lens, All-Spherical Cassegrain Catadioptric Telescope Design, *Advanced Telescope Making Techniques, Volume 1*, by Allan Mackintosh, Willmann-Bell, Inc., Richmond, Virginia, pp. 192–198.

4. H.G.J. Rutten and Martin van Venrooij, *Telescope Optics, Evaluation and Design*, Willmann-Bell, Inc., Richmond, Virginia, pp. 126–28, 276, 348.

5. H.G.J. Rutten and Martin A.M. van Venrooij, The Houghton Telescope, An Optimum Compromise, *ATM Journal #l*, Amateur Telescope Makers Association, Seattle, Washington, pp, 5–9.

> **Ed Note:** Since this article was written, I have come up with a number of superior designs. The two designs presented here were created for this book. Also, Houghton telescopes are growing in number—usually in their Lurie configuration. I know of instruments that have been completed in Texas, Oregon, Hawaii and Illinois in the U.S. and in other countries as well. From what I have heard, three Houghtons were present at Riverside 2002.

4.4 Astrofest 1993

By Richard Walker

The beginning of my annual five and a half hour drive to Astrofest dawned early. Past experience has shown that the sooner one arrives on Friday, the closer to the north side of the field one can set up.

Astrofest is hosted each year by the Chicago Astronomical Society at Camp Sbaw-Waw-NasSee 4H Camp, near Kankakee, Illinois. This year, Astrofest 14 was held on Friday, September 10 through Sunday, the 12th.

After the usual false starts (someone always forgets something), we took off. Seven members of our local club were going this year, in three different vehicles, leaving at three different times, with the idea that we would meet and set up together once we arrived. I spent the trip in the back of a motor home, reviewing star maps. The skies began to look better as we drew closer to Kankakee, and we were all looking forward to a good night of observing. We arrived around 2:30 pm and noticed that the telescope field was already almost half-full—"Early" is getting earlier every year. After finding the other members of our group, we pulled our vehicle into position and started unloading equipment.

I was not there more than 15 minutes when a friend asked, "You got $50?" It seemed that someone at the swap meet had a set of gears just perfect to replace the drive on the telescope we are putting into my observatory. The swap meet at Astrofest is always one of the highlights of the event, and this year's was one of the largest ever. It was also one of the earliest, with reports that it started as soon as the first vendor set up Friday morning. By Saturday morning tables, canopies, pickup truck beds and blankets on the ground stretched all the way from the dining hall past the first set of bunkhouses.

After check-in (560 people pre-registered this year, probably to take advantage of the significant discounts offered for early registration), I grabbed my camera and started doing what I really go to Astrofest for: talking with people and looking at telescopes. I did not take an instrument this year, so I had a lot of time to look at and through those brought by others.

The first telescope I looked at was pointed out to me across the field. Steve Lome from the NW Suburban Astronomers Group (Schaumburg, IL), ground and finished a 6-inch *f*/20 mirror, then realized that he did not know how to economically mount such a long telescope; see Figure 4.4.1. He settled on a design that he had seen at last year's Astrofest. Working entirely in wood, Steve built a very well-finished horseshoe-cradle mount for his square wood-tubed Newtonian. An interesting addition was green indicator lights on the corners of the base to let people see where it was in the dark. Steve's efforts received a Merit Award for Overall Design and Craftmanship.

Just a short distance from us was Al Woods with another interesting looking telescope atop a familiar fork mount; see Figure 4.4.2. I had seen his fork mount before, at my first Astrofest about six years ago, but not the telescope. As I got closer I noticed the letters "SPT" on the front end! Sure enough, Al confirmed that he had built a 6-inch *f*/12 Stevik-Paul. He started it in July after getting an advance copy of the article that appeared in *ATMJ* #3 (see article beginning on page 71).

Al said that he had only finished the final alignment of the four mirrors the night before, and was hoping for a clear night to try it out. Baffling the telescope so as to block stray light, yet not obstruct the light path, was the trickiest part of the assembly.

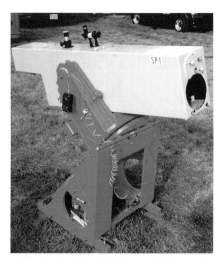

Fig. 4.4.1 *Steve Lome's award winning 6-inch f/20 Newtonian.*

Fig. 4.4.2 *Al Woods' Stevick-Paul telescope won a merit award for Groundbreaking Exotic Optical Design. This design was the cover story for ATMJ #3.*

With the sun still up, a few properly equipped telescopes were directed at it in hopes of finding sunspot activity. There were none. One of these belonged to Steve Sands of the St. Louis Astronomical Society. Steve built his 6-inch *f*/9 solar scope with an unaluminized primary, a 5% transmission one-way mirror, which serves as both a filter and secondary mirror, and a #9 welder's glass at the focuser. Steve also brought a 20-inch, *f*/4 (Galaxy Optics) Dobsonian.

After dinner I put the camera away and headed out into the darkness to look through some scopes. They ranged from 60 mm refractors up to Bob Ross' 36-inch *f*/5 Obsession; see Figure 4.4.3. This telescope, built by Dave Kreige, was the largest ever to appear at Astrofest. The eyepiece is just over 14 feet above the ground, when viewing at the zenith. The lines to view with this scope were about 45 minutes long all night. The next day, I got a chance to test its mount a little; it moved as smoothly as any of Dave's other telescopes.

Later in the evening, I returned for a peek through the SPT. I found the instrument to show a slight amount of astigmatism. However, Al suggested that much of the problem could be due to a slight misalignment of the mirror tilt angles. When the brief seconds of good seeing allowed, the SPT showed a clear, sharp image of Saturn. Al won a Merit Award for Groundbreaking Exotic Optical Design for his SPT. I expect to see a few more of these instruments in the future.

New at Astrofest this year was the "electric area". With the increase in computer usage and CCD and video imaging, a small area near the hall was filled with the glow of computer monitors. CCD imaging, along with conventional photography, was also going on at several locations throughout the telescope field. Richard Berry was showing his Merit Award-winning prototype for a very affordable CCD

Fig. 4.4.3 *This 36-inch f/15 Obsession was the largest telescope at this year's Astrofest. This instrument belongs to Bob Ross and was built by Dave Kreige.*

Fig. 4.4.4 *Winning the Merit Award for Outstanding Accessory Systems was the 20-inch f/4.2 built by Steven Aggas of Washington, Michigan.*

Fig. 4.4.5 *Another Merit Award winner was Thane Bopp's "Travel Telescope."*

Fig. 4.4.6 *Jason Kauffold of Sun Prairie, Wisconsin won a Merit Award for Excellent First Attempt by a Junior ATM with his 8-inch Dobsonian.*

camera kit, which hopefully will get to the market sometime before the end of the year. As suggested before, CCD imaging will change amateur astronomy.

Even though the skies never got really dark and the dew was quite troublesome, I was able to look through a lot of telescopes at many different objects—even finding a few objects myself. The rising third quarter Moon gave me a perfect excuse to go to sleep. It had been a long day.

Everyone was up early on Saturday, and by 7:30 the swap meet was going full strength. My first stop of the day was the dining hall. Then, donuts in hand, I

Fig. 4.4.7 *Drake Damerau received a Merit Award for Excellent Design & Craftsmanship. His Crayford focuser uses Teflon bearings.*

Fig. 4.4.8 *(Above) A close-up of Jim Carrol's TV camera hook-up.*

Fig. 4.4.9 *(Below) Jim Carrol's 18-inch telescope features an on-axis TV camera. This instrument has a mirror made by Dan Joyce and a modified JMI mount.*

Fig. 4.4.10 *(Left) Chris Engelhorn's 6-inch off-axis reflector. This instrument was featured in* Sky & Telescope *for April 1993, pages 93–95.*

Fig. 4.4.11 *(Right) José Sasián with another of his off-axis telescopes.*

was off to the swap meet, looking for bargains. With commercial and non-commercial vendors selling everything from full telescopes to all of the pieces you would need to build one; magazines, books, t-shirts, eyepieces and a lot of stuff that I had to ask "What exactly is this?", it was possible to buy almost anything. In addition to the gears mentioned earlier, I picked up a mirror blank, some books, Teflon and some stuff I still have not unpacked. After browsing for a couple of hours, I headed back to the telescope field.

At my first Astrofest, instruments built around Coulter 17.5-inch mirrors were the most numerous large telescopes, with only two larger. One was Richard Berry's 20-inch Dobsonian, and the other was the first showing of David Kriege's original 20-inch Obsession. Now with their widespread availability, 20-inch scopes seem to be the rule. Most of the innovations seen this year had more to do with making the telescopes easier to use than with the basic design.

Steven Aggas from Washington, Michigan, brought his 20-inch $f/4.2$ (Pegasus Optics) Dobsonian; see Figure 4.4.4. The mirror for this instrument is supported by a 36-point cell equipped with micro-switches with indicators to show if the mirror is not exactly centered in the cell. The focuser cage includes a built-in filter slide moved by a control wheel near the focuser. Again, there are indicators which light up to let you know which filter is in place, and that it is properly centered in front of the focuser. Steven won a Merit Award for Outstanding Accessory Systems.

In the Junior ATM category, Jason Kauffold of Sun Prairie, Wisconsin, won a Merit Award for Excellent First Attempt by a Junior ATM. Jason's 8-inch Dobsonian uses a Meade primary, but otherwise was built completely by him (Figure 4.4.6).

Thane Bopp of St. Charles, Missouri, brought his 10-inch $f/4.5$ Travel Telescope. Transport is easy as the entire telescope, including optics, weighs just 24 lbs and is designed to break down and fit into a moderate sized suitcase. This telescope was built for a trip to Australia and won a Merit Award (Figure 4.4.5).

A Design and Craftmanship Merit Award went to Dave Steven of Bartlett, Illinois, for his 10-inch, G5 fork mount with a ribbon drive.

Another large Dobsonian was built by Don Dewitt of Green Bay, Wisconsin. His 18-inch $f/5.5$ (Galaxy Optics), constructed of a foam and fiberglass composite was not only light yet strong, it also had a distinctive appearance. The telescope included an internal filter wheel and won a Merit Award for Overall Design and Craftmanship.

Drake Damerau (and friends) brought his 20-inch $f/5$ (Galaxy Optics) Dobsonian with an 18-point cell, a 3-arm spider, and a Crayford focuser with Teflon bearings. Its tube is a ½-inch plywood frame, skinned in dark blue Formica. A Merit Award for Excellent Design and Craftmanship was awarded for the focuser. In use this focuser is very stable and very smooth, even at high powers, (Figure 4.4.7).

The biggest change that I saw in the telescope field this year was the relatively small number of homebuilt optics. This is especially true for anything larger

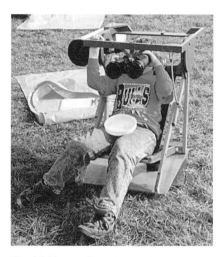

Fig. 4.4.12 *(Left) This 20-inch Newtonian belongs to Dan Joyce. Dan is a superb mirror maker whose work has found its way into a number of prominent instruments including Don Parker's 16-inch.*

Fig. 4.4.13 *Now this is living! This arrangement (builder unknown) allows the observer to scan the sky with a large binocular and then move in on his celestial prey with a 4¼-inch Newtonian.*

Fig. 4.4.14 *We have no names to go with these beautifully crafted instruments.*

that about 12.5 inches. It just shows how successful certain commercial optics suppliers are at providing good products at reasonable prices.

After dinner in the hall, the Astrophoto awards were announced. At Astrofest, the winners are determined by a ballot of the attendees. This year's winners were: Deep Sky, Detleff Schmidt for his picture of the Cresent Nebula; Solar System, Gordon Garcia for his photo of Solar Prominences.

The Telescope Making Merit Awards were then given out, followed by the last of the door prizes.

We left the hall to find the sky completely overcast, with lightning to the north. As it got darker, the thunder and lightning grew in intensity. Antennas for weather radios sprouted up, and the forecasts were not favorable: thunderstorms with a slight chance of some partial clearing much later. This prompted many people to pack up their equipment and leave. By Sunday morning, over half the telescope field was empty.

Despite the rain on Saturday night, this year's Astrofest was a success for the 588 people in attendance. Seeing old friends and making new ones, seeing the newest telescopes and ideas, and looking through them at many different objects on Friday night—that is what makes Astrofest an event to look forward to. It just gets bigger and better each year.

4.4.1   Astrofest 1993 Merit Awards

- Drake Damerau, Roseville, MI, Excellent Design and Craftsmanship, Home-made Crayford focuser.

- Ed Jones, Cincinnati, OH, 12½-inch f/5 Newtonian with home-made zero-power reflex sight.

- Don DeWitt, Green Bay, WI, Overall Design and Craftsmanship, 18-inch f/5.5 Newtonian/Dobsonian with low profile and composite materials.

- John Pratte, Charleston, IL, "One-man" Transport System, 10-inch f/6 Newtonian on German Equatorial Mount.

- Ray Minnich, Oak Lawn, IL, Technical Advance of Photography, Electronically Cooled Film Camera.

- Steve Lome, Overall Design and Craftsmanship, 6-inch f/20 Planetary Newtonian, English Yoke Polar Disk Mount with ½-inch diagonal.

- Dave Stevens, Bartlett, IL, Design and Craftsmanship, 10-inch f/5 with Fork Mount (bright yellow telescope!).

- Thane Bopp, St. Charles, MO, Largest "Carry-All" Telescope, 10-inch f/4.5 "Suitcase" Altazimuth Newtonian.

- Jason Kauffold, Excellent First Telescope Attempt by a Junior ATM.

- Richard Berry, Cedar Grove, WI, Technical Advancement of Electronic Imaging, Prototype of Affordable Cooled CCD Camera Kit.

- Al Woods, Kirkwood, MO, Groundbreaking Exotic Optical Design, 6-inch, Stevick-Paul Telescope.
- Steven Aggas, Washington, MI, Outstanding Accessory Systems, 3-filter Slide, Electrical Primary Adjustment and more.

4.5   The Krupa Collimator

By Jordan D. Marche II

The device illustrated here is a simple collimation tool that can be made by practically any ATM. It is useful for performing a check on the alignment of either refractor objectives or Newtonian reflectors. Easier to construct than the Cheshire eyepiece, it places a small bright "point" source of illumination (such as a low-voltage LED) on the axis of the telescope in place of the usual eyepiece. Light returned from the objective is viewed through the clear Plexiglas "window" holding the source. With a properly collimated instrument, one sees the reflected light returned exactly to the center of the device. By placing one's eye close enough, the two images may be superimposed. The simplicity and elegance of the tool permit a fast and easy check of alignment in these optical systems, even at night.

The device consists of nothing more than a piece of 1¼-inch o.d. metal tubing (this can be enlarged for 2-inch focusers), which holds a larger diameter piece of ³⁄₁₆-inch thick Plexiglas, machined at one end to fit the inside diameter of the tubing. Obtain the smallest possible diameter LED and drill a precisely centered hole through the Plexiglas to snugly accommodate it. This operation should preferably be done on a lathe. Paint the walls of the hole black to prevent light from escaping through its sides. Connect the LED (the back surface of which should also be blackened) to a 100-ohm resistor in series and power both items from a pair of AA batteries (an on-off switch is optional).

When testing a refractor objective in the manner described by Richard Berry (*Build Your Own Telescope* (1985), upper illustrations, p. 160), first cover the objective completely with black cloth or a lens cap. Install the collimation device in the focusing tube and light the LED. One looks to see that the reflections from all four lens surfaces appear merged into one. You may also find it useful to have a second hole, drilled near to one edge of the device, to place the LED intentionally off-axis. In this case, all reflections should appear in a straight line with a properly collimated objective.

When used on Newtonian reflectors, especially short-focus Dobsonians, a small center dot must first be placed on the primary mirror. Thereby, one can check to see that the source, center dot, and the reflected rays of light all coincide precisely. This technique works even if the diagonal mirror is intentionally "offset." Cassegrain-type reflectors cannot be collimated with the device, because there is no return of the light after twin reflections from the secondary and primary.

I am not the originator of this tool; it was invented by Mr. Jack Krupa of Wil-

Fig. 4.5.1

low Street, PA, during his refurbishing of a secondhand Unitron refractor. I had the pleasure of answering many of Jack's questions about telescopes during the six years that I taught astronomy at nearby Lancaster. He is a retired watchmaker, expert machinist, and amateur astronomy enthusiast. Although I have encouraged him for some time to submit a report of his collimation device, he has been reluctant to do so. Recently, I obtained his permission to publish this article so that others may enjoy the benefit of his keen insight. I hereby designate this tool, the Krupa Collimator, in his honor. It offers a superior method of optical alignment that is readily adapted to the two most common types of telescopes constructed by amateurs.

4.6   The Eye and the Use of Telescope Optics, Part II
By Dr. Richard A. Buchroeder

Q. When scanning across objects like Jupiter, a friend reports he can see finer detail.

A. I find this hard to believe, because as he moves his eye, he doesn't see anything because his brain filters out the blur from his consciousness. This is called scotopic suppression. Saccatic motion is a different thing. A possibility is that by moving the object to a part of the retina which hasn't been bleached yet (by prior exposure to the image), he may be able to see something better.

Q. When I observe Mars with a telescope with one eye, and then change to the other eye, I notice a differently colored image. Is this an effect of bleaching?

A. Yes.

Q. When I go to a dark site and first see the Milky Way, it is clearly outlined against the dark sky. But after awhile, sky brightness seems to rise and the Milky Way is no longer conspicuous. Is this because cone vision does not dark adapt, and

Fig. 4.6.1 *Dr. Ted Thall.*

is the principal star sensor, while rod vision does adapt and so eventually senses the background of unresolved stars and sky glow?

A. A dark adaption effect is the most likely explanation I can think of. Try readapting your eyes to light again, and then repeat the observation.

Q. Does the cornea regenerate itself, whether from an accident or RK?

A. Scars are formed, which hopefully become transparent. Scarring causes glare. Radial keratotomy can cause glare, with laser or knife. The matter is still being studied. There is no agency to approve RK. But there is an agency, FDA, which has to approve excimer lasers, and they have not yet been approved in the US.

Q. What about corneal transplants, if the RK procedure fails?

A. That's a real problem. Sewing it up causes irregular astigmatism. Your vision will not be as good as before. Surgical procedures work approximately 80% of the time, but there is risk. Anytime you have therapy, whether what you want to do is good or bad depends on what else is available. RK costs between $500 to $2000, hazardous, while spectacles are cheap and always a winner.

Q. The reason why we are interested in RK or contacts is because nearly all eyepieces are designed for non-eyeglass wearers.

A. Right, the eye relief stinks, and, how come, for a bunch of smart guys, the guys that make slit lamps don't put a plastic ring around the eyepiece so your glasses don't go 'ding, ding' against the metal? The eye relief is never long enough! That annoys the heck out of me.

C. I'm glad I'm not the only one that feels that way about it! There is no technical reason why eyepieces cannot have longer eye relief, and have soft rather than metallic eyecaps. The issue is apparently one of tradition and appearance. To fit the 95th percentile, spectacle wearers require at least 23 mm eye relief in an eyepiece.

Q. Intraocular implants that you've used were spherical and made from homogeneous material. Doesn't that suggest the lens does not rely on gradient index material and asphericity for good vision?

A. I've run this on CodeV, and looked at it in great detail; this is my own research. When you look at aberrations of the whole eye, it turns out that it doesn't make much difference what you do in the lens to correct aberration. You are not going to affect the spot diagram—so a pretty lousy lens can pass a test. But some shapes do not work very well and can cause distortion. But plano convex or double convex, doesn't make much difference.

Q. I wore hard contacts for a while, and I presume they were nominally spherical. They gave me very acute vision, when they were at their best, but of course, they float up and down each time you blink your eyes. This suggests that a spherical cornea can produce good acuity. Therefore, need we invoke anything other than spherical curves to model the optical behavior of the human eye?

A. I'm not convinced that's true. You get a sharper spot diagram when you aspherize the model. My contention is that aspherics and gradient index materials are an unnecessary complication in explaining why our eye works so well.

Q. My contention is that aspherics and gradient index materials are an unnecessary complication in explaining why our eye works so well.

A. That's being tossed right now, and there are two schools of thought. The FDA, which is reviewing the intraocular lens, is of the mind that it doesn't make much difference; but on the other hand, they don't know where to get good aspheric optics. The tip of the cornea can be measured, and is an ellipsoid.

Q. Is there any way to reverse the effects of aging on vision, say with exercise or diet?

A. If I find the fountain of youth, I'll let you know. No.

Q. Are certain individuals gifted with vision quite superior to the rest of us?

A. There's no question that you can train yourself to be a better observer. But you can't change your genes. Yes, there are definitely individuals who have acuities as high as 20:10. Normal vision, however, is 20:15. Snellen, 150 years ago, got a group of locals together with whatever vision they had, and he did a statistical average on this group. What he did at the time was perfectly reasonable, but "normal" had not been sufficiently established.

Q. Do people with high sensitivity have poor acuity, and vice-versa?

A. A normal person has both high sensitivity and high acuity.

Q. The photoreceptors lie at the bottom of an approximately 500-micron stack of cells, terminated with ganglia that lead out the optic nerve. Sounds like this would lead to a seriously blurred image, but apparently it does not. Does the light focus at the top or the bottom of this stack?

A. That has been the topic of much research. We believe it focuses on the bottom, on the photoreceptors, which have a waveguide-like structure. This is believed to cause the Stiles-Crawford effect, which states that as the rays come in at greater angles, the light is not conducted efficiently. The 500 microns of stuff is so transparent that it is hard to measure where the light is focusing. In the fovea it's all pushed out of the way and reaches the photoreceptors immediately, so you don't have the scattering problem.

Q. Is it possible the different photoreceptors are at different heights, to compensate for the primary color of the eye?

A. No, they are not. They are at the same level.

Q. Wouldn't the acuity of our eye-telescope combination improve if we designed our telescopes to compensate for this residual primary color, which may be one or two diopters difference in focus from the F to C wavelengths?

A. Hasn't that been tried before?

C. Yes, color compensating optics were designed and reported in *JOSA* decades ago, and measured acuity did improve. However, a diopter or more of primary chromatic aberration would be judged chromatically imperfect and would probably not be accepted by viewers. Again, an aesthetic judgment. Curiously, we perceive nothing to produce a whiter image than an all-reflecting telescope, as even fractions of a wavelength of tertiary spectrum in an apochromatic refractor can easily be detected with our eyes. It is interesting that the Airy disk is not perceived as being colored, since its rings are wavelength dependent.

Q. I recently broke my single-vision close-up reading glasses and had to use bifocals, which I despise, while waiting for replacements. As I read, I found my eyes becoming strained, and eventually blurred, perhaps caused by my eyes slipping across the dividing line on the bifocals. This condition persisted for more than an hour, and 'star testing' on city lights showed more intricate patterns than could be caused just by defocus. What was going on here?

A. It's your lens. You're making an accommodative effort, but the elasticity is less as you get older. In straining the lens, you get odd patterns, and there is considerable hysteresis. Bifocals need to be fitted much more carefully than single vision lenses. As an aside, gradient power lenses, which don't have that 'telling' dividing line that bifocals do, aren't satisfying to critical users.

C. Yes, I bought a pair of the Varilux brand, and found that my lateral field of view is highly aberrated, making walking difficult. Reading a book requires almost constant head-turning, since swiveling eyes affects the image so badly,

Q. From time to time, when viewing Venus, I can see multiple images within one eye.

A. This is called monocular diplopia. Monocular diplopia is physiologic, and we all have multiple reflections from the interfaces in our own eye. Certain conditions can bring it out.

Q. My tabby cat's eyes have a greenish retroreflection, whereas my Siamese cats show a red retroreflection in flash photographs, just like people. Please comment.

A. People only have redeye reflections. Redeye is a defocused reflection from the major retinal blood vessels that are on the surface of the retina, not the back of the eye. The back of the eye has a heavily pigmented layer, called the retinal pigmented epithelium, that has the job of absorbing the light not absorbed by the photodetectors, whose quantum efficiency is only about 10%. If you didn't have it, scattered light would wipe out spatial acuity. All those blood vessels you see on flash camera photographs are in front of the retina.

The cats have something different. Instead of the retinal pigmented epithelium we have, they have a reflective layer. That is intentional; they want a double pass because they are nocturnal, and this doubles their quantum efficiency. The price they pay is reduced spatial acuity, but they see moving objects very well. We have better spatial resolvers; they, motion sensors.

Q. Since these layers are so reflective, why don't you get serious reflections against your eyepiece lenses?

A. Although it looks bright on your redeye photographs, it isn't as bad as it looks.

Q. Some observers deliberately turn on the dome lights when observing bright, extended objects like the Moon or planets. Doesn't this lower contrast?

A. It's painful to observe bright, extended objects, and I frankly don't want to be dark adapted when I do. Your average level is set low by being predominantly surrounded by darkness; you need to set this level near that of the object under view by raising the ambient light level. This puts the dynamic range where you want it to be.

Q. Staring at the sun will produce blind spots. Since our field of vision is some 120°, why don't we have spots burned all over the periphery of our retina?

A. Solar retinopathy is a rare bird. We see a couple of cases after eclipses now and then. Even in the foveal region, where your optics are best, it takes a few minutes to cause damage, because there is a cooling layer of blood vessels beneath the retina. A temperature rise of just 5° will cause damage. Most cases we see are in subjects that are druggies.

C. Needless to say, any apparatus used to view the sun must be carefully thought out, especially in regard to failure modes. Never trust reputation or assurances; test anything you propose to use before you look through it. The government doesn't have time to check out all the fool devices put on the market, especially in optics.

Q. What kind of telescopes do you prefer?

A. There are a lot of things that go into choosing a telescope, one of which is convenience. I like to take pictures, so I need adaptability to 35-mm, and I'd like to do large format like 4 x 5. Some people talk about great optics, and never test them. The best design, if not well made, is not going to work very well. I'm taking lessons in optical fabrication from Bob Goff, a master optician here in Tucson.

Q. What are the long and short term effects of cigarettes and alcohol on night vision?

A. Not good. There are things called tobacco optic neuropathy and alcohol neuropathy. Although not bad for your retina, they are for your nerves. Alcoholism will fry your optic nerve. We can only pick it up when it is very gross; the optic nerve turns pale. However, there is no physiological evidence that moderate smoking and drinking affects your vision.

Q. What would be the effect of atropine on the eye to obtain a 10 or 12 mm pupil to see greater limiting magnitude threshold?

A. First, you don't want to self-administer atropine without a doctor's instructions since it has a lingering effect, ruins your accommodation, and has many possible side effects, including death. Second, it will increase light scatter and make things worse. The Stiles-Crawford effect precludes any significant advantage, even under the best of conditions.

Q. What about "floaters" when using a small exit pupil?

A. Everybody gets floaters. They are easier to see with a small beam. What you are seeing is a cell shadow cast onto the retina. But if one day you see a whole bunch of them, then you could have torn your retina, and if accompanied with sensations of flashing light, you should get to your ophthalmologist immediately. If caught early, it will only be a minor inconvenience. Floaters are in the aqueous humour.

Q. A friend asks, "the pseudoscopic effect in beamsplitter binocular, why does the mind interpret this as three-dimensional?"

A. One image split into two identical copies? First, pseudoscopic actually means reversed 3-D, so it isn't the right term. Disparities in cues are what give true and pseudoscopic vision. What you describe may be a couple of things. Ten percent of the population doesn't have stereovision. There are a lot of people out there who don't know they lack it. Unlike colorblindness, unless you've experienced true stereovision, you don't know what you're missing. Depth perception is not the same thing as 3-D. There are lots of monocular cues as to depth perception. But some people look with only one eye, and don't know they lack true stereovision. Let's assume that this is not what is going on here. Perhaps he is looking at one image from two different angles. But, other than that, I have no idea!

C. You can actually get a curved 3-D image from two identical monocular images viewed through aberrated eyepieces in which your two eyes are not perfectly aligned on the optical axes, nor positioned at the exit pupil. Field curvature and astigmatism in the eyepiece produce a curved image to be viewed by each eye. Your perspective in each eye is different when the IPD and exit pupils do not match those of your eyes. Seen from different perspective points, you can indeed see stereoscopic, flat, or pseudoscopic images, depending on the state of adjustment of the binocular device. Eyepieces like Naglers and Meade UltraWides do not produce this false curvature, because they are sensibly free from field curvature and astigmatism. Poor eyepieces, such as Kellners, are best at producing the "three dimensional" images. The impossibility of obtaining 3-D images by viewing two collimated, sensibly flat, identical images can be verified by studying a Don Parker Planetary Video on your television screen through two pieces of tubing. The 3-D effect with binocular attachments, often incorrectly and misleadingly termed "Stereo Viewers", is not an illusion, but is a curved virtual image produced by eyepiece aberrations and maladjustment of the binocular attachment,

Q. What advances are expected in cataract removal in the future?

A. A whole bunch. The surgery is already good, and will get better. Stay away from multifocal lenses, though, for reasons that should be apparent. An accommodating lens will be here eventually, but it's years away. Improvements in prosthetic devices are ones that proceed toward more physiologic devices.

C. Our tape ran out, but the conversation continued without further documentation. Dr. Thall saw my 18 x 120 modified TOKO binoculars that have soft ABS eyecaps to obviate eyeglass damage, and long eye relief (23 mm) to allow eyeglass wearers to see the field stops of the eyepiece. He was impressed and asked why designers did not implement such improvements in virtually all forms of optical instruments, from lavishly expensive funduscopes to common binoculars. I said that vendors try to shave costs everywhere, that styling is a big factor in product design, and long eye-relief optics appear bulky and awkward and, therefore, unfashionable. Consumers determine what will sell, and if it doesn't sell, then it disappears. I've seen some good products discontinued for lack of consumer interest.

4.7 Letter to the Editor

An editorial clarification from David Nagler concerning our interview with Dr. Thall in Issues 3 and 4.

4.7.1 Eye Relief vs. Field of View

Dear Bill: I believe some editorial clarification is due on Dr. Thall's comments regarding eyepiece construction and eye-relief. First, rubber eyeguards are standard on many eyepieces such as all the Tele Vue models shown on the back cover of Issue #4. On the models most recommended for eyeglass wearers, these eyeguards fold down leaving a non-marring rubber ring on which one can rest a pair of glasses.

Second, it seems Dr. Buchroeder believes there is untapped eye-relief that manufacturers are keeping from customers. Well, maybe that is true. But, practicality and economics are the reasons, not "tradition and aesthetics."

How can we tap into that eye-relief? The easiest way is to narrow an eyepiece's apparent field. Result: gain in effective eye-relief at the expense of a loss of true field.

At Tele Vue, the approach is to try to obtain the widest apparent field while maintaining image quality and eye-relief. A strong negative power close to the image plane results in greater eye-relief; however, aberrations become more difficult to correct. This necessitates designs with a greater number of larger diameter and thicker elements made of more expensive materials. Result: "space-walk" views and tack-sharp images, at higher cost.

The original series Nagler and new Panoptic designs are the results of our approach. The original Nagler series has eye-relief of about 1.4x each eyepiece's focal length. They are also big! Take a look at a 13 mm Nagler and ask yourself, "Is it practical to make it bigger, heavier and more expensive than it already is?"

The 35 mm and 27 mm Panoptics were designed to extend Nagler-like image quality over a larger true field, and with enough eye-relief for eyeglass wearers. A balance of aberration correction, eye-relief, size, weight and cost was achieved with a quite large 68° apparent field.

Adding a high quality Barlow (another strong negative lens) will increase the magnification while extending the ocular's eye-relief. The additional negative power moves the eye-relief out, creating a dramatic shift in pupil position and resulting aberrations.

The Panoptic Barlow Interface is the solution. It repositions the pupil back to eliminate the aberrations; and in fact, an additional amount of correction was added to make the Panoptics equal to the Naglers in edge-of-field sharpness.

In conclusion, I would agree that there are no technical reasons why eyepieces can't have longer eye-relief, as long as cost, size, weight and apparent field sizes are balanced for a given focal length.

David Nagler
Tele Vue Optics, Inc.

4.8   Instant Wavefront Analysis

By Philip Moniot

A. Danjon and A. Couder, in *Lunettes et Télescopes* (Paris 1935), specified a double criterion for a good telescope mirror:

1. The circle of least aberration should be comparable in size with the theoretical diffraction disk, and the mean of the transverse aberrations should not exceed the diffraction disk radius.

2. The maximum wavefront error must not exceed ¼-wavelength of light, and for most of the mirror surface, the defects should be considerably less.

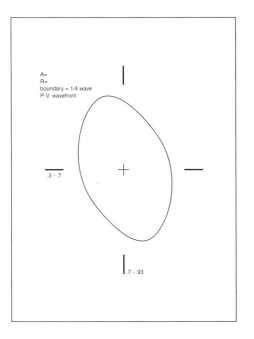

Fig. 4.8.1 *This illustration should be scaled to a width of 4.6 inches (190%).*

With Foucault test data, the first criterion can be checked quickly and accurately by using the Millies-Lacroix graphical analysis, described in the February, 1976 issue of *Sky & Telescope*. Its preparation requires only simple arithmetic, and the worker can consult the graph repeatedly to check the transverse aberrations in any desired number and location of zones throughout the figuring process.

For the second criterion, the present paper describes a graph that directly shows the wavefront error of telescope mirrors when tested in three zones using the Foucault test. It has the advantage of requiring only simple arithmetic in its preparation, but limits the worker to testing only three zones. Three zones, of course, are adequate for evaluating the small to medium-size mirrors (of moderate focal ratios) usually made by amateurs.

The wavefront analysis requires that the zones tested be located at 30%, 70% and 93% of the mirror's radius. The author computed the graph's ¼-wave boundary using the equations Roger Sinnott presented in the August 1977 *Sky & Telescope*.

The diagonal line that equally divides the graph into two sections describes plots of all conic sections in the vicinity of the paraboloid. In order along this line from lower left to upper right are found the oblate spheroids, the sphere, the ellipsoids, the paraboloid, and the hyperboloids. In addition to these shapes, the graph enables smooth curves that are not conic sections to be analyzed. Knife-edge test plots for such mirrors would fall to either side of the diagonal line.

To make the graph, first enlarge Figure 4.8.1. using a photocopy machine until the outside border measures exactly 4.6 inches wide (190%). After consulting

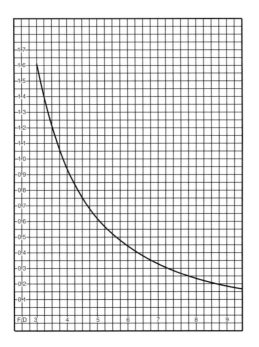

Fig. 4.8.2 *Periodic line spacing (inches) representing 0.005 inch intervals of longitudinal knife-edge travel, as a function of focal ratio.*

the graph in Figure 4.8.2 for the proper line spacing, draw the lines onto the blank and insert the mirror's theoretical values. The graph is now ready to use.

Example 1:

Figure 4.8.3 shows a graph prepared for an 8-inch $f/6$ mirror. The x-axis represents the immediate range of Foucault knife-edge differences between the 70% and 93% zones, and the y-axis represents the 30%–70% range. The intersection of the two ideal coordinates for this mirror describes the location of the ideal paraboloidal plot in the central region inside the oval ¼-wave boundary. For this mirror, the formula $r^2/2R$ prescribes, respectively, 0.031 inch and 0.033 inch for these differences.

The circle in Figure 4.8.3 marks a plot from a hypothetical set of knife-edge readings that fall a short distance inside the ¼-wave boundary. Its readings are x = 0.026 inch and y = 0.046 inch. This point is ⅔ the distance from the ideal plot (at the cross) to the boundary. The mirror's wavefront quality is, therefore, ⅙-wave, because decreasing wavefront quality in this graph is directly proportional to the distance from the ideal plot.

Example 2:

Figure 4.8.4 shows a graph prepared for a 6-inch $f/8$ mirror. In it we see the plot of knife-edge readings x = 0.027 inch and y = 0.012 inch. The plot falls 40% of the distance from ideal to ¼-wave. Thus, the wavefront error of this mirror is ⅒-wave.

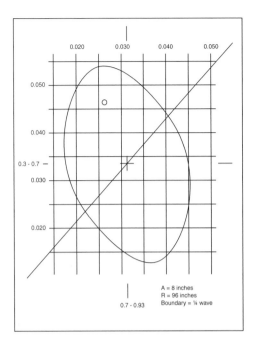

Fig. 4.8.3 *Sample test graph for 8-inch f/6 mirror.*

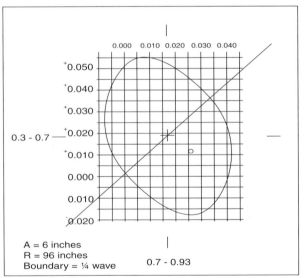

Fig. 4.8.4 *Sample test graph for 6-inch f/8 mirror.*

4.9   A "Sidereal Hinge" Equatorial Platform
By Albert Kelly

Almost a decade ago, the articles by White, Olson, and Barclay (*TM*'s #14, 15, 17, and 21) caused me to start thinking about "sidereal hinge" designs for equatorial platforms. In the interim, I have been closely associated with Andy Saulietis in the construction of several large equatorial platforms based on Andy's elegant "virtual axis" designs. I have retained the notion that a nested box platform would be the least expensive and simplest design for the average amateur astronomer to build. After all, most amateur telescope makers do not have metal lathes, milling machines, major wood-working tools, and advanced machining skills at their disposal.

The main advantage of any sidereal hinge design is the potential to harness gravity as the motive force. The primary drawback has been in developing a method to control the motion imparted by gravity. This motion must be controlled very smoothly, and must allow the sidereal hinge to rotate about its axis at as constant a rate as possible. The use of a deflatable bladder has been proposed and commercially developed. Bob Barclay, in *TM* #17, demonstrated a very clever adaptation of hydraulics as a control mechanism for a heavy-duty platform. His article outlined the problems associated with using a threaded rod, nut and a motorized drive to push against gravity, and concluded that hydraulic escapement may be the best answer.

In this article I propose a new solution that overcomes the basic difficulties outlined by Mr. Barclay, in the form of a simple, inexpensive design which amateurs of all levels of skill can implement.[1]

As in Mr. Barclay's platforms, gravity accomplishes most of the work required for tracking movement. In the prototype (see drawings), the rate at which the nested box is allowed to descend is controlled by the gradual movement of a wing-nut up a threaded rod which is turned at a controlled rate by an inexpensive DC motor powered by a 9-volt battery. Two of the good things about this particular design are: it requires very little electrical energy, and a potentiometer (variable resistor) can be added in series with the motor to provide a variable speed control or drive corrector.

If constructed properly, the "wing-nut escapement" platform is inherently very linear, in that the drive rate does not vary significantly from beginning to end of travel. Three sources of non-linearity must be considered:

1. the degree to which the suspension cables are not parallel to the threaded rod as they rise from the wing-nut to the Teflon support blocks;

---

[1] **Ed. Note**: This design has not been optimized to be all things to all people. This is a simple and inexpensive design, which will offer approximately one hour of tracking. With a number of modifications, the basic design could have its tracking time extended considerably.

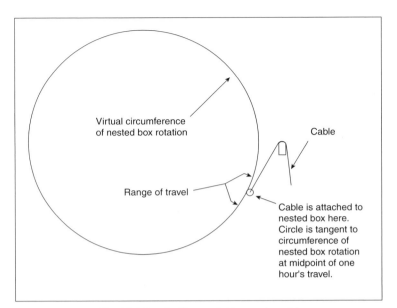

Fig. 4.9.1 *View looking south (down polar axis).*

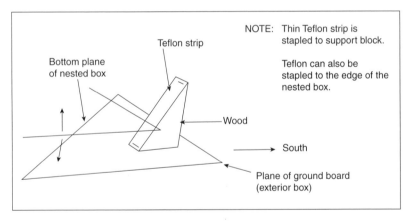

Fig. 4.9.2 *Support slide schematic.*

2. the constantly changing tangent error at the point of contact of the suspension cable to the nested box; and

3. the steady drop in available current from a battery-powered DC system.

The first of these is minimized by putting the edge of the Teflon block directly above the wing-nut attachment points. The second is minimized by assuring that the Teflon blocks are at the height and distance from the attachment point to the nested box which places the cable tangential to the circumference of nested box rotation. As shown in the accompanying schematic, the exact tangent point

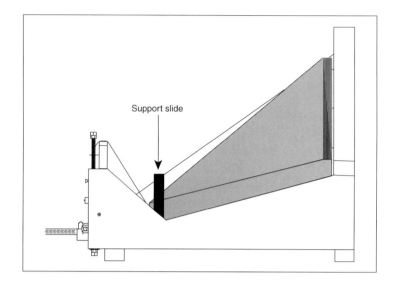

Fig. 4.9.3 *View looking north.*

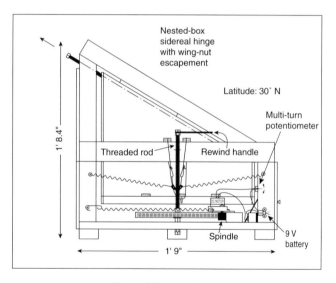

Fig. 4.9.4 *View looking east.*

should be at the mid-point of desired travel. The third is negligible, except when a battery is near the end of its lifetime. This is not much of a problem as the 9V battery used in our prototype lasted for many hours.

At first, 0.080-inch lawn trimmer line was used to connect the wing-nut to the nested box. This worked well, but was slightly springy. We are currently using a limber strand of metal cable of similar diameter, and it appears to be the best so-

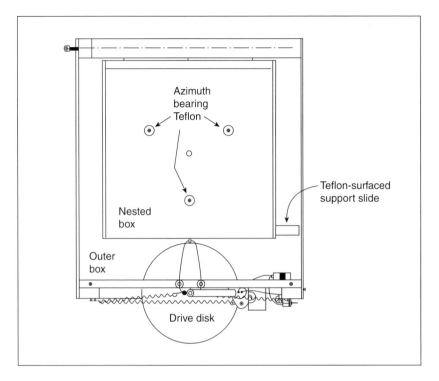

Fig. 4.9.5 *View looking straight down onto the base.*

lution thus far. Such cable is available at most hardware stores. Both the trimmer line and the stranded cable slide smoothly over the Teflon support blocks,

As with most prototypes, things were not done with great emphasis placed on economy and efficiency. The total cost of the prototype platform was about $35, which is about 40% more than we estimated. The list of materials was as follows:

- 6 sq. ft. of ¾-inch exterior grade plywood
- 1 sq. ft. of ½-inch exterior grade plywood
- 4 sq. ft, of ¼-inch light-weight lauan plywood
- 4-ft. length of 2 x 2 (pine)
- ¾-inch length of ⅝-inch wood dowel (for drive spindle)
- Drywall screws (mostly 1-inch)
- Teflon (a ¼-inch cylinder, about 3 inches long)
- ⁵⁄₁₆-inch x 36-inch threaded rod (18 threads per inch) and assorted nuts and washers
- ⁵⁄₁₆-inch wing-nut
- Lapping compound (small amount to lap threaded rod and wing-nut—auto parts store)
- ½-inch eyehooks
- Assorted springs (for tensioning the wing-nut and motor)
- 33 rpm low-current DC gearhead motor (used—$3.95)

- 2-position switch ($1.50)
- 5K-ohm multi-turn potentiometer ($0.95)
- One 9-volt battery (used to power the 12-volt motor)
- Small diameter stranded steel cable (36 inches)
- Small diameter insulated electrical wire to connect motor/potentiometer/switch/battery
- Small, surplus metal slat (3 inches long) for rewind handle
- Non-skid deck tape (for edge of large drive disk)
- Short lengths of stiff wire for assorted hooks

Construction was accomplished over one weekend by my 14 year-old son and me, using no specialized tools or special skills—any of my friends can vouch for that! We had the basic idea in mind and made the details up as we went along. As can be seen in the drawings, we neither worked for nor achieved the lowest possible platform. Obviously, the whole east side of the nested box arrangement could have been lowered by at least 6 inches to allow better swing clearance for a larger or more compact Dobsonian rocker box and tube assembly. This was not necessary for our 6-inch *f*/5, which has a tall rocker box to allow viewing at a comfortable height. A much lower platform, for a larger/heavier telescope, could be constructed along the lines of those discussed by Mr. Barclay in *TM* #21. A wing-nut escapement could still be used to provide smooth tracking.

Although the prototype looks a bit like a "Rube Goldberg special," it operates very simply and efficiently. After the platform hinge has been aligned with the axis of the earth, the telescope rocker box is attached to the nested box (which has three Teflon bearings) as if the platform were the ground board of a standard Dobsonian mount. With the motor switched off, the motor tensioning spring is swung to the southwest corner of the platform and secured to a convenient screw head to keep the drive spindle in rewind position, away from the drive disk. The rewind handle is then used to turn the threaded rod counter-clockwise, positioning the wing-nut and the platform at the start of travel. This requires one finger and a few seconds. After the motor tensioning spring is swung back to the north and re-attached to assure that the drive spindle maintains constant contact with the drive disk, the motor is switched on and observing may begin. At any time, the potentiometer may be adjusted to optimize the drive rate.

For photographic guiding, the potentiometer may be made remote from the motor through the use of an extension cord made from phono jacks and a little extra insulated wire. My son has rigged this arrangement for the prototype, and my experiments with illuminated reticle eyepieces indicate that this could be quite an inexpensive and effective astrophotographic setup if the hinge is well aligned with the earth's axis. A little declination adjustment can be provided, as suggested by Richard Berry in *TM* #14, to make a fully flexible system for long exposures with long focal lengths.

4.9.1   Construction Notes:

1. The large drive disk was jigsawed from ½-inch plywood and the non-skid tape (coarse grade) was affixed to its edge with quick-drying

epoxy. This provided all the texture necessary to eliminate slippage between the wooden drive spindle and the large disk.

2. The drive spindle ($\frac{5}{8}$-inch wood dowel) was press-fitted to the half-round steel rotor shaft extending from the motor. The low torque required to turn the large disk assured the durability of this low-tech spindle.

3. Holes were drilled through the "wings" of the wing-nut to accommodate the wire hooks which were fashioned as end attachments for the stranded cable and the tensioning springs. These springs prevent the wing-nut from rotating with respect to the nested box.

4. The pitch of the threaded rod (18 per inch), the motor RPM, the radii of the spindle and the large disk, and the distance of the screweye attach point of the nested box from the "sidereal hinge" played key roles in determining the tracking rate of the platform. Large variances in the RPM of available motors can be accommodated by changing the ratio of the aforementioned radii. Since the calculations were straightforward, no detail is provided; however, it is worthwhile to note that the prototype's final radii were based on the midrange of motor RPM available from the 9-volt battery and the potentiometer adjustment. We plan to standardize these values for other platforms to be built by the telescope-making group in the Johnson Space Center Astronomical Society.

5. A thin Teflon washer was used between the flat washer at the bottom of the threaded rod (underneath the ground board) and the bottom of the ground board. This, along with assuring that the bottom board was not clamped tightly by the nuts on the threaded rod, allowed the large disk to rotate smoothly and easily.

6. After initial construction, the threaded rod and wing-nut were lapped with both coarse and fine lapping compound by running the wing-nut up and down the threaded rod several times, using the rewind handle. The interface was under normal working tension with the telescope in place. This method, an afterthought, improved system performance significantly (after the lapping compound was completely removed!) and nearly eliminated the need for the tensioning springs seen extending outward from the wing-nut in Figures 4.9.4 and 4.9.5.

7. The nested boxes were hinged together by a length of threaded rod inserted through eyehooks. The threaded rod was supported at either end by insertion through the pre-drilled corner posts, and along its length by the eyehooks extending downward from the angled cross-piece of the outer box. This method allowed easy separation of the two boxes and was stronger and cheaper than most plated hinges.

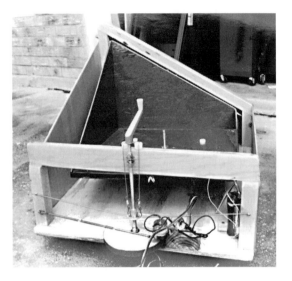

Fig. 4.9.6 *Albert Kelly's sidereal hinge equatorial platform with "disk drive" and wind-up handle in the foreground.*

4.9.2 Conclusion

Although findings are somewhat preliminary, I believe the practical limit for this prototype platform to be a 10- to 12-inch telescope, depending on weight. We have found that cylinder-based, virtual-axis designs can be easily built, and make much sturdier and lower profile platforms for large Dobsonians. Chuck Shaw of the JSCAS has designed and built several excellent virtual-axis platforms. Nevertheless, this nested-box, captured-nut escapement is probably the least expensive, most easily built equatorial platform for small-to-medium Dobsonians. Its smoothness and true sidereal accuracy make it a serious project for amateurs with not-so-serious skills.

I would be pleased to respond to any comments or questions. Readers should write to my home address (322 Gleneagles, Friendswood, TX 77546) or call me at (713) 482-5190 evenings or weekends.

4.9.3 Addendum

Since this article was initially composed, several members of the JSCAS and I have built 16 Dobsonian telescopes which incorporate my Gravity-Released-and-Battery-Assisted-Screw-System drives much as described here. We have had an a lot of fun and success with this little design. Fourteen of these are 8-inch *f*/6 Newtonians, and the other two are 10-inchers. Three have homemade mirrors.

We have learned that cost can be kept as low as I expected. We have also learned to use coupler nuts rather than wing-nuts. These provide more thread contact and can be ground round on the outside of one end and then press-fitted into large flat washers to provide the same functionality as a wing-nut. To make system performance extremely smooth perform a final lapping with Brasso; this polishes the threads on the threaded rod and the coupler nut.

A final, most important design addition has been to position a Teflon-sur-faced "support slide" at the south edge of the southwest corner of the nested box. This is made from a triangular wedge of ¾-inch thick hardwood which is screwed to the ground box (Figure 4.9.2 on page 145). The support slide provides constant, solid contact with the descending edge of the nested box, eliminating the pro-longed vibration or "swing" (6–8 seconds) previously experienced when the tele-scope was re-positioned, bumped, or focused. With this addition, all telescope vibrations dampen rapidly. The proper angle for the support slide (from the hori-zontal ground board) is the complement of the latitude for which the platform is built. Of course, the platform's feet can be shimmed for use at latitudes plus or mi-nus 5° or so from the design latitude, but this does not adversely affect the utility of the support slide.

The 14 platforms for the 8-inch f/6 "Standard JSDC Scopes" use 6-volt lan-tern batteries, ⅜-inch diameter hard rubber drive spindles, and window cranks for rewind handles.

One last addition: I have written a simple Lotus 1-2-3 spreadsheet which pro-vides design solutions when given variable parameters (threads per inch, motor rpm, disk diameter, spindle diameter, radius of attached point rotation). This makes up-sizing, down-sizing, or otherwise changing the design very simple.

4.10 From the Bench: Binoculars and Optical Coatings
By William J. Cook

Until recently, I never really considered a binocular to be of much use for astron-omy. My mind changed in a hurry after scanning the sky with a 20 x 100 that I borrowed from a friend. I don't have the money for a binocular that big, but I have been giving strong consideration to a 20 x 80. My question concerns the special anti-reflective coatings. I want to get the "biggest bang for the buck", but I am con-fused by all the claims and performance ratings that I see in the magazines and sale brochures…Can you make any recommendations?

Few features in binocular ads are touted more than the quality of anti-reflec-tive optical coatings. Some companies try to illustrate the benefits of their coatings by showing a photograph, half of which is crystal clear and the other half looking like it was shot through a piece of frosted glass. This is not really a fair compari-son. Some of the photographs show a much greater difference in clarity than one would find when comparing a non-coated lens system to the finest multi-coated system available, let alone comparing two similar coatings.

When you move from magnesium fluoride coatings, which have been the in-dustry standard for about 50 years, to multi-coatings, you pick up about 9% to

14% greater light transmission. However, when comparing the multi-coatings of two manufacturers, you might as well be trying to split hairs with an axe; and you should certainly not weigh the quality difference in the coatings too heavily when making a buying decision.

One crude way to perform an in-store test of the anti-reflective coatings of instruments you are considering would be to place your hand over the ocular and look down at the objective lens as if you were looking for dust. In an instrument with the standard magnesium fluoride coatings, you will probably be able to see your face quite well. In other instruments sporting multi-coatings, such as Leica, Zeiss, Steiner, Swarovski and Optolyth, you will see only a very dim image, illustrating that more of the light is going through the instrument and is not being reflected back out into space. In some cases, you may not be able to see your face at all!

There are those who profess to tell the quality of coatings by the color of the film. This is simply not a valid gauge. Case in point: In the 1960's binoculars with beautiful, deep blue tints became quite popular, and many supposed that they were seeing better through optics thus coated. Many of these beautiful coatings were the by-product of magnesium fluoride being deposited at a temperature which fell short of the ideal due to a production shortcut.

Some have also speculated that the coatings are to "protect" the glass surfaces. This, too, is an erroneous assumption. Magnesium fluoride is harder than the glass beneath—575 on the Knoop hardness scale vs. 520 for Bk7 glass. However, at a thickness of 4 millionths of an inch, it offers little in the way of protection.

One of the more recent trends in advertising is to claim that certain coatings cut out "100% of harmful ultra-violet," or that they greatly reduce infra-red transmission. These claims are true. However, let's consider a few things that the manufacturers might not tell you. First of all, most types of uncoated glass stop UV at a rate of about 90% per millimeter. Glass types which do transmit UV, such as Ultran 30 by Schott Glass Technologies, are rare enough to cause manufacturers to take out full-page ads in the trade magazine to promote the product.

Yes, infra-red does get through. However, just by putting a binocular to your face, you cut your exposure to infra-red to a small fraction of what you would get merely walking down the street on any given day. The bottom line: buy a binocular with some type of multi-coatings. Beyond that, make your purchase based on optical quality, structural strength, dealer reputation and price.

Issue 5

5.1 Building a Split-Ring Porter Equatorial
By John Shelley

I was very pleased with my 13-inch Dobsonian when it was new, but the trunnion bearings became worn after one season. This made the friction noticeably different between azimuth and elevation, and it became difficult to follow objects. I was jealous of those having polar mounts. It also occurred to me that the various Poncet mounts are challenging to build, and offer only limited tube rotation. The inadequacy of existing Dobsonian mirror cells for use with an equatorial mount became a growing disappointment. Therefore, I was happy when Coulter Optical told me that improved cells for thin mirrors made it possible to use them with equatorial mounts; no problem with tube rotation.

I rebuilt my 13-inch based on Gary Walker's elegant adaptation of the Porter split-ring displayed at Stellafane '87. However, I made several modifications to his design, such as replacing the trunnions with Teflon face bearings.

Formerly, I rejected the Porter as a home project because of the large size of the required horseshoe bearing. Who would have thought that wood, with its variation in grain and density, could be stable enough to serve in this application? Well, a fine-grained plywood can be stabilized with multiple coats of exterior polyurethane varnish, and a metal band can enhance the smoothness and durability of the rim.[1] A lathe big enough to turn a large disk is not needed. A sawed-out circle can be turned smooth and true with a sanding drum, while mounted on a plank with a large bolt. Having found the solutions to these problems, I decided to build a lightweight and versatile instrument. The present configuration contains some novel and radical features.

Although a person having a 13-inch f/4.5 mirror and living near 42° latitude can use the dimensions in the accompanying drawings, others will have to do some design work. In either case, one should start by making a full-size drawing of the light path and telescope tube with a tube inner diameter 2 inches greater than the diameter of the primary (Figure 5.1.2). I laid out the primary mirror, the reflected rays and the focal point. Then, I used paper patterns for several sizes of diagonal mirror to determine the optimum diagonal position and the focal plane

---

[1] Thin aluminum has been used, but stainless steel is available in the form of ¾-inch wide chimney straps for TV antennas. These can be found at Radio Shack or other electronics distributors.

Fig. 5.1.1 *John Shelley transports his 13-inch split-ring Porter equatorial with ease. (Note wooden cover over mirror.) Photo by Paul Pompa.*

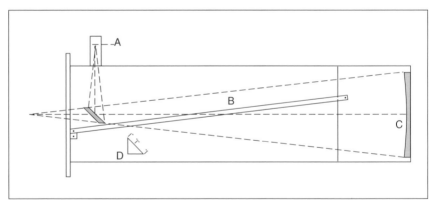

Fig. 5.1.2 *Tube Layout: A) Eyepiece mount & focal plane, check for vignetting. B)Truss tube. C) Primary mirror. D) Diagonal offset layout using 2.6-inch diagonal. Drawing by John Shelley, Jr.*

position at the proper height in the eyepiece holder.[2]

The lightweight, open frame or truss tube design has been criticized by Texereau[3]; he objects to the turbulence caused by uncontrolled air movements,

---

[2] The catalog for Kenneth Novak & Co. has a comprehensive explanation for determining diagonal size, and gives specifications for the quality, lightweight components used by the author.

[3] Texereau, Jean. *How to Make a Telescope*, 2nd English Edition. Richmond, VA: Willmann-Bell, Inc., Chapter 4, p. 119.

body heat and use of metal rods. However, modern plastics have made it possible to avoid these problems without detriment to tube weight and strength.

A lightweight vest of polyethylene bubble-pack can eliminate these first two sources; and the use of PVC water pipe, instead of metal tubing, can take care of the third. My experience has been that, while a single piece of the PVC pipe sags considerably under its own weight, the triangular makeup of the truss tubes limits distortion to that from extension and compression.[4] When Teflon bearings are used, only rapid movements will cause a backlash effect. One should not move the tube except by the top ring.

The truss tube assembly is very simple to build. A vertically-centered side view can be added to the full size drawing, and it will show the tubing in near true length. After cutting to this length, the mounting holes need to be drilled. But first, push-fit 2-inch long wood plugs into the tube ends. These get drilled with the tubes and prevent their collapse when the hardware is tightened. A drill jig should be made to drill all tubes vertically on one end and for the right- or left-hand 45° angle on the other. Two carefully drilled wooden blocks can be fastened to a board, slightly longer than the tubes, to accomplish this.

The tube frame (Figure 5.1.3) and upper tube ring (Figure 5.1.4) have to support the tube assembly and allow it to rotate freely for eyepiece positioning. This rotation feature will more than double the number of people who can use the scope at a star party. First thoughts were to make only the top end rotatable, but the added weight at the top end of the tube and the need for counterweight positioning made this undesirable.

Also included on the tube frame are the declination bearing disks. These are 1¼-inch thick and have flat Teflon rings attached to them. The rings have a much larger surface area than the traditional trunnion pads. In the customary arrangement, tube weight deforms and wears down pads; in this design, tube weight is transferred to threaded bolts which also allow friction adjustment. Furthermore, mating the fork and horseshoe to the tube frame, through the Teflon rings, yields an integrated assembly that is quite rigid.

Anything attached to the top end of the scope will have to be balanced by nearly five times its weight at the bottom. The best place to add weights is in the bottom of the mirror cell enclosure at the side opposite the eyepiece. This is important to maintain balance at all angles of the tube, to minimize stresses and to eliminate sudden movements by the top end. The weights are chosen to balance the usual accessories, but minor differences can be compensated for with the adjustment nuts. Steel was chosen for the cell enclosure both for its strength and its weight to help achieve a low center of gravity. Of course, insulation is required

---

[4] An experiment was performed by clamping the bottom ring of a PVC truss tube assembly to the side of a heavy, metal frame so that it projected out parallel to the floor. The distance from floor to a point on the edge of the top ring was measured and a 1-lb weight attached. Another measurement was taken and followed by three more at one hour intervals. The temperature remained within 1° of 85° throughout the test. Initially the added weight caused an estimated ⅓ of a ¹⁄₃₂-inch deflection. No further deflection was observed during the test period.

A) Right and left declination bearing disks, 1¼-inches thick by 8-inch O.D., use ½-inch and ¼-inch plywood.

B) ⅝-inch wide by 1/16 wide Teflon® ring, 8⅝-inch O.D. countersink wood under Teflon and hold with 4 screws.

C) 1-inch wide by ¼-inch deep notches, 120° apart on top and bottom rings to clear cell mounting bolts.

D) 2¼-inch by ¾-inch Teflon pads, 120° apart, recessed ⅛-inch at each end to lower screwheads.

E) 5/16-inch threaded rod, countersunk to clear nuts.

F) Tube frame, upright supports; F₁) 8 each, 5⅞-inches long by 2½-inches wide by ¾-inch deep, grooved ⅜-inch wide by 3/16-inch deep; F₂) Glue 4 pairs; F₃) Hollow one side of assembly to a depth of ⅛-inch.

G) Top frame ring, double ¾-inch plywood.

H) Bottom frame ring, identical to top ring except O.D. is circular with a 9-inch radius.

I) Counterbore with 1-inch bit, ⅜-inch deep, then drill ⅝-inch bolthole, bolt heads to be pressed in.

J) ⅝-inch by 3-inch bolt with large washer and double nuts to control friction.

K) ⅝-inch, ¾-inch long sleeve, use ½-inch conduit (which is actually ⅝-inch OD).

L) Truss tube mounting holes, 4 pairs ~1¼-inches apart, located ½-inch from edge of cell enclosure.

M) ¾-inch plywood ring, 17¼-inch O.D., ~15¼ inch I.D.

N) 1/16-inch thick steel mirror cell enclosure formed around 15-inch diameter wooden disks and welded. 1-inch pan head wood screws spaced at 3-inch intervals.

O) Mirror cell mounting holes, 3 each to fit cell located ⅝-inch from bottom of cell enclosure.

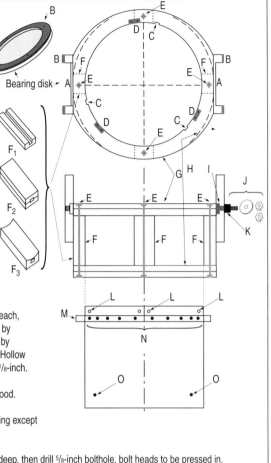

Fig. 5.1.3 *Top & bottom tube rings, tube frame support and mirror cell enclosure. Drawing by John Shelley, Jr.*

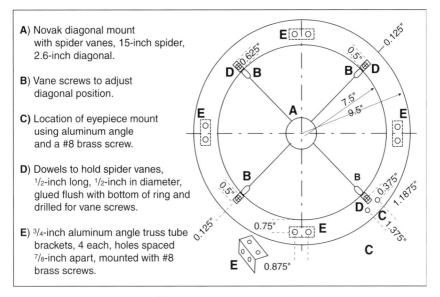

A) Novak diagonal mount
   with spider vanes, 15-inch spider,
   2.6-inch diagonal.

B) Vane screws to adjust
   diagonal position.

C) Location of eyepiece mount
   using aluminum angle
   and a #8 brass screw.

D) Dowels to hold spider vanes,
   1/2-inch long, 1/2-inch in diameter,
   glued flush with bottom of ring and
   drilled for vane screws.

E) 3/4-inch aluminum angle truss tube
   brackets, 4 each, holes spaced
   7/8-inch apart, mounted with #8
   brass screws.

Fig. 5.1.4 *Upper tube ring. Use ⅜-inch quality plywood for ring. Drawing by John Shelley, Jr.*

around it.

Excess weight is avoided at the top. To achieve this, a lightweight PVC drawtube was designed (Figure 5.1.5). The drawtube bearing pressure may have to be changed to adjust for temperature by using paper shims under the Teflon.

Tube imbalance can affect the mount's overall balance in both right-ascension and declination. Since the horseshoe is inherently unbalanced due to its being an incomplete circle, that is enough for a drive system to contend with without adding more load that varies as the tube changes its angle.

Next, a diagram of the horseshoe/fork assembly is made using the local latitude for the angle of the fork (Figure 5.1.6). Tube length below the declination axis determines horseshoe diameter and, therefore, the length, width and weight of the base. A single circle serves as the limit for the bottom corners of the cell enclosure in the superimposed views in Figure 5.1.6. These cell corners should come within 1/2 inch of the horseshoe; but can be further away from the fork-bow, if needed. The need arises in reaching down to a short, firm rear footpiece; but another limit is reached as the fork shoulders approach the side rails of the base. The horseshoe and siderails may have to be sculpted to make up for cumulative errors.[5]

The wooden parts of this assembly are made from birch plywood. While the horseshoe and tines are made from a single thickness, the fork-bow is doubled. Temporary tines should be made from inexpensive plywood and will have to be tailored, along with parts of the base, for the lowest, most compact configuration.

---

[5] **Safety Note**: When clearance is reduced below 1/2 inch, an effective shear mechanism is created that can be a danger to the fingers of children and adults.

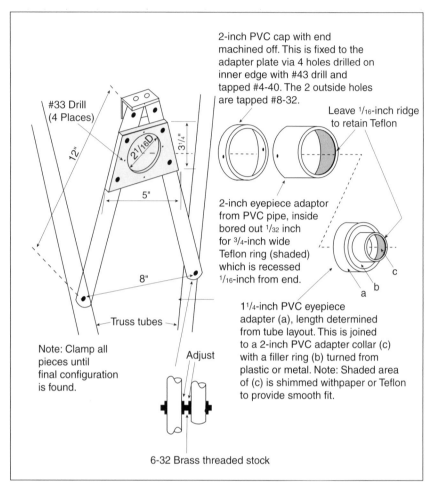

Fig. 5.1.5 *Eye-piece mount. Drawing by John Shelley, Jr.*

The final ones could be made from the cut-out center of the horseshoe.

I learned some hard lessons after repeated visits to the lumber company. Plywood may not be the exact thickness that is ordered. If full thickness is specified, it will be more expensive. The number of gaps in the plys also varies with price. Generally, fir is of fair quality. Yellow pine, however, is terrible to work with, and it is dimensionally unstable. Once you have tried these, you will appreciate the birch. Even then, care is needed to prevent edge-splitting of the outer plys.

The above leads to the fact that everything must be measured, and not assumed. It is very important that the inside, flat edges of the horseshoe fit exactly or be shimmed to the combined width of the tube frame, bearing disks, Teflon and aluminum rings and the fork tines (Figure 5.1.6). These tines should be fitted to the others and made exactly parallel and symmetrical before gluing. All of this en-

A) Fork tine, 4-inch outside radius, 3/4-inch slot to accept steel sleeve on declination bearing, faced one side with thin aluminum by sanding and contact cement which rides against the Teflon bearings, edges are beveled to prevent wear, 14⁵/₈-inches overall length, 4-inch cap to accept fork-bow.

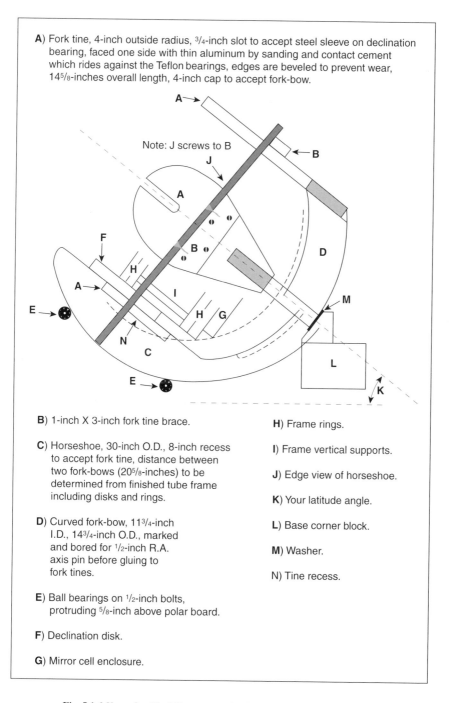

B) 1-inch X 3-inch fork tine brace.

C) Horseshoe, 30-inch O.D., 8-inch recess to accept fork tine, distance between two fork-bows (20⁵/₈-inches) to be determined from finished tube frame including disks and rings.

D) Curved fork-bow, 11³/₄-inch I.D., 14³/₄-inch O.D., marked and bored for ¹/₂-inch R.A. axis pin before gluing to fork tines.

E) Ball bearings on ¹/₂-inch bolts, protruding ⁵/₈-inch above polar board.

F) Declination disk.

G) Mirror cell enclosure.

H) Frame rings.

I) Frame vertical supports.

J) Edge view of horseshoe.

K) Your latitude angle.

L) Base corner block.

M) Washer.

N) Tine recess.

Fig. 5.1.6 *Horseshoe/Fork/Base composite. Drawing by John Shelley, Jr.*

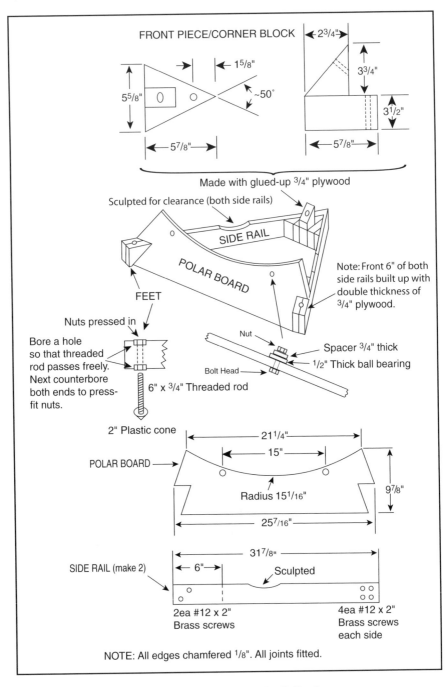

FRONT PIECE/CORNER BLOCK

Made with glued-up 3/4" plywood

Sculpted for clearance (both side rails)

SIDE RAIL

POLAR BOARD

FEET

Note: Front 6" of both side rails built up with double thickness of 3/4" plywood.

Nuts pressed in

Bore a hole so that threaded rod passes freely. Next counterbore both ends to press-fit nuts.

Nut

Spacer 3/4" thick

1/2" Thick ball bearing

Bolt Head

6" x 3/4" Threaded rod

2" Plastic cone

POLAR BOARD

Radius 15 1/16"

SIDE RAIL (make 2)

Sculpted

2ea #12 x 2" Brass screws

4ea #12 x 2" Brass screws each side

NOTE: All edges chamfered 1/8". All joints fitted.

Fig. 5.1.7 *Base. Drawing by John Shelley, Jr.*

Fig. 5.1.8 *Horseshoe and fork assembly. Photo by John Shelley.*

sures that the horseshoe stays round, and the bearings have constant friction through all angles of declination. To further ensure this, a temporary brace should be made and fastened across the open end of the horseshoe as soon as it is cut out.

The dimensions of the base (Figure 5.1.7) come from those of the horseshoe/fork, and it is here that some serious fitting and juggling will allow a good, low configuration. You can use scrap, 1 x 4-inch lumber to make patterns for the side rails. These should be made longer than needed and laid down, on edge, to form two sides of a triangle. The third side, which can also be an inexpensive board, will face the polar axis; one edge is sawed to fit the contour of the horseshoe. Placing the pieces together and starting with an equilateral pattern, the side rails are positioned to touch at their corners. A wooden cleat is fastened to the side rails, on top, near the apex, with one screw into each. The polar board is temporarily clamped with two, foot-long square sticks which are then clamped to the inside surfaces of the side rails. Each stick should be equidistant from the apex, and the rail corners should remain in contact or adjacent at the apex

The temporarily assembled horseshoe/fork is placed into the above setup and blocked up to hold it in its required attitude. This may require narrowing the polar board to allow better positioning. The general layout in the diagram will be a guide; but each builder should seek the lowest, best-fitting scheme for his needs. This will vary for different scopes and will require a trial and error approach. When a satisfactory arrangement is reached, angles and dimensions can be measured and used to make the final parts from birch plywood. Do not forget that the

Fig. 5.1.9 *Paul Pompa, a friend of John's, stands 5 ft. 7 in. and has no difficulty in reaching the eyepiece when the telescope is pointed straight up.*

polar board will be displaced, and the ball-bearings will occupy its position.

The cell enclosure (Item "N" of Figure 5.1.3) can be made from a discarded well tank having a $\frac{1}{16}$-inch wall. Many of these show up at dumps in good condition, except for the ruptured air bladder inside. Most of the cutting can be done with a fine-toothed hacksaw or blade wrapped in cardboard to save fingers. Marking the end cuts accurately will require a strip of thin, straight-edged material that will go around the tank twice for good alignment. After cutting out a strip to reduce the diameter, the new seam can be welded or fastened with screws and square cleats. The latter may not seem so undesirable after an amateur welder gets through with the job—unless he used a shielded arc welder. Rounding the final sleeve will be easier if welding has not distorted the metal around the seam. Insulation needs can be satisfied by an internal sleeve of one-sided, corrugated cardboard, painted flat black, along with fixed panels in the sides of the tube frame.

During my four-year quest to make a more convenient scope from my original Dobsonian, I decided to go for the lightest possible instrument—yet keep to all of the rules. I thought of different ways to implement a right-ascension drive; but I became well satisfied with a simple, Teflon friction block instead. The new mirror cell (Novak), combined with the face bearings and the PVC truss tube assembly, have helped achieve a fully articulated (equatorial with rotating tube) system. At 70 lbs and close to perfect balance with 1¼-inch eyepieces, it is a far cry from some of the massive, artillery-like mounts of the past.

5.2   A 200 mm *f/4* Houghton Telescope
By F.T.A. Hoenderkamp

In their April 1991 article about the Houghton Telescope[6], Harrie Rutten and Martin van Venrooij expressed surprise at the lack of interest shown by ATM's for this type of instrument. This may be due to the fact that until recently the Houghton design has remained relatively obscure. Apart from an article in *Sky & Telescope* back in 1979[7], little has been published about it—certainly not about the variant designed by Lurie.[8]

It is this design which, in my opinion, lends itself most favorably to realization by amateurs. True as this may be, it does not mean there are no hurdles to be surmounted in the process. Of these, the prospect of grinding the two corrector lenses will be the main cause of hesitation for most ATM's. It was for me! Even though several useful guidelines have appeared over the years, lens making remains a somewhat uncanny exercise for most of us. In the case of failure, it may even involve a painful financial setback, considering the price of high quality optical glass. In this respect, the H.T. has proven itself to be amateur-friendly. It requires no special type of glass, and both lenses may be made of the same material. They must, of course, be reasonably free from internal strains, air bubbles, etc. I was lucky to obtain two blanks of B 270 "Superwite," a product of Deutsche Spezialglass AG. Their product information included the refractive index, so that corrections could be calculated for the radii of curvature, which proved to be only minor.

For readers who would like to learn more about the theoretics of telescope optics, and want to learn how to evaluate or modify existing designs, or even create designs of their own, I would recommend *Telescope Optics, Evaluation and Design*[9] by Harrie Rutten and Martin van Venrooij, along with its accompanying computer program. I found the booklet "Spiegeloptik" by Kurt Wenske[10] less theoretical than *Telescope Optics*, but rich in practical guidance for the ATM. It contains concise and clear explanations of the various skills involved in making mirrors and lenses.

---

[6] Rutten, Harrie and Martin van Venrooij. "De-Houghton-telescoop, een optimaal compromis?," *Zenit*, April 1991, p. 138.

[7] Turco, Edward. "Gleanings for ATM's—Making an Aplanatic 4-inch Telescope," *Sky & Telescope*, Nov. 1979, p. 473.

[8] See Chapter 17-18.6 in Volume II of *The Best of ATMJ*, Richmond, VA. Willmann-Bell, Inc. 2003.

[9] Rutten, Harrie and Martin van Venrooij. *Telescope Optics, Evaluation and Design*. Richmond, VA: Willmann-Bell, Inc., 1988.

[10] Wenske, Kurt. *Spiegeloptik, Entwurf und Herstellung Astronomischer Spiegelsysteme*. Verlag Sterne und Weltraum Dr. Vehrenberg GmbH. Muenchen.

Fig. 5.2.1 *A simple grinding machine facilitates the centering of the lenses, but is not indispensable.*

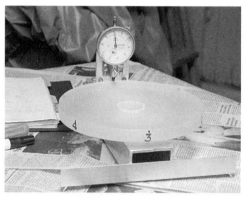

Fig. 5.2.2 *Monitoring wedge error by comparing edge thicknesses along the perimeter.*

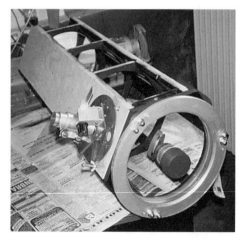

Fig. 5.2.3 *The corrector and diagonal assembly in place. Photo also illustrates the positioning of the threaded rod used in collimation, the hexagonal cross-members and Mr. Hoenderkamp's intricate Crayford focus mechanism.*

Fig. 5.2.4 *The Houghton telescope mounted temporarily on a Dobsonian-Poncet combination.*

5.2.1 Making the Corrector Lenses

I would like to pass on a few things I learned while making a Houghton telescope, in hopes that my comments might inspire enthusiastic TM's to follow suit. It will not be necessary to go into much detail about making a spherical 200 mm *f*/4 mirror, as this should not present a major problem for prospective lens makers. Fortunately, no parabolizing will be required; thus, checking the mirror with a simple knife-edge tester will show when you have produced a smoothly curved specimen with something like ⅛-wave precision, depending on the amount of effort you are willing to invest in this part of the project.

Regarding grinding the corrector lenses, it greatly encouraged me to find the dimensions to be familiar, as they are comparable to mirrors I have made for other projects. There are, however, a few important differences. First is the fact that lenses have two optical surfaces, both of which have to become, and remain, highly polished and scratch-free. This requires great care in handling compared to a mirror, especially in the later stages of grinding, polishing, and figuring. Furthermore, the relative thickness of the lenses is less than that of the average telescope mirror. This increases the risk of creating a less than perfect sphere which could result in astigmatic images. I reduced the potential for causing scratches and deformation by using plaster casts for support and protection of the lens face opposite the one being worked. The casts were made immediately after each radius had been roughed in. As a base for pouring the plaster, I used ½-inch thick disks of MDF (medium density fiber, obtainable in hardware stores and do-it yourself shops). At the same time, these disks served as reinforcement for the castings, to which they became firmly attached using tacks, halfway driven in, so that the nail-heads were embedded in the plaster. After waterproofing with several coatings of shellac, each casting could be taped to one of the lenses to form a temporary unit. Naturally, a thin layer of a resilient material should be inserted between lens and plaster before taping them together. I used ⅛-inch foam rubber. The total thickness of the sandwich was approximately 1¼-inch, making it very similar to an average telescope mirror.

A matter of great importance with lenses is the centering of their two surfaces. Defects in this respect—called wedge error (or simply wedge)—must be kept to a minimum. This condition may be controlled by frequently comparing the edge-thickness in a number of places along the edge of the lens. In the present case, a wedge error of 0.1 mm may be considered acceptable. A complicating factor lies in the fact that the two lenses are being ground, one against the other. Logically, having equal radii makes this work-saving method possible. At the same time, however, centering will become more difficult by using this method. While trying to improve the centering of lens A by grinding it against lens B, the amount of wedge in lens B will also be affected—and may even get worse.

To avoid this problem, I decided to use a simple turntable-type machine for rough and fine grinding (Figure 5.2.1). It was with parts from a discarded grinding machine and a washing machine motor. It had only one speed, for which I chose 100 r.p.m. Adding a horizontal swing-arm over the turntable, the top lens was made to spin around a point where the arm rested. By choosing a point off-center, the lens would not only spin, but it would also make a swinging motion. Slightly more glass would be removed from that part of the surface where its edge was nearest to the center of rotation (and pressure). Meanwhile, the spherical shape of both surfaces would be preserved. The bottom lens, nicely centered on the turntable all the time, did not swing at all. Glass was removed evenly along its circumference, keeping the amount of wedge unchanged. By alternating their position, both lenses could be corrected. In this way, I succeeded in keeping the edge thickness of both lenses uniform within 0.05 mm.

For monitoring the edge thickness, I used a dial gauge mounted on a section of rectangular aluminum extrusion on which three ⅛-inch steel balls were fixed to support the lens (Figure 5.2.2). Horizontally, its position was defined by resting it against two vertically placed steel pins. The gauge having $\frac{1}{100}$ mm divisions, the slightest change in wedge could be noticed immediately. The same dial gauge proved very useful in a simple spherometer, made from a piece of heavy aluminum plate. It rested on three steel balls, as above, fixed to the plate in a circle, with the gauge placed vertically in the center. The device, although not extremely accurate, enabled me to determine the lens radii to within 0.5%.

Finally, I will offer some explanation for going to the trouble of perforating the corrector. The reason I decided not to avoid the extra effort and the risks involved, is the extreme convenience of being able to adjust the diagonal from the outside. The risks associated with cutting holes are not very great, provided the procedures described by Texereau[11], among others, are carefully followed. How important this is, may be illustrated by what happened when I overlooked what seemed to be only a minor detail; i.e., cementing the glass plug back into the blanks. Although I had not forgotten to bevel the edges of both the lens opening and the core, I realized later that the deeper of the two concave surfaces of the negative lens requires an extraordinary wide bevel. Otherwise, it will wear away before the final depth of the curve is reached. The depth of that curve being as much as 4 mm, the bevel should be at least 6 mm wide! In my case, it gradually disappeared, and I was compelled to increase its size with the core in position. Although I succeeded in the end, this method is certainly not to be recommended. The lesson I learned from this experience is not to start cutting out the center before the curve has reached at least ⅔ of its final depth.

5.2.2   Construction of the Telescope

Only simple tools being available, I decided to go for a wooden tube. It had a hexagonal cross-section and a number of diaphragms. Six reinforcing members, one in each corner, running lengthwise inside the tube, together with the diaphragms, constituted a rigid frame to which the external panels could be nailed and glued. The mountings of the mirror and of the corrector were interconnected by means of three threaded rods (M6). The diagonal was then attached to the corrector. In

---

[11] Texereau, Jean. *How to Make a Telescope*, Richmond, VA: Willmann-Bell, Inc., 1984.

Fig. 5.2.5 *Fork mounting, soon to be complete, has electronic drive on both axes.*

Fig. 5.2.6 *(Above and Below) The complete telescope. Concrete platform constitutes compromise between garden table in the daytime, and solid base for the telescope at night. Carriage, partly visible below, lightens the job of placing the telescope on the table. The big hour wheel rests on two small wheels, one of which is driven by an electronically controlled motor.*

Fig. 5.2.7 *Test photo showing southern region of constellation Lyra with Ring nebula. 10 minute exposure on Kodak TP 2415 covering entire* mm *frame in prime focus. Coma fully corrected!*

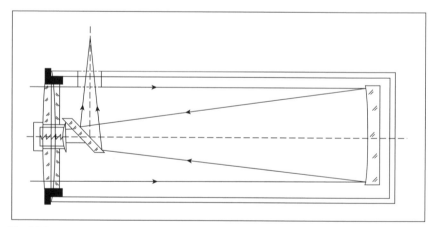

Fig. 5.2.8 *Layout for a 200 mm f/4 Lurie-Houghton telescope as described in* Zenit. *Design specifications may also be found Table 1.1.1, on page 6.*

this way, all optical elements were firmly held together, so that only temperature variations might create a problem in keeping the image plane exactly in position during photography with long exposures. Given the focal ratio of $f/4$, the margin for critical focusing was only 0.2 mm. Any existing curvature of field further reduced this margin. In this case, the sag of the image plane would only be 0.07 mm diagonally across a 35 mm film frame. Still, 0.13 mm then remained to deal with.

If, in addition, we must allow for a temperature change, of say 10° Celsius during a picture taking session, the focal plane would shift another 0.07 mm; because of this, the whole question of securing sharp images became critical. I, therefore, decided to use Invar—(low thermal expansion steel) for the connecting rods. At this point, I encountered a well-known obstacle for amateurs; namely, minimum charges for small orders. I now looked for an opportunity to combine orders with a fellow ATM.

Finally, the tube was lined internally with black velvet to reduce stray light. The dewcap was treated in the same way. It was also cut off at an angle so as to reduce the problem of street lights and the like coming from one direction.

5.2.3 Provisions for Collimation

The threaded rods between the mirror and corrector provided an opportunity for adjusting the mirror from the front-end of the telescope, while watching the effect through the eyepiece-tube. For this purpose, the rods were taken through the lens mount flange and grooved at the end to accommodate a screwdriver. After adjustment, they were locked in position by a capped nut. This provision allowed the tube to be completely closed at the mirror end. Therefore, in the case of faulty images, tinkering with adjustment screws and wing nuts by inquisitive admirers could be ruled out as the cause.

The diagonal is adjustable from the outside—this was the reason for perforating the corrector. The principle of this provision was described by Texereau (in his Chapter 14). The main point is that it will keep the center of the diagonal exactly in place when angular adjustments are being made.

5.2.4 Mounting and Drive

Various telescopes I have made over the years were adapted for use on Dobsonian mountings. For automatic following, they could be placed on a Poncet platform (*Zenit*, June 1989). For visual observing, this combination was great considering its unsurpassed stability. Lunar and planetary photography presented no problems. However, long exposures were a different matter. Lacking a declination axis, guiding on a star became troublesome. An exception to this was viewing objects near the meridian, for which turning the southern supporting bolt of the platform achieved something similar to declination control. Therefore, as a temporary measure for testing the photographic capabilities of the completed telescope, I motorized that supporting bolt. Of course, as this Houghton telescope was definitely meant for astrophotography, a proper equatorial mounting was indispensable. In fact, at the time of this writing, I have one nearing its final stages. From two small-scale models, one a split-ring and the other a fork construction, I chose the latter. The main reason for not choosing the split-ring with its inherent stability, is the position of the center of gravity. The relatively heavy front-end, with the full size corrector lenses and mount, causes the point of balance to move away from the mirror; therefore, in order to keep the diameter of the split-ring within reasonable limits, a quite hefty counterweight would be required.

With all of the apparatus nearly finished, it was time to tackle the really big problem: finding a dark sky not too far away, where $f/4$ exposures in excess of four minutes may be made that would yield fine nebula images—instead of those depressingly gray negatives of city skies.

The Houghton telescope described in this article is, in principle, an improved Newtonian. Its superior characteristics are obtained by employing a twin-lens corrector, placed in the entrance of the tube. This corrector has zero power (it does not influence the focal distance of the mirror), but eliminates both the spherical aberration and coma. Consequently, the mirror only needs to be spherical, making testing and figuring relatively easy. The coma correction dramatically improves the image quality away from the optical axis, so that pinpoint star images are produced to the extreme corners of a 35-mm frame and even slightly beyond. These advantages, and several more, make the Houghton design an attractive proposition for the amateur (and for manufacturers), and were discussed in the April 1991 issue of *Zenit*.

5.2.5   A Summary of the Main Characteristics:

- Compact design
- Closed system (reducing internal turbulence)
- Good image quality over a wide field
- Fast $f/4$ system
- All surfaces spherical
- Common glass types, the same for both lenses
- Diagonal attached to the corrector (improved contrast)
- Suitable for visual and photographic application
- Relatively simple construction
- Equal radii of curvature permit grinding and testing one lens against the other

5.3   A Reclining Chair Binocular Mount
By Randall Wehler

Holding up 5 pounds with both hands, arms bent at the elbows for minutes at a time, can get quite tiring, even for well-seasoned giant binocular observers. Just when you have located that elusive deep-sky object you have been hunting, muscle strain really starts to set in. When that happens, the fatigue usually leads to more shakiness. It's time to put your arms down to rest! Later, you work your way back to the object you previously had in view. Soon, fatigue sets in again. Sound familiar?

Of course, your giant binoculars could be placed on a photographic tripod. That eases things quite a bit—no more muscle strain, and the binoculars stay where you aim them. But some terribly awkward positions are in store for you at times, especially when viewing near zenith. Backache sometimes sets in as does

Fig. 5.3.1 *Carla Wehler demonstrates the comfortable viewing position available with her reclining chair and binocular mount.*

eventual neckache. Is there no solution to these discomforts?

Yes, there is. I have read a number of descriptions of building binocular mounts. Some are better than others, and they all have their own merit. I personally enjoy lying in a chaise lounge when I use binoculars to gaze at the night sky. Maybe some would call this a "lazy" way to view, but the comfort keeps me observing longer. Resting my arms on the chaise arms works as long as the altitude of the objects is within a certain limited range. But what about the rest of the sky? How could I construct a special reclining chair mount to help me keep objects fixed in view, and let me observe with as much comfort as possible? An idea took shape.

Why not construct a box-shaped support base with two arms attached to its top sides that bend somewhere near the middle and hold a device that moves the binoculars up or down in altitude and left or right in azimuth? Many motions to position oneself for comfortable viewing could be possible—up and down for altitude, left or right for azimuth, up and down in elevation, forward or backward as I sat in the chair—aside from moving the recliner itself forward and backward on the ground and with respect to changing the angle of the back-rest.

Other than several pieces of 1 x 4's and 2 x 8's, most of the mount was built out of 2 x 4's which are inexpensive and readily available. The hardware required was wood screws, bolts, nuts, washers, eye screws, clevis pins, floor flanges, pipe nipples with end caps, and metal strips (mending plates). Rubber tarp straps were also part of the design to assist with counter-balancing and securing the binoculars

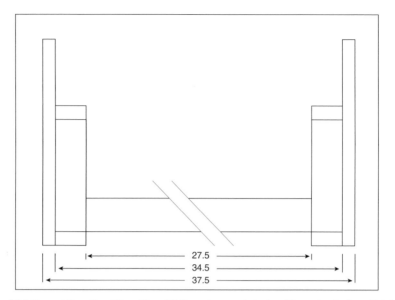

Fig. 5.3.2 *Support Base-Front/Rear View. All dimensions are in inches. Illustration by Steve Strickland*

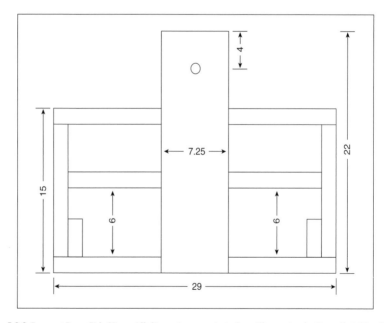

Fig. 5.3.3 *Support Base-Side View. All dimensions are in inches. Illustration by Steve Strickland.*

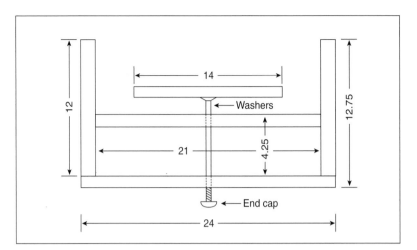

Fig. 5.3.4 *Binocular Holder/Yoke. All dimensions are in inches. Illustration by Steve Strickland.*

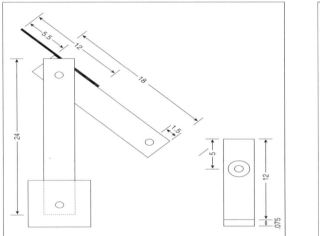

Fig. 5.3.5 *Arm System-Side View. All dimensions are in inches. Illustration by Steve Strickland.*

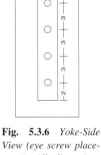

Fig. 5.3.6 *Yoke-Side View (eye screw placement). All dimensions are in inches. Illustration by Steve Strickland.*

in place.

The first part of the project was construction of the box-like support base out of 2 x 4's. The overall dimensions of this were 29 inches in length, 15 in height, and 34½ in width. The main thing was to keep the 2 x 4 pieces either parallel or at 90° to one another. Wood screws of 3- or 4-inch length solidly secured the pieces.

Next came the attachment of the two arm supports. These were made of

2 x 8's 22 inches in length. Each was attached midway on the outward side of the support base with 4-inch wood screws. At 4 inches from the top of each arm support, I drilled a ⅜-inch hole. Through the outside, a 4-inch carriage bolt (½ inch in diameter) was firmly attached.

Two 2 x 4's of 24-inches length were cut for the lower arms. About 2 inches from each end of these, I centered and drilled ⅜-inch holes. The bottom holes were reamed out with a round rasp file to ½ inch diameter to be fitted over the carriage bolts that were firmly in place. One-half inch flat and lock washers were put in place as well as ½-inch wing nuts. The lower arms were now moveable. Nine equidistant holes were drilled through the bottom holes in the lower arms into the arm supports with a ⅜-inch drill bit, defining 90° of a circle. Three-eighths inch clevis pins held the lower arms securely and could be removed to tilt the arms forward or backward, as needed.

The holes at the top end of the lower arms were enlarged slightly to firmly screw in another 4-inch carriage bolt. Two 2 x 4's, 18 inches in length, would be attached to the tops of the lower arms by means of these carriage bolts, the bolts being pushed through ½-inch holes drilled at the ends of the second arms. The holes were placed several inches from the end of each second arm. Squares of emery cloth (3½ inches in width), with a hole in the center for the bolt, were cemented to the 2 x 4 ends on the inside of the "elbow joint" for added friction. One-half inch flat washers, lock washers and wing nuts completed the joint for the arm system.

Metal strips, 12 inches long, with six holes were attached to the top of the second arm at the "elbow" end by ⁷⁄₁₆ inch, 4-inch long bolts inserted into drilled holes and held by ⁷⁄₁₆-inch flat washers and nuts. Three holes in the metal strip were now available to attach rubber tarp straps that were hooked into the bottom part of the lower arm via eye screws—five of them in a series, 3 inches apart. Rubber straps of 9 and 15 inches on each side would serve as tension "counterweights" to offset the weight of the yoke holding the binoculars. This was designed to supplement elbow joint tension provided by tightening the upper wing nuts.

At about 1½ inches from the other end of each second arm, ½-inch holes were centered and drilled. Simple bearings were made by attaching ½-inch (inside diameter) wide-faced flat washers to each side of the arms to define the hole that had been drilled into the ends of the 2 x 4's. Four such washers were used for this. Small holes were drilled into each washer to allow small wood screws to pass through and into the 2 x 4's. These "bearings" would later accommodate ½-inch pipe nipples, 4½ inches long, inserted into the floor flanges that would hold the binocular yoke with the addition of nipple end caps.

Before this could be done, the "yoke" or binocular holder had to be assembled. Two 2 x 4's (12-inches in length) made up the sides of this in addition to 1 x 4's (21 and 24 inches in length), put together in a rectangular shape by wood screws. One-half inch floor flanges were placed on the outside of the 2 x 4's secured by wood screws, centered at 5 inches from the top ends of the studs to balance the yoke and binoculars properly. One-half inch holes were drilled in the

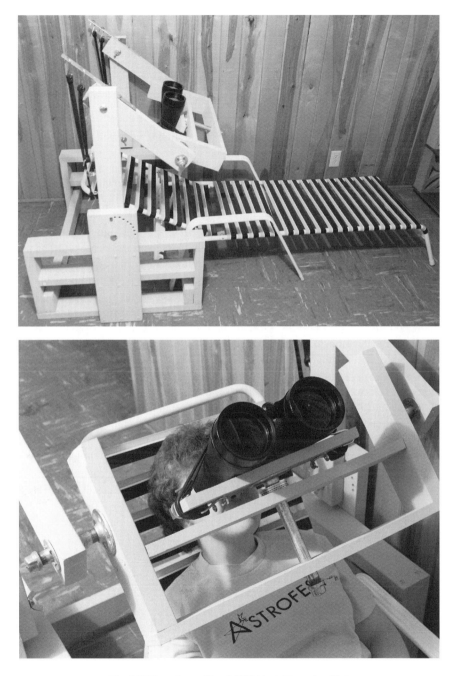

Fig. 5.3.7 *Two views of Randall Wehler's Binocular Chair.*

centers of the two 1 x 4 pieces to accept an 8-inch pipe nipple. This would be screwed into a ½-inch floor flange holding a 14-inch long 1 x 4, which would carry the binoculars with the attachment of eye screws and rubber straps 15 inches long. Four ½-inch (inside diameter) wide-faced flat washers served as spacers between the top 1 x 4 and floor flange. An end cap on the 8-inch nipple would prevent the T-shaped azimuth axis component from accidentally falling out of the yoke support.

The last task to complete the mount was a paint job. I chose a gray—neutral, but pleasant—color, and used latex "one coat" exterior, flat finish paint.

The mount was now finished. I slid the chaise lounge under it and positioned myself. It worked like a charm! Facing in any direction, I could view from the horizon to just past the zenith. The azimuth axis allowed flexibility, by angling the binoculars to the sides, to take in a fairly large section of the sky. When I wanted to observe in another direction, it was no problem to move the mount and chaise. Very comfortable views were possible reclining in the chaise. There would be no more bending, crouching, squatting, or contorting my body in awkward positions as when using a photographic tripod. The binocular mount permits four easy motions: (1) up and down in altitude, (2) left and right for azimuth, (3) up and down in elevation via the "elbow joints", and (4) forward and backward via the swivel at the bottom of the lower arms. I found that I seldom needed to take out the clevis pins to move the binoculars forward or backward. Simply pivoting them at the elbow joint and changing the position of the backrest was usually sufficient. The yoke stayed in place very well due to the rubber strap "counterweights" and tension at the elbow joints, but could still be moved easily up or down without typically having to adjust the straps or wing-nuts. I found that if the azimuth axis loosened significantly with use, a tighter fit could be achieved by applying a little interior wood finish "liquid plastic" to the ½-inch holes of the 1 x 4's.

I have thoroughly enjoyed using this binocular mount. It handles my 5-pound giant binoculars very well, and could easily support models ranging from 7 x 35s to the 100 mm "super giants" that weigh about 7½ pounds. The cost was not a "budget buster", and parts were easy to obtain. Of course, I always think of refinements. One current idea is to attach some hard rubber wheels at the bottom of the support base for more ease in movement.

5.3.1   Parts List

Hardware

     two ½-inch pipe nipples, 4½ inches long
     one ½-inch pipe nipple, 8 inches long
     three ½-inch pipe nipple end caps
     three ½-inch floor flanges
     four ½-inch carriage bolts, 4 inches long
     four ½-inch wing-nuts
     four ½-inch flat washers

four ½-inch lock washers

two ⅜-inch clevis pins, 3 inches long

two mending plates, 12 inches long

four ⁷⁄₁₆-inch hexagonal head bolts, 4 inches long

four ⁷⁄₁₆-inch hexagonal nuts

four ⁷⁄₁₆-inch flat washers

fourteen eye screws

eight ½-inch (inside diameter) wide-faced flat washers

wood screws, variety of sizes

Wood

—2 x 4's

two, 35 inches long

six, 29 inches long

two, 24 inches long

two, 18 inches long

six, 12 inches long

—1 x 4's

one, 24 inches long

one, 21 inches long

one, 14 inches long

—2 x 8's

two, 22 inches long

Other

four, rubber tarp straps—15 inches long

two, rubber tarp straps—9 inches long

two, 3½-inch squares of emery cloth

5.4   An Illuminated Viewer-Magnifier for Star Charts
By Randall Wehler

The ability to view a star chart field that corresponds to that seen in your binocular or telescope would make an evening's observing go much better. I have spent enough nights fumbling with flashlight and star charts, fingers numbed and cold, or hands mosquito-bitten, to know that there had to be a better way to find my way around the heavens with a celestial map. My solution to this problem proved rather simple.

Magnitude 6 to 8 star charts are ideal for typical observing sessions with binoculars and small telescopes. The 52 atlas charts found in *A Field Guide To The Stars And Planets* by Menzel and Pasachoff cover the entire sky, are drawn with high precision by Wil Tirion for Epoch 2000.0, and contain about 25,000 stars down to visual magnitude 7.5 and some 2,500 deep-sky objects.

Fig. 5.4.1

Fig. 5.4.2 *The foundation for the illuminated viewer-magnifier.*

Fig. 5.4.3 *Projector lens with ceramic magnet f/stop in place.*

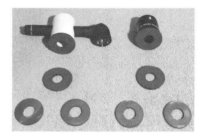

Fig. 5.4.4 *Projector lens (now with felt in place), flashlight and under-sheet magnets.*

Fig. 5.4.5 *Swiveling mount for the star chart viewer.*

Fig. 5.4.6 *The illuminated star chart viewer on a binocular mount.*

However, they are printed in a small format, and you need fairly bright illumination (such as normal room light) to read them. Deciphering them in the dark with a dim, red light poses somewhat of a challenge.

I decided to construct a device that would magnify and illuminate these charts and which could be placed relatively close to my binocular's eyepieces. That would make it easy to check out what I was seeing in them.

A 2 x 8 board (actually measuring 1½ by 7-inches) was cut 8 inches in length to provide a base measuring 7 by 8 inches. At the four corners of this base I drilled holes about ¼ inch inward from each of the sides to insert ⅜-inch dowels which were secured in place with carpenter's wood glue. (Figure 5.4.2.) These would be used to attach two sheets of Plexiglas. One such 7 x 8-inch sheet was carefully drilled out at the corners to the diameter of the dowels to serve as a template-like cover for the star charts to hold and protect them. The other sheet was attached to the top of the dowels by small wood screws. Atop this I would place a magnifying lens and flashlight for viewing the charts.

The f/3.5 anastigmat projection lens (Figure 5.4.3) which I chose as my magnifier has a working distance of about 2½ to 2¾ inches. This is just right to allow upward movement of the bottom sheet of Plexiglas for changing the star charts which I had removed from Menzel and Pasachoff's book. Ceramic permanent ring magnets, ⅛ inch thick and 1¾ inch diameter with a ¹³⁄₁₆-inch center hole, were purchased from Edmund Scientific Company[12]—two packages, four rings to a package, at $11.00 for all 8 of them.

One ring magnet was cemented to the front (bottom) end of the projection lens and covered with a protective felt-like material to prevent scratching the Plexiglas as the lens moved across its surface. Three of the rings went underneath the sheet. The top ring (felt-covered), magnetically holds the lens in place, even when the device is turned upside down and allows free movement of the lens across the Plexiglas surface. Various fender washers of differing inside diameters placed on the bottom of this stack of magnets control the size of the lens' field of view making it correspond to that of the fields of view of my two binoculars (7½° for my 10 x 50's and 4½° for my 11 x 80's).

To provide illumination for the charts I purchased a Streamlight revolving head, multi-angle flashlight from Wal-Mart.[13] This light has an adjustable focus from "spot" to "flood," thus providing just the right amount of light for any situation.

I fashioned a simple holder (to position the flashlight body parallel to the Plexiglas) out of a short section of 1½-inch PVC pipe. (Figure 5.4.4) The remaining ring magnets and felt protectors were used in a fashion similar to that for the magnifier assembly—to hold the flashlight securely in place, yet allow it to be re-positioned easily.

A piece of gray-black picture-framing matt between the base and lower

---

[12] Edmund Scientific Company, 101 E. Gloucester Pike, Barrington, NJ 08007-1380.
[13] Manufactured by Streamlight, Inc., 1030 W. Germantown Pike, Norristown, PA 19403.

Plexiglas sheet kept my star charts snugly in place with the assistance of rubber bands.

Attaching the device to the yoke of my "saw horse" binocular mount[14] proved to be fairly easy. At the center of the back side of the viewing base, I attached a ⅜-inch floor flange with wood screws. Threaded into the floor flange was a ⅜ x 4-inch pipe nipple. This shaft was then inserted through the closed ends of a pair of U-bolts (holes having been drilled into the bottom cross-piece of the yoke) to fasten them in place with the included plates and wing nuts. (Figure 5.4.5) An end cap for the nipple prevented the viewer and shaft from falling out. This shaft rode smoothly on the corner braces and U-bolt plates with negligible back-and-forth movement since the distance from flange to end cap was the same as the width of the yoke cross-piece between corner braces.

The viewing device could thus be rotated 360° so that the orientation of the charts could match the sky. Adjusting the wing-nut tension allowed such rotation as well as tightening the viewer in place.

Being left-eye dominant, I put the viewer in the lower left corner of the yoke and found that to be a very convenient location for a compact star map that (1) doesn't have to be held, (2) can be oriented correctly to the sky, (3) is illuminated with flashlight in place that you don't have to handle, (4) shows a magnified view for easier readability at low light levels, and (5) has a viewer with a defined circular field roughly corresponding to your binocular field of view.

Illuminated viewer-magnifiers based on this design could be used with telescopes on other types of mounts as well. Where and how to position them are considerations left to the particular user. I know that my device works very well for me and complements the reclining chair mount that I had designed and built. I'm almost looking forward once again to those cold, crisp Minnesota winter nights when the stars are at their crystal-clear best. Warm in my sleeping bag, I will only occasionally have to move my hands and arms out of its coziness to search for objects and confirm them on my atlas charts.

5.5   Telescopes For CCD Imaging
By Richard Berry

As any type of observing places a special set of demands on telescopes, so CCD cameras place new demands on the telescopes intended for imaging. This is nothing new. Similar changes have, in fact, happened many times before in the history of amateur telescope making. The move away from the small refractors of the Fifties occurred because the long-focus Newtonian of the Sixties satisfied the aperture and resolution needs of lunar and planetary observers; and the growth in popularity of highly-portable Schmidt-Cassegrain telescopes in the Seventies occurred at least in part because car-portable telescopes became a necessity for many urban amateurs. The remarkable spread of the Dobsonian in the Eighties

[14] See *Sky & Telescope*: April, 1993, pp. 90–92.

Fig. 5.5.1 *For CCD work, bigger is not always better—unless you're talking mounts. In his observatory, Richard Berry uses a 4-inch f/5 Genesis and a 6-inch f/5 Newtonian on a beefy Byers mount. The accompanying "photo" was taken as a CCD image. This printing does not do justice to the quality of the original image in which detail was clear enough to read Genesis on the focus mechanism.*

happened because it answered so perfectly the need for a low-cost, large-aperture telescope for visual deep-sky use. As new types of observing have become popular, each has placed new demands on telescopes, and the types of instrument that amateurs build and use have changed in response.

Today, we face another great change brought on by the needs of observers who do CCD imaging. These needs are not, for example, the same as the requirements of visual deep-sky observing. In this series, I will discuss these new requirements and how we, as amateur telescope builders, can address them. This series will, hopefully, prove useful for amateurs who build telescopes optimized for astrophotography, portability, deep-sky observing, planetary observing, and other semi-specialized purposes even if those amateurs do not intend to try CCD imaging for themselves. If nothing else, the series may help everyone understand what the CCD types are trying to accomplish.

CCD imaging closely resembles astrophotography in its fundamentals. Factors such as field of view, exposure time, and focal ratio are just as important for digital imaging as they are for film imaging. The major differences are that CCDs are generally much smaller than the standard film formats, they require shorter exposure times, and the image is read out and stored digitally, which means that the CCD image can be enhanced more easily than film images.

Another major factor that determines what sort of telescope you need for CCD work is what type of imaging you wish to do. The instrument characteristics required for taking color pictures, shooting faint galaxies in black and white, tracking comets, or shooting high-resolution planetary images are as much different for CCDs as they are for conventional photography. It is, therefore, necessary that you decide what type of imaging you want to do before you start designing a telescope.

Finally, it helps to recognize that it doesn't take a specialized telescope for CCD imaging. We are still talking about refractors, reflectors, and various types of catadioptric systems. The big surprise comes from the fact that CCDs do not call for small telescopes or big telescopes in the same ways that visual observing and astrophotography usually have, and that means that CCDs may offend our traditional sense of "what's right." In a nutshell, for many types of deep-sky observing, the best CCD telescopes are small, fast Newtonians; and for imaging detail on the planets, large Newtonians with good optics outperform everything else. Small, fast refractors with good color correction in the near infrared part of the spectrum also do well for deep-sky imaging.

If there is any one area where home built telescopes have trouble meeting the demands of CCD work, it's their mountings. For every type of observing, a solid, shake-free equatorial mount is crucial. Visual observing tends to forgive the vagaries of the mount, and in astrophotography, sheer perseverance and careful guiding can overcome a clock drive with annoying periodic errors. But for CCD imaging, where individual exposures seldom run over ten minutes, it is hardly worth the time needed to find and set up on a guide star. That means, however, that the clock drive must be good enough to track without guiding; and to make matters worse, because the pixels on the CCD chip are so small, you can see tracking errors $1/3$ to $1/5$ as large as in conventional astrophotography.

Largely because of the tough tracking demands CCDs place on the telescope and mounting, small, fast refractors and Newtonians tend to be easier to use than other types of telescopes. For my CCD work I use a 4-inch $f/5$ Genesis and a 6-inch $f/5$ Newtonian with a Byers 812 mount. This mount, designed for an 8 to 12-inch reflector, does a great job with the 4-inch and 6-inch telescopes. On a good night, I can count on the drive to perform flawlessly during four-minute exposures; and if the drive hiccoughs during one of them, I simply repeat it.

However, CCDs do offer an important way to minimize the impact of a drive system with tracking problems. Using a track-and-stack technique, a CCD camera shoots a series of exposures short enough for almost any clock drive to handle; for example, making 20 exposures each 15 seconds long. The camera software determines the positions of one or more tracking stars in each individual image, and then shifts the image to register the stars and adds it to an accumulating master image. Although track-and-stack with a poor drive takes a long time compared to one-shot imaging with an excellent drive, it makes CCD imaging possible on many more telescopes.

5.6 Baffling Baffles

By Robert Bunge

During the early 1980s I was lucky enough to have occasional access to a beautiful 6-inch $f/15$ Alvan Clark refractor located in Charlottesville, Virginia at Leander McCormick Observatory. This telescope, dating from 1982, was as beautiful to look at as it was to look through. Inspired by the outstanding planetary views it gave (even though it suffers from some astigmatism), I decided to embark on the journey of making a similar telescope.

To lower the cost of the project, I planned make the lens myself. That is another story. In a perfect world, I would be able to make an exact replica of the Clark telescope; but materials and cost would not allow that. However, I could study the designs of the masters and try to learn from their experience.

5.6.1 Enter the Baffle

Over the next five years, as the making of the lens progressed, I managed to examine no less than two or three dozen 12-inch or smaller refractors, ranging from old to new. One area of interest was the baffling used in these telescopes. I noticed that older refractors made by companies like the Clarks and Warner & Swasey often used sheet metal that was somehow folded—much like the petals of a rose—to form a circular opening. These metal baffles were then either wedged up the tube, on smaller telescopes, or riveted in place on larger models that had steel tubes.

With newer instruments, the gauntlet ranged from no baffles—surprisingly common in homemade refractors—to beautiful machined units bolted or pressed into place. Conversations with builders of home-assembled refractors revealed that more often than not baffles were forgotten in the haste to get the telescope operational, and then further neglected once the telescope was in normal use.

As the project advanced, I started to collect parts for the tube assembly. I kept my eyes peeled for ideas or articles that would show me an easy way to decide what size to make the baffles, then how make them (in particular, make perfectly round cuts), and finally how to hold them in place without the use of fancy machine tools that I did not have.

5.6.2 A Simple Solution

During the course of the project, as I tackled the problem of making pieces that I could not find, the problem often solved itself. I was describing the need of securing the baffles in the tube to my uncle, a mechanic and an all-round handy type of man, when he suggested running a series of three .22 caliber gun cleaning rods, $120°$ apart, up the inside of the tube, with the rods supporting baffles at the threaded joints.

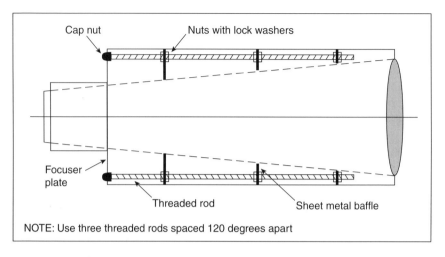

Cap nut
Nuts with lock washers
Focuser plate
Threaded rod
Sheet metal baffle

NOTE: Use three threaded rods spaced 120 degrees apart

Fig. 5.6.1 *A cross-section of Robert Bunge's telescope, illustrating how his baffles made of aluminum flashing are held in position by three threaded rods. When the baffles are correctly spaced and affixed to the threaded rods, the assembly is fitted into the telescope as a unit.*

An investigation into cleaning rods proved them to be awfully expensive, but 72-inch long ¼-20 threaded rods were only a couple of dollars. For smaller 3-, 4- and 5-inch refractors, smaller threaded rods should work.

The threaded rod allowed me to place the baffles wherever I wanted, and had the added benefit of enabling me to move the baffles for cleaning purposes in the future. You would be surprised to see just how dusty baffles are in hundred-year-old refractors.

5.6.3 Cutting the Baffles

As assembly of the scope progressed to the point that the baffles were next, necessity proved to be the mother of invention since I could not find an easy way to make the circular cuts in thin sheet metal.

A walk though a hardware store produced a turntable bearing for a Lazy Susan for $1.50. With the bearing mounted on a ½-inch piece of pressboard, another board made a nice turntable. Spinning the turntable, I used a pen to mark the center of the top board.

I chose lightweight aluminum flashing (made for furnace ducts and available in any hardware store) for the baffle material. The key is that the flashing needs to be thin enough to cut with scissors. Since my tube has a 6.5-inch inside diameter, I cut out several 8 x 8-inch squares using a photographic paper cutter (of course, scissors will work here, too). Drawing a line from corner to corner marked the center of each piece of metal. A compass was used to draw in the position of each baffle's opening and its outer edge which must fit the inside diameter of the tube. The flashing was then fastened to the turntable—one tack through its center and another off to one side to keep it from slipping as it was turned.

With the turntable base weighted for stability, I placed a medium duty photo tripod over the table with the tripod's center post near where I wanted to make the cut for the baffle opening.

Armed with a sharp utilityknife I grasped it and the center post of the tripod in one hand, lowered the knife onto the flashing, and lightly scored a circle in the surface of the metal by rotating the turntable with the other hand.

Next, I repeated the process, pressing harder as I retraced the score from the first rotation. Generally, I found that if I pressed down too hard on the first cut it would not be sufficiently round. Sometimes I did this twice, or until I felt that I had made a deep enough score around the circle.

Finally, I took the flashing off the turntable. When the scoring was done correctly, I could see light through it when I held the metal up to a 100-watt light bulb. Grasping the baffle with both hands, I carefully bent it back and forth a couple of times to break the score, and the center circle popped out in my hand without any trouble.

The result was a clean, round hole centered in the baffle. Some light sanding took off any burrs that remained. I then made the outside cut with a pair of scissors.

5.6.4   Placement of the Baffles

As for the placement of the baffles, not being much of a computer programmer, I resigned myself to a full-scale drawing. Gee… where do I get paper 7 inches wide and 90 inches long? From the continuous forms that I feed into my computer printer, of course!

Another thought took it a step further. A simple CAD program provided an easy and accurate way to make a full-scale drawing. It does not have to be a highly detailed drawing of the tube. It only has to show the inside diameter of the tube, the true light path leading to the area of full illumination at the focal plane (in my case, $1\frac{3}{4}$ inches since that is the field stop of my largest 2-inch eyepiece). For a more involved discussion of this, see Chapter 19.2 of Rutten and Venrooij's *Telescope Optics, Evaluation and Design* published by Willmann-Bell, Inc. Please note that a larger diameter tube will require fewer baffles than one that is just a bit bigger than the diameter of the objective.

5.6.5   Assembly

With the proper holes drilled in each of the baffles and the focuser plate of the telescope, I spent an evening in front of the TV putting the assembly together, using six nuts and lock washers to hold each baffle in place. It took about two hours for me to roughly position the baffles and the nuts. A lot time was spent spinning nuts along those rods!

Careful measurements were made from the focuser plate to align each baffle up and make it square with the tube. Finally, two wrenches were used to tighten the nuts down, after which the assembly was surprisingly strong, showing very little sag.

With a loud screeching noise that sent the cat running under the bed, the baffle assembly was slid up inside the tube. With the focuser in place and at the point where most of my eyepieces focus, a careful visual check was made to make sure that the lens was not vignetted from anywhere in the focal plane.

One note: you will have to remove anything attached to the tube with bolts (i.e., the finder) since the baffle assembly will not slide up it properly if bolts protrude into the tube. Another option would be to notch the baffles for any screws that are in the way.

With the whole thing painted black, I was ready to go…almost. The baffle closest to the lens was only about ½-inch wide. Under the weight of the threaded rods, it refused to hold its shape (perhaps an argument to use smaller, lighter threaded rods joined together). To solve this problem, I formed a ring from a coat hanger to serve as a stiffener to help the baffle hold its shape.

In this manner, I was able to place four baffles in the tube of my refractor without using a lathe or drilling holes.

Once at a star party, I saw a refractor with baffles that had been forced up the tube creating a tight fit. Under the watchful eye of mother time and the steady forces of thermal expansion, one of them had turned sideways in the middle of the tube during the drive to the observing site. The owner was painfully trying to use a long stick to force it back into position. . .

5.7   The Maksutov Telescope: Past and Present

By Thomas A. Dobbins
with Yuri A. Petrunin and Eduard A. Trigubov

5.7.1   The Invention

As German armies prepared to lay siege to Leningrad in August of 1941, Soviet authorities began the evacuation of those fortunate inhabitants considered non-essential for the impending defense of the city. Aboard one of the refugee trains rode a very tall 45-year-old optician, Dmitri Dmitrievich Maksutov. During the long eastward journey, Maksutov's thoughts were occupied by an assignment now interrupted by the war—the design of a portable telescope suitable for mass production, intended for use in schools throughout the Soviet Union.

Initially, he envisioned a conventional Newtonian or Cassegrain reflector augmented by a plane-parallel optical window. The window would serve to support the diagonal or secondary mirror, thus eliminating the diffraction effects of conventional mechanical supports, prevent loss of collimation in transport and handling, and minimize the internal convection currents that plague open-tubed instruments. Most important, the window would protect the telescope's delicate reflective coatings from dust, dew, and tarnish. However, a nagging problem remained: the added expense of fabricating a window of sufficient optical quality to not detract from the instrument's performance.

Maksutov's thoughts turned to the meniscus lens, essentially a "bent" window, as an alternative. Such a lens suffers from virtually no chromatic aberration, but does exhibit considerable spherical aberration. For several hours Maksutov struggled in vain to conceive of a meniscus lens free of spherical aberration. He was on the verge of abandoning the idea when he suddenly realized that the spherical aberration of a meniscus was capable of canceling out the spherical aberration of equal magnitude but opposite sign of an easily manufactured spherical mirror. Maksutov became convinced that the use of meniscus lenses would make possible a new generation of extremely compact Newtonian, Cassegrain, Gregorian, and Herschelian telescopes and wide-field astrographs.

The first example of a Maksutov telescope, constructed in October of 1941, was an f/8.5 Gregorian variant of only 100 mm aperture on a tabletop pillar and claw stand. The most modern optical system of its day was supported by a very antiquated mounting, indeed! On November 3, 1941, Maksutov applied for a patent on his invention, which was announced to a Western audience in a detailed 15-page article in the May 1944 issue of the *Journal of the Optical Society of America*.

It should be noted that the Maksutov was independently discovered in February of that same year by the Dutch optical scientist, Albert Bouwers of Delft. Since Holland was occupied by the Germans at that time and the potential military applications of the invention were obvious, Bouwers patiently waited until the end of the war to publicly disclose the invention, although with the cooperation of the Dutch authorities, he had been secretly awarded a patent on July 7, 1941. Not surprisingly, the design is often referred to as the "Bouwers-Maksutov", particularly in Dutch literature!

5.7.2 Theory Becomes Reality: A Chronology

In the October and December, 1944 issues of *Scientific American*, Norbert J. Schell published the complete optical parameters of an 8-inch f/4 Maksutov-Newtonian. Schell's design permitted the Newtonian diagonal mirror to be removed so that a film holder could be substituted, converting the instrument to an astrograph. Albert G. Ingalls, editor of the magazine's "Telescoptics" department, organized a buyers' club to reduce the cost of molding and casting the thick blanks of crown glass required for molding the meniscus corrector lenses. The Corning Glass Works constructed a temporary mold and cast 24 of these special blanks, which were quickly sold.

Suffering none of the off-axis coma characteristics of conventional "fast" Newtonians, the performance of these Maksutovs as rich-field telescopes and comet-seekers, delighted their builders. The July 1946 issue of *The Journal of The British Astronomical Association* contained the description of a 6-inch Maksutov-Newtonian, based upon data in Schell's *Scientific American* article, built by three Australian amateurs, C.J. Tenukest, R. Schaefer, and H. Pinnock. "Remarkably small and sharp images of stars were obtained, free from coma and color," they reported. "The image of Jupiter was as sharp as if viewed through a first-grade refractor, yet the bluish halo visible around the disk with even the best glass was entirely absent."

With the assistance of Dmitri Maksutov, mass production of the "TMS-70", a 70 mm f/11 Maksutov-Cassegrain began shortly after the end of the war. Initially manufactured in Leningrad, production was later transferred to Novosibirsk, where it continued until Maksutov's death in 1964. These telescopes featured oculars providing magnifications of 25x and 75x in a revolving turret and were sold for 50 rubles, a typical month's salary during this period. They were found in virtually every secondary school and university in the Soviet Union.

1952 Limited production of the "AZT-7", an 8-inch f/15 Maksutov-Cassegrain, began in Leningrad. Principally employed by professionals for evaluating atmospheric conditions at potential observatory sites, some of these instruments were also used for instructional purposes in universities.

1953 Alma-Ata Observatory in Kazakhstan received a wide-field Maksutov camera of 500 mm aperture with a speed of f/2.4. This instrument was employed by astronomers, V.G. Fesinkof and D.A. Roskovsky, to compile a photographic atlas of galaxies. A similar instrument with a speed of f/4 was later installed in the Crimea. With 50-minute exposures, it proved capable of recording 21.5 magnitude stars over a 4° field.

Wollensak, a firm renowned for photographic optics, offered the "Mirroscope", a 20x Maksutov-Gregorian terrestrial telescope that sold for $59.50. This optical system provided an erect image without the Porro or Amici prism required by refractors and Cassegrains.

1954 The American inventor and entrepreneur, Lawrence Braymer, introduced the West's first commercial Maksutov-Cassegrain telescope, the Questar. Weighing only 7 lbs, this revolutionary equatorially-mounted, motor-driven 3.5-inch aperture telescope provided an effective focal length of almost 4 ft in a tube only 8 inches long. For many years, Questar placed the aluminized secondary spot on the front, rather than the rear surface of the meniscus corrector, protected by a layer of black paint, to avoid violation of Bouwers' patent.

As portable as a microscope and made to the most exacting optical, mechanical, and cosmetic standards, the elegant Questar soon earned the reputation as the "Rolls Royce" of small telescopes. These beautifully crafted instruments continue to be manufactured in a form that is virtually indistinguishable from the first examples produced over four decades ago, testimony to Braymer's genius as an engineer. However, the high prices commanded by Questars (in excess of $1,000 [U.S.] per inch of aperture, at present) have largely restricted their use to the most affluent hobbyists.

1955 The world's largest Maksutov, of 700 mm aperture and featuring f/2.9 photographic and f/14 Cassegrain foci, was commissioned in Abastumani, Soviet Georgia. Equipped with an objective prism, this instrument recorded low-dispersion spectra of 17th magnitude stars.

1957 In the United States, the superb reputation of the Questar soon became inextricably intertwined with that of the Maksutov itself. For many years the "Mak" was coveted to a degree rivaled only by the popularity enjoyed by apochromatic refractors today. In the March, 1957 issue of *Sky & Telescope*, Perkin-Elmer Corporation engineer, John Gregory, published a design for a Maksutov-Cassegrain suitable for construction by advanced ATMs. Embodying the rare combination of highly skilled practical optician, machinist, and theoretician, Gregory fabricated a modest number of 8.2-inch and 10.8-inch $f/15$ instruments for sale to amateur astronomers. His optical design was employed in the largest Maksutov in the Western Hemisphere. Completed in 1965, this 22-inch instrument, located at the Stamford Observatory in Connecticut, featured $f/15$ Cassegrain and $f/3.7$ photographic foci.

1958 So enthusiastic was the response to Gregory's *Sky & Telescope* article by both courageous and foolhardy "glass pushers," that the Maksutov Club was founded by Allan Mackintosh for the purpose of disseminating construction tips, and making it possible to purchase molded meniscus lens blanks in a variety of sizes.

1963 A 6-inch $f/4$ Maksutov-Newtonian, the "Vega Six", was introduced by the Vega Instrument Company, founded by Robert T. Jones. Supplied with a rugged German equatorial mounting and a 3x Barlow lens for high-power work, several dozen of these versatile instruments were produced. Late in 1993, the "Vega Six" was resurrected by the F.J.R. Manufacturing Company of West Bend, Wisconsin, complete with a state-of-the-art servo-driven mounting.

1964 Thermoelectric Devices Corporation of Massachusetts advertised a 4.5-inch $f/23$ Maksutov-Cassegrain. The product was soon discontinued due to inadequate mechanical performance.

1965 A Questar accompanied the Gemini astronauts into orbit, becoming the first astronomical telescope to be employed above the Earth's atmosphere.

Tinsley Laboratories of Berkeley, California, noted for excellent large telescopes for professional use, offered a 5-inch Maksutov-Cassegrain. During the next several years, about thirty units were produced.

1967 Johannes Haidenhain of Traunreut-am-Traunstein, Bavaria, offered a 12-inch Maksutov-Cassegrain for permanent observatory installation. By rotating a Nasmyth tertiary flat mirror 180°, the focal position could be alternated from one tine of the equatorial fork mounting to the other via hollow declination axes.

1968 Ernst Popp, an optician of Zurich, Switzerland began to produce a series of fork-mounted $f/15$ Maksutov-Cassegrains in apertures ranging from 6- to 12-inches.

1969 Questar Corporation introduced the "Questar 7", essentially a scale-up of their now legendary 3.5-inch Maksutov-Cassegrain.

1970 The world of amateur astronomy was forever changed by the introduction of a very affordable Schmidt-Cassegrain, the "Celestron 8". Tom Johnson, founder of Celestron Pacific (later Celestron International), developed a proprietary process for economically producing the complex aspheric curve of the Schmidt corrector plate. Interest in the Maksutov waned drastically, for the Schmidt promised comparable compactness and portability at markedly lower prices.

1973 The Cleveland, Ohio firm Impex Optics introduced a 3-inch $f/11$ Maksutov-Cassegrain dubbed the "Copernicus", in honor of the 500th anniversary of the birth of the Polish astronomer. Imported from Poland, this $200 instrument featured a rudimentary altazimuth fork mounting intended for use with a medium-duty photographic tripod, a detachable solar projection screen, and two achromatic oculars providing magnifications of 50x and 92x. Until 1986 the Copernicus was marketed by the New York firm, A. Jaegers, a major supplier of optical and mechanical components for ATMs.

1976 Questar Corporation introduced the "Questar 12". Unlike its smaller predecessors, this instrument was supported by a German equatorial mounting rather than a fork, and was available only on a custom order basis.

1977 Two former Questar executives founded Optical Techniques, Incorporated of Newtown, Pennsylvania. The firm produced the "Quantum" 4-inch and 6-inch $f/15$ Maksutov-Cassegrains on single-arm fork mountings, a design that had been prototyped by Questar 25 years earlier, but rejected for aesthetic reasons. Shortly before its untimely demise in 1981, O.T.I. introduced an 8-inch $f/15$ optical tube assembly. Quantums were highly regarded and now command high prices on the second-hand market.

1978 The "C90", a 90 mm $f/11$ Maksutov-Cassegrain, was introduced by Celestron in both astronomical and terrestrial versions. Focus was achieved by moving the meniscus lens.

1981 Astro Works Corporation of White Rock, New Mexico introduced the "Astromak", a 12-inch $f/5$ derivation of the Maksutov-Cassegrain based upon the "Simak" optical formula published in 1980 by designer Mike Simmons. With its generous image scale of 2.3 arc minutes per mm and ability to deliver pinpoint star images across a 90 mm diameter image circle, the Astromak proved to be a powerful tool in the hands of several skilled astrophotographers, notably Jim Riffle, the firm's proprietor.

1983 Celestron introduced a short-lived product called the "C65", a Maksutov-Gregorian spotting scope.

1984 Carl Zeiss (Jena) introduced the "Meniscas 180", a 7-inch $f/10$ Maksutov-Cassegrain.

1992 6-inch Maksutov-Cassegrains produced in Moscow by the Manufacturing

Cooperative "INTES", appeared on the market in the United States, Europe, and Japan.

The Italian firm, Costruzioni Ottiche Zen, introduced a 7.3-inch $f/15$ Maksutov-Cassegrain optical tube assembly.

1993 Canadian optician, Peter Ceravolo, took a novel approach to the design of Maksutov-Newtonians. His firm, Ceravolo Optical Systems, introduced $f/6$ instruments of 145 mm and 216 mm aperture, optimized for high-contrast visual work, by employing unusually small central obstructions of only 10% in order to achieve exquisite definition rivaling that provided by much less compact apochromatic refractors.

> **Ed. Note:** Telescope Engineering Company in Golden, Colorado now offers a quality line of 6" to 12" MAKS.

5.8 From the Bench Binocular Collimation, Part I
By William J. Cook

In From the Bench, Issue #3 (see Section 3.8 on page 108), we discussed the de-cementing, coating and re-cementing of binocular objective lenses. That column ended with the words, "Now, if your prisms are clean and properly affixed on the prism shelf, you are ready for collimation." Since then, some readers have expressed an interest in learning how to collimate their own binoculars. Ok, here goes.

The first thing to recognize is that you do not need a sophisticated test setup to determine whether or not the binocular is aligned—nature has provided you with all you need. Simply focus the instrument on a distant, high contrast target, and slowly move it away from your eyes, keeping the target in the field of view. When the binocular is about 10 inches from your face, open and close each eye several times; if you do not notice the image dancing around, the instrument is probably well within alignment standards for that particular IPD (interpupillary distance).

In the alignment of binoculars, two standards may be achieved. The first, "conditional alignment" means that the optical axes are parallel at a given IPD or at a predetermined standard IPD. More than likely, this is the condition that exists when an individual hands his binocular to a friend who immediately starts complaining about seeing a double image, or how the instrument "draws" his or her eyes. Unless a prism is mispositioned on the prism shelf, conditional alignment is not difficult to achieve and, unfortunately, it is the preferred alignment method for a large portion of repair facilities. The bittersweet situation here is that so many of today's low priced binoculars suffer from so many mechanical weaknesses and acute aberrations, that trying to achieve collimation is an effort in futility.

"True collimation" is considerably more taxing. This condition is only achieved when the optical axes are parallel at *all points* along the "swing" of the telescopes or at all IPD settings. The process here requires the optical axes of both telescopes to be parallel with each other *and* with the mechanical axis or axle. When these conditions have been met, the 6 foot man can hand the binocular to his 8 year old and know that the child will enjoy an image of equal quality.

With this in mind, the next step is to explain how collimation is achieved. This is not as easy as it might seem since there are a number of ways to collimate a binocular which will vary with instrument size, style, and manufacturer.

In Part II (see Section 6.9 on page 235) we will describe a number of the most common design conventions (relative to collimation), and then offer some insight into how to use the differing conventions to collimate your own binoculars.

5.9 Viewpoint

By R. A. Buchroeder

This writer has viewed with consternation the recent brandishing of MTF (Modulation Transfer Function) and Strehl Ratio as descriptors of telescope quality. Not only have authors tended to oversimplify the issue, but the reader is still left with only a vague feel for how his images will actually look!

The "old fashioned," and still completely effective, method for appraising a telescope's quality is to regard an extended image as being made of a myriad of points, each of which is smudged by the diffraction- and aberration-blurred star image in that area of the field of view. Put technically, we treat the telescope as being isoplanatic over a portion of the field, in which the perceived image is the convolution of the geometrical image and the diffraction PSF (Point Spread Function).

The trick, of course, is to know the diffraction point spread function. Two-dimensional plots are found in many optics texts, (e.g., *Principles of Optics*, by Born & Wolf); but three-dimensional representations are comparatively scarce, especially in comparing the same peak-to-valley forms of the different aberrations, measured in the wavefront.

Recently, Fred F. Forbes of National Optical Astronomy Observatories (formerly known simply as "Kitt Peak") showed me a preprint of a paper he wrote for professional astronomers, and it contained an excellent plot of diffraction point spread functions which may be of interest to the readers of *ATM Journal*. Therefore, it is reproduced here with permission. (Figure 5.9.1)

The diagram illustrates that as the image becomes increasingly degraded, the differences in the characteristics of contributing aberrations become more pronounced. Thus, when someone states that his telescope is "good to an eighth-wave," a serious telescope maker or lens designer might rightfully inquire, "An eighth-wave of what?" Note that in the row of 0.4-wave aberrations, the plot for the effects of pure spherical aberration more closely resembles the ideal than does the plot illustrating 0.4 waves of pure astigmatism. From this we may ascer-

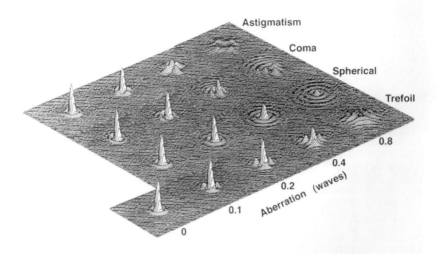

Fig. 5.9.1

tain (as one example) that a system suffering from pure spherical aberration would provide superior star images; more so than a similar system suffering from an equal amount of astigmatism.

Another issue that should be addressed someday is the comparative irrelevance of geometrical spot diagrams as compared with the fundamental value of the wavefront measurement of aberration.

5.9.1   Additional Reading

1. Schroeder, Daniel J. *Astronomical Optics.* San Diego, California: Academic Press, Inc., 1987.

Issue 6

6.1   Riverside 1994

By Cheryl Wilcox

With photos by Gary Hall, Kreig McBride and Bob Stephens

The Riverside Telescope Maker's Conference (RTMC) is held each year over Memorial Day weekend. The conference began in Riverside, and later found its current home at Camp Oakes, eight miles outside Big Bear City in the beautiful San Bernardino Mountains, at an elevation of 7,300 feet. This year marked the 26th Annual meetings, and was the first without RTMC founder, Clifford W. Holmes Jr., who passed away in September, 1993.

Originally created as a gathering at which telescope makers could show off their designs and get new ideas, RTMC also has many things to offer the non-builder. Other activities include commercial product demonstrations, a swap meet, talks on various astronomy related topics, and, of course, a star party each night.

This year the RTMC adopted a theme. Due to the prominence of the Moon at this year's conference and the impending collision of Comet Shoemaker-Levy 9 with Jupiter, planetary telescopes became the focus. Noted specialists Don Parker and Jeff Beish came from Florida to discuss planetary observing and modifications to telescopes which can greatly improve them for high-resolution applications.

The conference opened at 9:00 a.m. on Friday, and by Saturday morning, the camp was overrun with cars, trucks, trailers, RV's, tents, and telescopes of every kind, shape and size. Most of the homemade instruments were displayed on the main telescope field, in front of the Walker Observatory, or down "Telescope Alley."

On Friday afternoon vendors began selling their wares. Companies like Celestron, Meade, and Parks usually have a large assortment of telescopes, eyepieces, and other equipment on sale; this year was no exception. Other vendors offer everything from astronomy apparel to meteorites. Many good deals were there for those willing to seek them out. The first meal got underway Friday evening in the meeting/dining hall, and attendees had their choice of several meal plan options that go along with dorm or camping accommodations. The food is pretty tasty, and you always get plenty to eat. For those who chose not to purchase the meal plan option, a snack bar was open at each mealtime.

Following dinner on Friday, a few talks were given. This year, a CCD demonstration was conducted by John Sanford, a member of the Orange County Astronomers. His hands-on demonstration included telescope set-up, camera attachment, initialization, focusing, finding objects, exposure, and simple image adjustment techniques. It was a nice opportunity for observers who had been considering the purchase of a CCD camera. After the talks, attendees adjourned outside to enjoy the first of three nights of star parties.

The Merit Awards judging started Saturday, and is RMTC's way to recognize outstanding craftsmanship, design, and innovation in telescope making. The judges look for outstanding examples of metal and wood craftsmanship, well-made first telescopes, complete systems that are homemade, and generally innovative ideas. Two award categories exist—Honorable Mention and Merit Award. Winners of the latter receive a certificate, as do those given the Honorable Mentions; but, in addition the Merit Award recipient also receives a small metal plaque that can be fastened to their instrument.

Entrants for the Merit Awards sign up in one of four time blocks and anxiously await the judges. This year a panel of six judges and two photographers spent a tiring day looking over 31 entries. The judges set out in small groups, hunting down entries in order to talk to the owners. After the entrant met with all of the judges and had his or her entry photographed and videotaped, they were free to partake of other RTMC events. Unfortunately, every year someone is overly anxious to get to the swap meet or to hear a talk in the meeting hall, and abandons the competition after talking with only one judge.

This year the Saturday talks largely focused on CCDs. With these talks being held in the meeting hall and the hustle and bustle around the tables at the swap meet, there were those who preferred to just roam the telescope field and down Telescope Alley looking over Merit Award entries, past and present.

The main program on Saturday night this year was in remembrance of Cliff Holmes, RTMC founder. It was an open forum for Cliff's friends to come up and share a funny story or to tell about how he had touched their lives. We laughed and thought fondly of Cliff as we were reminded of his famous yodel, of how he liked to leave long messages on our answering machine, and of his passion for the "heavenly bodies" of our universe. His wife, Jackie, shared some memories with us, as well.

Following this, the door prize drawing began. Vendors donated prizes to be given away on either Saturday or Sunday night. While the winning numbers were called, a rivalry seemed to develop between those seated inside the meeting hall and those seated outside. When no one out back had won anything after a few prizes were awarded, the chant "Outback! Outback!" began. The same happened when those inside the hall hadn't won in a while.

After the lesser-valued door prizes were given away, this year's main one, sponsored by Meade Instruments, was shown off—a 12-inch LX200 Schmidt-Cassegrain. Following the door prize drawing, the Saturday night star

Fig. 6.1.1 *(Left) Merit Award winner Clyde Bone, of San Angelo, TX, describes his unusual telescope to Richard Berry. The 20-inch f/5 Newtonian has dual capacity. According to Clyde, the Newtonian can convert his 4-inch Tele Vue Genesis into a 20-inch instrument by sending parallel light rays directly into the refractor's objective. With this combination, Clyde enjoys the opportunity to use a wide range of eyepieces, and either a 35 mm or 200 mm lens for film plane imaging.*

Fig. 6.1.2 *(Right) Peter Hirtle of Seattle (in the light shirt) shows of his Tri-Schiefspiegler. The primary mirror, which is about 55% parabolized, is a 6-inch with a 124.8-inch radius. The telescope has found a home atop a mount that was started for a different project and which required the f/14.7 focal ratio to be increased to f/15 in order to increase the back focal length enough for the light path pass through the mount. The secondary is spherical and was tested with a test plate. The drive uses a stepper motor and friction disc. This instrument will be featured in our next issue as part of our coverage of Stellafane.*

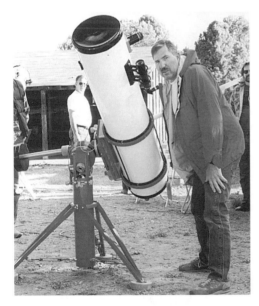

Fig. 6.1.3 *With "A little bit of sandpaper and some attention to detail can add a lot of pride of ownership." as his battle-cry, fireman Jim Hannum (of Orange County Astronomers) brings another winner to Riverside. This year it's a beautifully executed 10-inch f/6 Newtonian.*

Fig. 6.1.4 *A closeup of the mount for Jim Hannum's 10-inch f/6 Newtonian. The workmanship on this instrument really made it a show stopper. This bright red piece of machinery features knurled knobs, a removable shaft and was painted with DuPont acrylic auto-body paint.*

Fig. 6.1.5 *Wayne Kaaz of Costa Mesa brought his blue 10-inch f/8 Ritchey-Chrétien. This instrument has an extended axis mounting (torque tube) that allows him to cross the meridian without having to flip the tube around. This enables him to take longer exposures at higher latitudes. With its 3-inch diameter steel shafts encased in aluminum tubing, it will hold up to 80 lbs of instruments.*

Fig. 6.1.6 *Steve Swayze's 40-inch f/5 was the largest on the field and won the Merit Award for Outstanding Large Telescope. This instrument, built by Steve, his brother Bruce and Mel Bartels, features a 27-point flotation system, a 12½-inch finder and some excellent woodworking. In all, it took a year to complete; the primary alone requiring four months to finish, even though it came with a pre-generated f/5 curve!*

The line of people waiting to look through this instrument was long and moved very slowly. It took time to scale the tall ladder, take a look, and then plan your descent. The three-legged ladder worked well and, if you were patient, you eventually got a peek. However, once perched at the eyepiece, some observers noted that they had to make some critical choices—like "should I attempt to focus or just hang on for dear life."

Mel Bartels has developed a computer program that shows the curvature of lines produced in Ronchi testing large, fast optics, which was critical to the successful completion of the 40-inch f/5 mirror. For more information concerning this program, contact Bruce Swayze at 7635 S.E. Deardorf Road, Portland, OR 97236. Remember that a self-addressed, stamped envelope encourages a speedy reply.

Fig. 6.1.7 *Steven Overholt stands beside his 30-inch f/3.75 Dobsonian. This telescope can be disassembled and packed into a compact car, in about seven minutes. His book,* Lightweight Telescopes, *is loaded with information for those interested in making such instrument, as is available from Owl Books, Box 901, Santa Margarita, CA.*

Fig. 6.1.8 *Randy Johnson's 6-inch f/15 Tri-Schiefspiegler is one of six instruments made working with others from the Seattle Astronomical Society. With its unobstructed aperture, Randy enjoys what he refers to as the "poor man's apochromat." A good amount of frustration was encountered in testing the tertiary mirror using the Foucault test at a distance of 83 ft!*

Fig. 6.1.9 *Paul Jones of Flagstaff, AZ, stands beside an unusual dual telescope he built with Gene Purtick. The instrument consists of an 8-inch f/9 off-axis reflector and a 10-inch, f/15 refractor which has been stopped down to 8 inches so that a direct comparison can be made between the two telescopes. The refractor has a fold mirror at the base of its tube so that the focus mechanisms are close together. Paul says the off-axis reflector "compares favorably" to the refractor, but stops short of favoring the performance of one instrument over the other.*

Fig. 6.1.10 *Judging by the horror films of the 1950's, one would believe that piranha just swim from one feeding frenzy to another. The truth is, they only behave that way at certain times of the year. The same can be said of that rare species known as* Telescopium Grabacus Aparatacus. *The major difference in the two is that the piranha is considerably more docile and discriminating!*

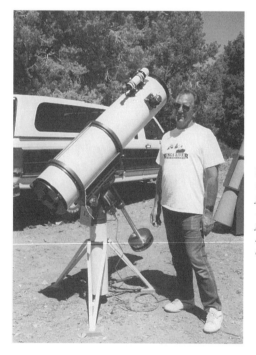

Fig. 6.1.11 *Larry Bohne of La Habra, CA received an Honorable Mention for his first effort at telescope making. His 10-inch f/5.5 Newtonian was recognized for Excellence in Metal Craftsmanship. All parts were machined by either Larry or Jim Hannum.*

Fig. 6.1.12 *Stephen Collett of Bakersfield, CA received an Honorable mention for his 14 ¼-inch f/5 Dobsonian. The instrument is easy to assemble due to special fasteners at the rear of each truss rod, which holds it in proper position to facilitate installation of the focuser section. To make life better still, collimation can be adjusted at the eyepiece.*

party began and, although tired from chasing down entries all day, the Merit Award judges joined the star party to look through the many entries that remained on the telescope field.

Talks continued on Sunday in the meeting hall while the Merit Awards judges met elsewhere to discuss this year's entries. After much thought and consideration, the winners were selected. Merit Award Director, Randall Wilcox, then started the lengthy job of putting together the awards program, which was presented later that evening.

Sunday afternoon, an RTMC "first" took place. Wayne Johnson, Vice President of the RTMC Board of Directors, married Arlene Thomas. Both are members of the Riverside Astronomical Society. A large crowd of well-wishers was present for the ceremony and reception that followed.

The Merit Awards program was held Sunday night—Chairman Randall Wilcox presiding—as judges Richard Berry, Rick Shaffer, Paul Zurakowski, Pierre Schwaar, and Keith Lawson made the presentations.

This year, there were several teacher-student entries. Jim Hannum of Fullerton and Steve Collett of Bakersfield, CA, passed on the art of telescope building. Jim's Merit Award winning entry, that he termed "the party scope"—a beautifully machined white 10-inch Newtonian with red and gold accents, portable enough for public star parties—was part of what became known to the judges as "10-inch Alley"—four 10-inch scopes in a row, all Merit Award entries. Jim's student, Larry Bohne of La Habra, CA, received an Honorable Mention for crafting a beautiful pearl-white Newtonian with blue accents. Jim showed Larry how to machine the parts, and Larry manufactured a duplicate. Steve Collett, who received an Honorable Mention for his 14.25-inch *f*/5 Dobsonian, gave Darren Bly the help he needed to build a 17.5-inch *f*/4.5 open frame Dobsonian—his first telescope.

Unique designs entered this year included a Tri-Schiefspiegler by Peter Hirtle of Seattle, WA, and a Nasmyth-Mersenne by Clyde Bone of San Angelo, TX, both of which won Merit Awards. Philip Alotis of San Francisco, CA had an interesting way of magnetically attaching the brass finder he had fashioned for his beautiful oak telescope (a 1992 Merit Award winner and work of art), winning him an Honorable Mention.

John Laborde of San Diego, CA garnered a Merit Award with his outstanding photographic instrument—an 8.6-inch *f*/3.7 Wright Schmidt. It has a 6-inch *f*/20 Cassegrain guide scope on an off-axis mounting. He was able to show the judges several stunning photographs of the objects he had photographed through his entry.

The largest telescope entered this year was a 40-inch, crafted by Steven Swayze of Portland, OR. He made the mirror himself, using a grinding machine in the living room of his one bedroom apartment. He had to go to an office building that had a long corridor just to test it. On Saturday night, the wait to look through it was about 45 minutes but the view was well it. Steve's entry was a Merit Award winner.

Following the Merit Awards program, the Sunday night door prize drawing began. The grand prize was a C-5 donated by Celestron. Then the final star party of the 1994 RTMC took place.

Monday morning following breakfast, the tear-down began, and the 1994 RTMC came to a close.

6.2 Impressions of Riverside 1994
By Kreig McBride

After three years without a vacation, my wife, Carol and I decided to explore the coast of Oregon, the valleys of Yosemite, and the sights of Riverside. The beaches of Oregon greeted us with the traditional monsoons, so we traveled inland to camp inside one of the caldera, one of many to be found in the lava landscape the state's central region. It snowed. We found millions of tons of obsidian and dreamed of building a very large furnace to cast a very large mirror. The signs stated, "Do not remove objects from the National Forest". We then moved on to Yosemite where we watched the full Moon rise from Glacier Point. The Moon appeared full as it "rolled up" the ridge of Merced Peak. On clearing the summit, the Moon was complemented by yellow Jupiter to the south and Venus peeking out from behind a glowing red and yellow thundercloud.

After five glorious days in the valley, it was on to our first Riverside; and we were looking forward to a unique and enjoyable experience. One of my goals was

Fig. 6.2.1 *The line waiting for Meade Instruments to start offering their telescopic goodies for sale.*

Fig. 6.2.2 *One of the many displays of non-commercial items.*

to view Omega Centauri, a globular cluster that is not visible from Bellingham, Washington's 49° north latitude. Omega was not visible from the coast in the rain, the caldera in snow, or from Yosemite with its oversized cliffs. But it would be from Riverside.

We arrived at Camp Oakes at 10 a.m. Friday; and after waiting for the dust to clear as the last buses carried screaming kids away from the camp, we drove in, located a campsite next to a small lake and planted ourselves for three memorable days of telescopes and astronomy stuff. The first scheduled event was the swap meet at 1 p.m. I took a walk over to the site and found people already lined up for the Meade and Celestron tables. For the next hour I checked out all of the dealers tables looking for those $5 Naglers and $100 apo-refractors trying to decide which one to be at when the clock struck one. I found myself standing in the center of the swap area when the official hour struck, and what a sight to behold. There were over 50 people in line at the Meade booth, and the first ten were let in to check out the good deals while the rest were held back desperately waiting their turn. Those first ten sure took their time! At the Celestron booth another 50 rushed forward all at once for a first-come, first-served opportunity. It reminded me of a celebration after a soccer match. Looking at the other tables, money was already exchanging hands where items had been pre-sold or scoped out (no pun here) ahead of time. I may have missed out on some great deals, but the entertainment was great. It was very hard to pass up all those disks of Pyrex calling out, "Take me home." However, I kept thinking of the four blanks at home that still need finishing.

I had always visualized RTMC as a large open field of grass and dirt where everyone camped and set up their scopes. I was surprised to find so many trees, nearly every one with a car, truck or camper under it. The only open area was a small grassy field next to the lake, reserved for telescopes only. This was definitely the place to be with the best sky views and the least dust. Most sites were limited to a smaller portion of the sky because of the trees; however when the sun was up, there was shade under dem der trees. At 7300 feet elevation one sunburns quickly and dehydrates readily. Next year, we camp in the trees. (You might read Richard Berry's article about the 1980 RTMC in *TM* #8 for a different point of view.)

As night fell, I had an opportunity to look through dozens of instruments. I was very impressed with a 12-inch $f/8$ reflector which gave stunning views of M13, and an 1888 6-inch Clark refractor which gave the best images of any scope I looked through the first evening. Most larger ones did not perform exceptionally well, as the atmosphere was quite turbulent. I was told by the locals that at the right time Omega Centauri is visible in a gap between two hills located south of Camp Oakes. I watched carefully all evening but never caught a glimpse. Maybe tomorrow night. An almost full Moon rose two hours after sunset and ended the night's observing for most.

Day two began hot and sunny (where is that tree?). At 9:30 a swap meet for non-commercial venders started and again almost every item related to the subject of astronomy was available for sale. If you did not see what you were looking for on your first pass, it would be there on your second one. I tried to purchase a pair of 25 x 100 binoculars, but the vendor would not take travelers checks. I saved a lot of dollars that day. Somehow I spent most of the day meeting very interesting people and learning the nuts and bolts of telescopes and their construction. Al Nagler of Tele Vue answered all my questions about eyepieces with one word, Panoptic. Edward Byers had all of his clock drives and worm gear units on display in a glass case. I watched them disappear for two days; and when I found out Mr. Byers was retiring and these were the last units to be manufactured, I purchased the very last 11-inch precision drive unit.

Ralph Aeischlemen was one of the founding fathers of our group in Bellingham and eventually landed a job with NSGS in Flagstaff. Ralph has the responsibility of preparing the maps that are drawn from information relayed by our fly-by spacecraft. It was a great pleasure to see him again. Ralph has great enthusiasm for the night sky, and it rubs off on everyone he meets. He had a table set up and was handing out free maps of the outer solar system objects, and also pictures of Venus generated by the Magellan spacecraft. Normally these materials are discarded, being out of date or rejects. I have spent a few hours exploring Venus with a magnifier and the views were…well, apochromatic!!

In the dining room/lecture hall, talks were being conducted all day and evening for three days covering subjects from CCD's to "Flexible Remote Telescope and Remote Astronomy". Richard Berry gave a very well-organized and informative talk about CCD's and basic image processing. His presentation and ideas are easy to follow and has convinced me that even I could build and operate a CCD camera. Earlier this year he took 99 separate exposures of M51 and combined them to produce an image that turned out to be a pre-discovery photo of supernova 1994 I.

One of the most impressive instruments on display was designed and built by Graham Flint, who is planning to publish a photographic sky atlas using a modified 36-inch $f/7$, 7-element Baker/PE camera. This camera was redesigned by Mr. Baker himself and is now color corrected. The central lens group is movable to compensate for distortion caused by changes in both temperature and atmospheric pressure! The plates are hypered 14" x 14" Tech-pan film. Images will be taken through a diffraction grid producing patterns that can define colors and spectral information. The atlas will consist of 118 plates each 20° x 20° and will be reproduced by a stoicastic printing process. This type of reproduction is affordable and is amenable to magnification.

On the second and last night I charted out exactly where and when Omega Centauri would be visible. I hiked up the hill to the north and waited. I estimated that it was about $\frac{1}{2}$° short of being visible. During the drive home I again attempted to find this now elusive object from the Central Valley but found that all stars within 15° of the horizon were obliterated by air pollution. RTMC was a great experience; maybe next year Omega will come out of hiding.

6.3 Telescopes for Observing Planets
By Jeff Beish

Observing planets requires a great deal of time and patience. It is not something most amateur astronomers like to do because of certain challenges involved—the two main ones being the derogatory effects of Earth's atmosphere on the tiny planetary images, and the equipment required to magnify those images. Because planetary observers use high magnifications, any imperfections in the atmosphere—and in the telescope itself—are amplified causing images to blur and move about. A planetary telescope must contain the highest quality optics and should be made with materials designed to reduce the effects of air turbulence within the telescope tube.

The planetary astronomer has a wide variety of telescopic equipment available in these modern times. You may wish to purchase commercially- made telescopes or build one of your own design. You may even want to ask for time on an instrument at an established observatory. In any case, one needs to consider several important factors about telescopes before getting started.

The most common instruments in use today are: the Newtonian reflecting optical system, the achromatic or apochromatic refracting lens system, the classical Cassegrain reflecting optical system, and the Schmidt-Cassegrain and other catadioptric instruments (which combine elements of both the refractor and the reflector). Telescopes are also classified for different uses according to the focal length of their optical systems.

Newtonian Telescopes: For Lunar and planetary work, these should have focal ratios from $f/6$ to $f/12$ and small secondary mirrors. Image contrast is an inverse function of the area of the central obstruction caused by the secondary mirror, its holder, and the spider support system. The simple rule is: the smaller the diameter of the secondary with respect to the primary mirror, the better the apparent contrast. An obstruction ratio between 10% and 15% will make an excellent planetary Newtonian telescope. The Newtonian reflector is completely achromatic and gives sharp images through all color filters.

Refracting Telescopes: Refractors usually have $f/8$ to $f/16$ focal ratios, are the most light efficient optical design, give maximum image contrast at high powers, are simple to use, and require little cleaning and maintenance. However, moderate size (10 inches or more) refractors are expensive, difficult to house, not portable, and usually not completely achromatic (thus requiring a new focus each time a different color filter is used for visual observation or photography).

Classical Cassegrain Telescopes: These have focal ratios from $f/15$ to $f/60$ with small secondary mirrors. They are comfortable to use, have folded, compact optical systems, require only medium size mounts, and are portable in modest sizes. The optical design is completely achromatic, gives excellent image contrast, and is a stable scope for photographic patrol programs. Focal ratios below $f/20$ require relatively large secondaries that reduce contrast significantly.

Schmidt-Cassegrain Telescopes: Schmidt-Cassegrains are usually a compromise between a planetary type and a deep-sky system, having $f/10$ or $f/11$ focal ratios. This design is extremely compact, easy to use, lightweight and portable. Little maintenance is required because of the closed tube. Image quality may vary from poor to satisfactory in commercially produced models. It is advisable to thoroughly test the quality of the optical system before the warranty expires. The use of a Barlow lens can turn this type of instrument into a fair planetary telescope by increasing the focal ratio two- or three-fold, to $f/20$ or $f/30$. A systematic photographic patrol of the planets is also possible by use of eyepiece projection.

6.3.1   Image Contrast

An image of a star formed by a perfect lens or mirror system under ideal conditions is seen as a tiny spot of light surrounded by several delicate bright rings separated by dark intervals. In theory 84% of the light falls in the central spot (or "Airy disk"), 7.1% in the first bright ring, 2.8% in the second, and 1.5% and 1% in the third and fourth rings.[1,2]

Since mainstream reflecting telescope designs require a secondary mirror to reflect the image to the eyepiece, it usually has to be positioned somewhere in the optical path. This obstruction in the optical path slightly reduces the light gathering power of the primary and more important, it adversely effects image contrast. If we scatter stray light throughout the image, it makes the dark areas of the object brighter and the bright areas darker; therefore, a loss in image contrast results. What really happens is the diffraction arising from the obstruction by the secondary tends to remove light energy from the center of the Airy disk and distribute it among the bright rings surrounding it, in the case of a stellar image, or the many points comprising an extended image. The space between the bright and dark rings is unaffected by the obstruction, but since extended objects are made up of innumerable overlapping Airy disks, this brightening of the rings will in effect "smear" the image points, thus causing a loss in contrast, see Figure 6.3.1.

Image contrast, as perceived by our eye, is the difference in brightness or intensity between various parts of the telescopic image; i.e., a star against the background sky or a planetary disk. A simple formula for calculating contrast is as follows:

$$c = \frac{b_2 - b_1}{b_2}, \qquad (6.3.1)$$

where b_1 and b_2 are the intensities, or brightness levels, measured in candle power/meter squared (cd/m^2) of two areas of the object and c is the contrast.

For example, light areas of Jupiter's disk have a surface brightness of around 600 cd/m^2. If we compare a dark belt of 300 cd/m^2, then the contrast between these areas would be:

$$c = \frac{600 - 300}{600} = 0.5 \text{ or } 50\% . \qquad (6.3.2)$$

If we scatter light from the bright area, say 50 cd/m^2, and add it to the dark belt, then the contrast between the two becomes:

---

[1] Hurlburt, H.W. "Improvement of the Image Contrast in a Newtonian Telescope," *J.A.L.P.O.*, Vol. 17 Nos. 7–8, July-August, 1963, pp. 153–158.
[2] Johnson, L.T. "Improving Image Contrast in Reflecting Telescopes," *J.A.L.P.O.*, Vol. 18, Nos. 7–8, July-August, 1964, pp. 142–146.

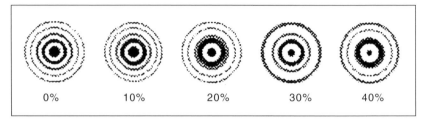

Fig. 6.3.1 *Computer graphics of Airy disks illustrating effects of various secondary mirror obstructions in reflecting telescopes.*

$$c = \frac{550 - 350}{550} \text{ or } 0.36 \text{ or } 36\%, \tag{6.3.3}$$

Thus showing that a relatively small amount of scatter may cause a significant decrease in image contrast. The Earth's daylight sky brightness has been measured at about 8000 cd/m^2.[3]

From the graphs and tables published in the referenced articles on Newtonian improvements in the *Journal of the Association of Lunar and Planetary Observers*, a general equation can be arrived at approximating the "contrast factor" value for your system:

$$CF = 5.25 - 5.1x - 34.1x^2 + 51.1x^3 \tag{6.3.4}$$

where x is the obstruction ratio or secondary/primary diameters, see Figure 6.3.2.

Table 6.3.1 Values for obstruction and contrast factor.

Obstruction (%)	CF
0	5.25
10	4.46
20	3.28
30	2.03
40	1.02
50	0.55

An image of a planet or extended deep-sky object may appear sharp and bright, but barely show any surface details in a telescope with a 35% obstruction. This same telescope can be made to show very fine surface detail and give refractor-quality, high contrast images if the secondary obstruction is reduced to say, 12%. This can be accomplished without perceptible vignetting of the image, as shown in the next section.

---

[3] Chapman, Clark R., and Dale P. Cruikshank. *Observing the Moon, Planets, and Comets*, Schramm and Groves, Laguna Niguel, CA, 1980.

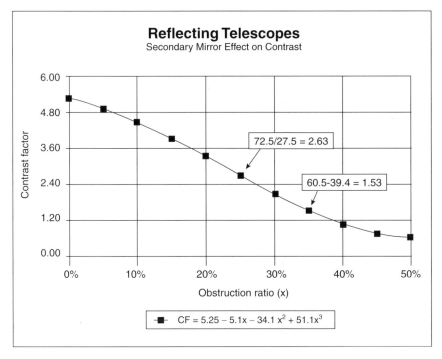

Fig. 6.3.2 *Plot of contrast factors as it relates to secondary mirror obstruction in reflecting telescopes.*

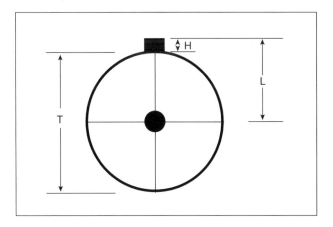

Fig. 6.3.3 *Cross section of reflecting telescope showing the distance between the fully racked-in focuser and the secondary mirror.*

6.3.2 Secondary to Focal Point

The Newtonian secondary mirror or diagonal is usually centered within the telescope at a point near the opposite end of the tube from the primary mirror. The distance from the secondary to the focal point (C) can be found by dividing the outside diameter of the tube (T) by 2 and adding the length of the fully racked-in focuser (H). This distance becomes important when selecting the final image size and size of the secondary mirror. The size of the secondary is a key parameter when attempting to optimize a telescope for maximum contrast. Some telescope makers add ½-inch (Figure 6.3.3).

The final choice for the tube diameter is predicated on the aperture and how much extra space is required for airflow. You will want at least 0.75 to 1-inch clearance between the primary and the inside of the tube, because heat waves (tube currents) from its walls will enter the optical path causing effects similar to bad "seeing." So, we have one parameter for selecting C—tube inside diameter should be the aperture plus 1.5 or 2 inches.

6.3.3 Secondary Diameter Determined by Image Size

To determine the absolute minimum size that the Newtonian secondary mirror can have and yet reflect the entire geometric cone of light received from the primary mirror to the focal point, let's start with this equation:

$$\text{Minimum secondary} = \frac{DC}{F},$$

where C = the distance from the secondary to the focal plane, D = the primary diameter, and F = the focal length.

However, this calculation only gives the minimum size; and in order to fully illuminate the image of the primary at the focal plane, we must increase the size of the secondary by a small amount. This can be found by the following:

$$\text{100\% illuminated secondary} = \frac{C(D-i)}{F+i},$$

where i is the linear image size at the focal plane.

If we use as an example a 10-inch $f/6$ in the above with a 12-inch O.D. tube and 1.6-inch focuser, where C = ½ tube diameter plus focuser height plus a ½-inch fudge, then C = $^{12}/_2$ + 1.6 + 0.5 = 8.1 inches, and the minimum secondary will be:

$$\text{Minimum secondary} = \frac{10 \times 8.1}{60} = 1.35 \text{ inches}.$$

For a fully illuminated image of 0.5 inch, the secondary would be:

$$\text{Fully illuminated secondary} = \frac{8.1(10-0.5)}{60} + 0.5 = 1.78 \text{ inches}.$$

In other words, the focal plane is not just a point—it is a circle or disk that represents the image of the primary mirror. But how large should this illuminated image be, and is it necessary to illuminate the entire field?

6.3.4 Reduced Secondary Mirror Size

Some telescope makers point out that a loss of a half-magnitude at the edge of the image field is barely noticeable to the visual observer. This is only one photographic f/stop, and they say this is acceptable for photometric purposes. An article illustrating a reasonable fall-off in the illuminated field of no more than 0.5 magnitude was published in the March 1977 *Sky and Telescope*[4] and gave a set of complex equations as follows:

$$M = 2.5 \log\left(\frac{1}{I}\right),$$

where "I" is found from:

$$I = \frac{\arccos A - x\sqrt{(1 - A^2)} + r^2 \arccos B}{\pi}.$$

Hence:

$$r = \frac{aF}{lD}$$

$$x = \frac{2b(F - l)}{lD}$$

$$A = \frac{x^2 + 1 - r^2}{2x}$$

$$B = \frac{x^2 + r^2 - 1}{2xr},$$

where:
M = Magnification
D = Diameter of primary
F = Focal length of primary
l = Distance from secondary to focal point
I = Fractional illumination
a = Diameter of secondary (minor axis)
b = Distance from center of the secondary to edge of field
π = 3.14159265

---

[4] Peters, William T., and Robert Pike. "Gleanings for ATM's—The Size of the Newtonian Diagonal," *Sky and Telescope*, March 1977, pp. 220–223.

Note: angles expressed in radians.

To prove the above theory, let's step through the equations. For a 12.5-inch *f*/7 (87.5-inch focal length), telescope with a 15-inch tube diameter, a 3.25-inch high focuser, and a 1-inch image, how small can we make the secondary before it causes 0.5 magnitude loss; i.e., it reaches the minimum size? After iterating down from the calculated secondary of 2.35 inches, we finally end up with 1.48 inches:

$$r = \frac{1.48 \times 87.5}{10.25 \times 12.5} = 1.011 \qquad (6.3.5)$$

$$x = \frac{(2 \times 0.5) \times (87.5 - 10.25)}{10.25 \times 12.5} = 0.603 \qquad (6.3.6)$$

$$A = \frac{0.603^2 + 1 - 1.011^2}{2 \times 0.603} = 0.284 \ (\text{arccos } A = 1.283 \text{ radian}) \qquad (6.3.7)$$

$$B = \frac{0.603^2 + 1.011^2 - 1}{2 \times 0.603 \times 1.011} = 0.315 \ (\text{arccos } B = 1.248 \text{ radian}) \qquad (6.3.8)$$

$$I = \frac{\text{arccos } 0.284 - 0.603 \times \sqrt{1 - 0.284^2} + 1.01^2 \times \text{arccos} 0.315}{3.14159265} \qquad (6.3.9)$$

$$I = \frac{1.283 - 0.578 + 1.276}{3.14159265} \qquad (6.3.10)$$

or

$$I = \frac{1.981}{3.14} \qquad (6.3.11)$$

or

$$I = 0.631 \qquad (6.3.12)$$

$$M = 2.5 \log\left(\frac{1}{0.631}\right) = 2.5(0.200) = 0.5. \qquad (6.3.13)$$

The sizes of secondaries in the above examples are calculated values and one should note that most manufacturers produce standardized sizes. You will have to pick a size near or slightly above the calculated one.

6.3.5 Telescope Modifications

Many telescopes, both commercial and amateur made, take excessive time to reach thermal equilibrium with the surrounding air. The most common reason for this is the material used in their construction. Materials that store heat, such as fiberglass, plastic, formica, or paper do not quickly transfer energy to the outside environment. Telescope tubes made of these materials will conduct heat into primary the mirror cell and secondary mirror support arms for long periods of time after being exposed to the cool night air. Heat transferred into the secondary will cause a continuous column of unstable warm air to rise into the optical path and give the same results as bad seeing.

Metals such as steel or aluminum conduct heat well and will cool down many times faster than the above materials. When exposed to the night air, metal tubes will immediately begin to radiate heat energy into space (straight up into the sky) and transfer heat into the surrounding air. The temperature of the telescope will equalize with the surrounding air many times faster if it is made of conductive materials.

A major factor in stabilizing tube temperatures is the color and thickness of the coating on the outside of the tube. A white or brightly colored tube may look very nice; however, heat (infrared radiation) is reflected by bright colors and will be reflected into the blackened inside of the tube and cause the air to become unstable. From the thermodynamic principle of black body radiation, it appears that the best choice for painting the outside of a telescope tube is black! The author has experimented with these concepts and found a marked improvement in the performance of his telescopes over the years.

6.3.6 Modifications for Schmidt-Cassegrains (SCT)

Years ago while observing under the very dark sky of the Everglades National Park, the author accidentally dropped his Dynamax-8 and rearranged the optics, so to speak! After attempting to set up and collimate the optical system, several ideas came to mind that might improve the performance of this telescope. The corrector plate had to be removed and replaced several times during this operation; and for convenience, the secondary baffle was left out. While observing Saturn to adjust my scope's optical alignment, a noticeable increase in image contrast was seen.

The author decided to reduce the size of the secondary mirror holder from 2.5 inches to 2 inches, and not reinstall the secondary baffle, which constituted a 2.75-inch obstruction. This reduced the secondary obstruction from 34% (contrast factor 1.54) to 25% (contrast factor 2.65) resulting in a 72% increase in contrast (Figures 6.3.4 and 6.3.5).

The design equations for Cassegrain optical systems indicate that the distance between secondary and primary is very critical. The equations for the typical Cassegrain optical layout are as follows:

$$a = \frac{F + b}{X + 1}$$

$$A = aX$$

$$B = A - b$$

where:

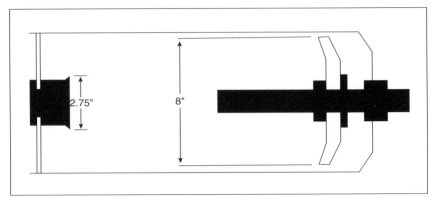

Fig. 6.3.4 *The layout of a typical 8-inch Schmidt-Cassegrain telescope. The 2-inch secondary mirror is attached to an aluminum plate and collimating screws with secondary baffle screwed to mounting plate. Total obstruction by flanged baffle was 2.75 inches or 34%.*

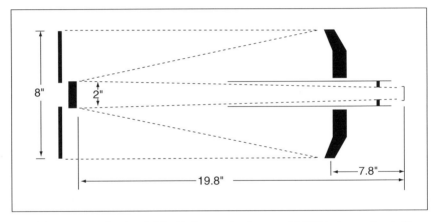

Fig. 6.3.5 *The layout of an 8-inch Schmidt-Cassegrain telescope with 2.75-inch secondary baffle removed. The 2-inch secondary mirror causes only a 25% obstruction, much improving the contrast factor.*

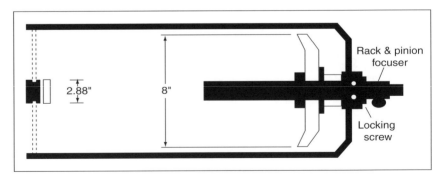

Fig. 6.3.6 *Layout of modified SCT illustrating relative optical layout, primary locking/collimating screws, rack and pinion focuser.*

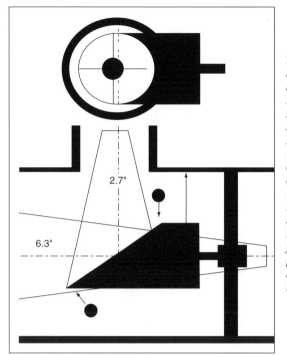

Fig. 6.3.7 *Typical fast Newtonian optical layout—no offset secondary mirror. Secondary is centered in tube and focuser. Secondary mirror image is centered in focuser. This example illustrates a 3-inch f/3, 3.4-inch O.D tube, 1-inch high focuser racked in, 0.5-inch linear image, and 0.9-inch secondary (minimum). A 1.25-inch secondary (100% image illumination) was used. Shadows of spider, secondary and reflection of primary are not centered in focuser, but are offset toward the primary end of the tube. Secondary misses optical path at top and bottom.*

Fig. 6.3.8 *Same as Figure 6.3.7, except the mirror is now offset to capture the entire light cone. Secondary is dropped away from focuser and is still centered in it. Primary image is larger and fills the entire secondary mirror. Reflection of the primary is centered in focuser. Shadows of secondary and spider are still offset toward the primary end of the tube. The secondary has been offset a distance of 0.158 inch.*

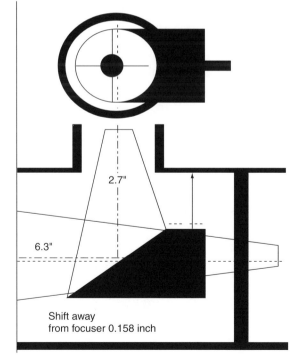

a = Primary intercept point
F = Focal length of primary
b = Back focus
X = Second magnification
A = Cassegrain back focus
B = Mirror separation .

The tolerance for mirror separation is critical and can be found by: separation limit (in millimeters) = 0.063 fr4, where fr = primary focal ratio. This limit can vary somewhat, as the separation of the mirrors can be as much as 45% more or 55% less and still give acceptable results. From this we find that for any Cassegrain system, the mirror separation should vary no more than 17 millimeters for an $f/4$ primary or 1 mm for an $f/2$ system. That is one millimeter for an $f/2$ primary in a typical SCT system.

The typical SCT focuses by moving the primary along the primary baffle tube to decrease or increase the distance between mirrors. In the author's SCT, this travel was about 12 mm, therefore violating the above design limits. Test results using my original SCT optical/baffle configuration indicated image degradation when mirror separation went beyond the limits. So, the correct separation was calculated, and the primary was locked into place. This was accomplished by drilling three holes 120° apart in the aluminum plate that holds the baffle tube, mirror cell, and other parts to the rear of the tube housing. Three 3-inch, #10 screws were used to lock the mirror in place and also provided a small adjustment for collimating it. Then a 1¼-inch rack and pinion focuser was installed to the outside of the plate. A glare stop was placed inside the primary baffle tube near the end which prevented direct light from entering the final image plane (See Figure 6.3.6).

6.3.7   Offset Newtonian Secondary Mirror

Fast Newtonian telescopes may require the secondary mirror to be offset from the optical path or dropped away from the focuser in order to fully illuminate the entire primary image in the secondary. From the optical layout of the typical fast Newtonian telescope in Figure 6.3.7, it can be seen that some portion of the secondary mirror's top and bottom misses the optical path.

To correct this, several equations are used to calculate the proper offset of the secondary as follows:

Mirror shift away from focuser = B1 − B2,

where:

$$B1 \;=\; \frac{i(F-SA)+L(D-i)}{2(F-SA)-(D-i)}$$

$$B2 \;=\; \frac{i(F-SA)+L(D-i)}{2(F-SA)+(D-i)},$$

and:

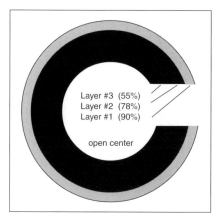

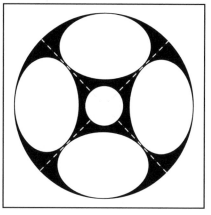

Fig. 6.3.9 *Typical apodizing mask with three layers of window screen material fitted onto a frame with center sections cut out. Three layers of screen, 30° rotation. Screen hole #1 is cut 90%, #2 to 78%, and #3 to 55% of aperture. For refractors the figures would be 88%,76%,52%.*

Fig. 6.3.10 *Bean/oval diaphragm. Circular disk of light material with bean shaped holes cut into it and fitted onto tube entrance. Dashed lines indicate spider.*

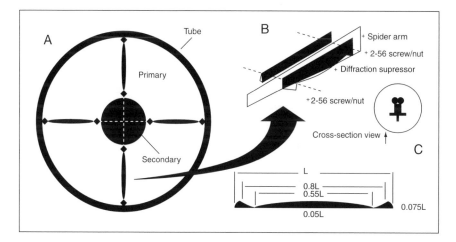

Fig. 6.3.11 *A) Flower box diaphragm. Individual sections are fitted onto each secondary support arm. Dashed lines indicate spider arms. B) Mechanical view of the Diffraction Spike Suppressor design from* Amateur Telescope Making Book 2. *From blow-up of drawing in book measurements of proportions were taken. Shaped from ½-inch aluminum angles (0.030-inch). Each half-section is wedged onto each secondary mirror support arm (spider) using ¼-inch 2-56 screws and nuts. C) Shown is the relative proportions for the 'Diffraction Spike Suppressors'. All dimensions are in relative lengths. Example: if L = 5.5 inches, then 0.8L is 4.4 inches, and 0.55L is 3.025 inches. Mid-section thickness = 0.05L or 0.275 inch, and end sections are 0.075L or 0.4125 inch.*

i = Image diameter (linear)

$$SA = \frac{r^2}{4R(\text{Sagitta})}$$

D = Diameter of primary

F = Primary focal length

L = Secondary to focal point

$$r = \frac{D}{2}$$

R = Radius .

The finished offset secondary configuration illustrated in Figure 6.3.8 is not necessary for Newtonian telescopes over $f/6$, because the difference will most likely be smaller than the usual centering error in these longer tube assemblies.

6.3.8 Apodizing Screens and Other Devices to Reduce Spider Diffraction Spikes

After testing several apodizing screens and other devices that are supposed to help improve "astronomical seeing" and/or eliminate secondary support arm diffraction spikes, this author has determined that for all practical purposes they are of little use. However, a very useful design was found that completely eliminates the apparent effects of diffraction spikes, and actually helps reduce the effects of bad "seeing" in some cases.[5] The author calls this device the "Diffraction Spike Suppressor", and is illustrated in Figures 6.3.9 through 6.3.11.

6.4 Techniques and Hints for Lunar and Planetary Observing

By Don Parker

Successful and enjoyable observing of the Moon and planets places some unique demands on both equipment and observer. This type of activity, often called high resolution work, requires that high magnifications be employed on fairly low contrast objects, making it essential that the wavefront presented to the eye or imaging medium suffer as little deformation as possible—that it does not exceed the ¼-wave Raleigh criterion. This is easier said than done, however, since many conditions exist that degrade the wavefront during the light's passage from the celestial object to the eye. While some of these factors are completely beyond our

---

[5] Couder, A. "Dealing with Spider Diffraction," *Amateur Telescope Making Book 2*, Scientific American, Inc., pp. 621.

control to influence, there are a number of things that the observer can do to min-imize their effects on his image delivery.

6.4.1 The Atmosphere

Aside from inferior optics, the single element most ruinous to ground-based astronomy is the Earth's atmosphere. When we view an astronomical object through a telescope, its light must travel through several miles of air. If this air were uniform and at rest, the only problem would be some loss of intensity of the light, or a so-called decrease in transparency. Unfortunately, the air above us is composed of rapidly moving layers of different temperatures and densities that refract the incoming starlight. It is as if the atmosphere was composed of millions of tiny lenses, constantly drifting across the light path and deforming the wave-front. If the size of these lenses, or "seeing cells," is larger than the telescope's aperture, there will be moments when the image will be sharply defined, albeit in motion. Classically, it has been taught that these cells are 4 to 6 inches in diameter, explaining why images are often better in smaller telescopes than in large ones. More likely, this "rule" has been masterminded by those who sell small tele-scopes! In general, such instruments will indeed present pleasing views of the Moon and planets; but in moments of atmospheric steadiness, they cannot hope to compete with larger telescopes.

This brings up one characteristic which is found in all successful lunar and planetary observers—patience. Serious observing, whether deep-sky or high res-olution, is not fast food: the observer must spend time at the eyepiece waiting for the fleeting moments of steadiness. This extra time spent at the telescope provides an additional bonus of training the astronomer's eye. While the exact mechanism of this training is uncertain, it has been recognized for well over a century that re-petitive viewing of an object is necessary for one to appreciate its most subtle fea-tures. An example is the planet Mars. Jeff Beish and I have made thousands of detailed observations of this object over the years, but when we start observing it at the beginning of the apparition, it appears totally unfamiliar! It often requires three or four observations before we can feel at home with the planet.

The trained observer does not necessarily have to possess exceptional vision; rather he or she combines past visual experiences with judgment to produce the perception of the current observation. Boston greats, Steve O'Meara and Ted Wil-liams, do not have superhuman vision—what separates them from mere mortals is their ability to concentrate and to correlate past observational experience with a present situation, whether it is a view of Halley's Comet or a ball breaking over the plate.

6.4.2 Site Selection

When selecting an observing site the amateur cannot avoid the Earth's atmo-sphere, but he or she can attempt to avoid places where the air is turbulent, and rather seek out sites where there is a laminar flow pattern. On a global scale, the middle latitude jet stream produces turbulence. "Expert" thinking has maintained that the best locales for astronomy are in mountains, since there is less atmosphere above a mountain than there would be at sea level. This is not necessarily true, since the seeing is often abominable when cold air cascades down lee slopes or warm air rushes up the weather sides of mountains. Worse yet are the mountain valleys, where these turbulent air masses often mix. These conditions are adverse to deep-sky observers and astrophotographers as well as to planetary astronomers; the atmosphere, while clear, is so agitated that faint extended objects and star images will appear so distended that their light will be spread out over a large area, diminishing their visibility.

In mountainous regions, laminar air flow is often found on the lee ends of gently upward sloping plateaus, such as Lowell Observatory's Mars Hill in Flagstaff, Arizona, and in maritime districts, such as Pic du Midi in the French Pyrenees and Mauna Kea in Hawaii. In general observatories located in tropical and subtropical zones, where one finds laminar flowing gentle trade winds, and sites near large bodies of water will experience the best seeing, although atmospheric transparency may be less than optimum in these locales.

Transparency is, however, of minor consequence in planetary observations; in fact, seeing is often exquisite just before the arrival of a cold front when the sky is hazy and high cirrus clouds ("mares'" tails) prevail. High altitude ice crystals that produce rings about the Sun and Moon likewise advertise a stable atmosphere. Foggy nights usually have the very best seeing because the conditions that allow fog to form require a very tranquil atmosphere. The presence of puffy "fair weather" cumulus clouds often discourages observers from setting up; however, the seeing is often superb at the leading edges of these clouds. We have these little clouds as nearly constant companions in South Florida; and while they hamper long exposure astrophotography, they have provided excellent planetary images!

While the foregoing discussion may be of interest for the very few amateurs who can erect observatories wherever they choose, most of us do not have the option of leaving our jobs and homes to move to an ideal observing location. We are stuck with what we have! Even if one lives where the seeing is usually poor, this does not mean that quality planetary work cannot be done. Even the poorest sites have an occasional fine night, and very often many nights have good seeing during certain hours. Many observers have noted that less air turbulence occurs after midnight, with the best seeing at sunrise. At certain seasons, lunar and planetary detail is extremely sharp just at sunset, when there is a brief pause in the heat balance between day and night, and the atmosphere has not yet had time to radiate its heat into space. At this time the atmosphere's temperature gradient (called the "lapse rate") varies more uniformly with altitude. Later, as heat is lost to space and to terrestrial conductors, air masses begin to move and seeing deteriorates. As dawn approaches, thermal equilibrium is re-established and seeing improves.

Regardless of his geographical position, the amateur astronomer can do much to optimize his situation by simply choosing an observing site that has the best conditions. Placing the telescope out of the glare of streetlights is an example.

When one seeks the most stable local airflow, he should avoid concrete parking lots or buildings, as concrete retains the Sun's heat and radiates it throughout the night. Asphalt is even worse in this regard. If the telescope is set up permanently on a concrete pad, try to shade the pad during the day. Covering the deck with indoor-outdoor carpet, or Astroturf, protects the concrete from heat buildup and is much easier on the feet. In addition, the astronomer's disposition is usually greatly improved because his incredibly expensive eyepieces no longer shatter on the concrete when dropped! As an aside, if the observer spends considerable time on a stepladder, gluing heavy shag carpet remnants to the steps will reduce foot pain and permit longer, more comfortable viewing. (After experiencing Riverside-94, I am convinced that the principle factor limiting the size of Dobsonians will be ladder technology!)

Ideally, the telescope should be set up on a grassy area. Trees, especially pine trees and large-leafed tropicals, emit carbon dioxide during the night and may cause turbulence if the telescope is sighted directly over them. Also avoid looking over chimneys and air conditioning units of neighboring houses. If the telescope is permanently mounted, the observatory should not be constructed of brick or cement block, but rather of wood or aluminum. A simple, economical roll-off roof structure usually has far better thermal qualities than the more elegant and expensive dome. If possible, the observatory should be elevated a few feet above the ground. While this is often impractical, it does eliminate many of the problems caused by convection occurring near the ground. Hurricane Andrew permanently rolled off the roof of my observatory, and I now protect my instrument with tarps and bungee cords. This works very well, but care must be taken to shade the telescope from the afternoon Sun so that it will not take several hours to cool down at night. Shiny covers, like the "Desert Storm" blankets help greatly in keeping the telescope cool.

6.4.3 The Telescope

A telescope's optics must be of as high a quality as possible if one is to truly enjoy observing. There are many other factors in the telescope itself that contribute to wavefront degradation. Many of these can be easily minimized by the observer. The following discussion will concentrate on Newtonians, since this is the design with which I am most familiar. In addition, I feel that a well-designed Newtonian is hard to beat for lunar and planetary observing.

6.4.4 Tube Currents

The presence of turbulent air in the telescope's tube can be ruinous to the image. Newtonians suffer from this more than other designs, since the light must traverse the tube twice. Worse still, incoming light advances along the sides of the tube, where eddy currents are most prevalent. This is why the inside diameter of the Newtonian tube should be at least 2 inches greater than the diameter of the primary mirror. Some have advocated capping the mirror end of the tube, preventing warm ground air from entering. In all but the shortest tubes this practice should be avoided, since cold air still enters from the top. With no way for air masses of differing temperatures and densities to exit the telescope tube, the effect is like filling the tube with millions of tiny lenses! The solution is to allow air to flow freely through the tube in a laminar fashion. This is easily accomplished by placing a fan at the mirror end. The flow need not be great (in fact, high flows create turbulence as air flows past the primary mirror.) A small muffin fan suspended by rubber bands works well on a small scope. On my 16-inch Newtonian, I employ three 12 volt DC, 24 cubic feet per minute (CFM) fans attached to a three-legged piece of plywood that is screwed directly to the mirror cell. These fans run continuously, even during high-resolution imaging. To avoid unwanted vibration, purchase new (not "rebuilt") fans, preferably running on ball bearings.

Another source of unwanted air currents is the tube itself. The tube must be rigid enough to hold the optical elements in collimation and yet be reasonably light. It should not allow transfer of heat into the optical path. Small instruments successfully employ fiberboard composites, such as Sonotube. However, Newtonians should be reinforced in apertures much over 10 inches. Fiberglass has been used in many telescopes, but in large instruments its poor strength-to-weight ratio becomes a problem. In addition, fiberglass, while a good insulator, retains heat well and is thus a poor choice for large telescopes.

Metal tubes, such as aluminum, provide very high rigidity per unit weight. They are excellent heat conductors and so should be lined with an insulator like cork so that heat will not conduct into the tube from the ground or from warm hands. Open truss tubes provide superb rigidity and are relatively lightweight. They can also be disassembled for transport. While open tubes avoid the problem of eddy currents creeping along the upper side of a closed tube, they do not prevent air currents, such as those produced by the observer's body heat, from crossing the optical path. Metal struts should be wrapped with insulating material, lest they bleed heat into the light path. In general, it is best to cover the truss tube with a lightweight insulator like Styrofoam or even cardboard.

Objects in the light path, like large metal Cassegrain secondary-mirror cells, can retain heat and then slowly bleed it into the atmosphere, destroying fine definition. Jeff Beish experienced this problem with his fine 12.5-inch Cassegrain. When viewing a slightly out-of-focus star image, one could actually see the heat coming off the secondary holder, imparting a teardrop shape to the image. Beish solved this problem by simply clamping a heavy wire to one of the spider vanes. The other end of the wire was fastened to a small heat sink and allowed to dangle from the tube's mouth. The image improvement was dramatic and immediate.

6.4.5   Focal Ratio

Despite our most careful efforts at minimizing the effects of air turbulence, it will still play a major role in high magnification observing. In addition to image excursion and blurring, shifts in the focal plane occur as the wavefront is deformed. If the telescope has a fairly high focal ratio, there will be sufficient depth of focus so that the object will remain reasonably sharp. If, on the other hand, the instrument

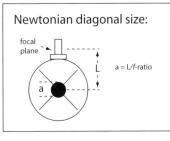

Fig. 6.4.1

has a low *f*-ratio (like many of today's "deep-sky" scopes), the depth of focus will be so small that the observer will have to refocus almost continuously. The depth of focus for a ¼-wave wavefront in an *f*/7 system is 0.01 inch, but falls to only 0.001 inch for an *f*/5! This is why many suggest that the minimum focal ratio for a planetary telescope should be *f*/6. This is not to say that serious high-resolution work cannot be done with short scopes—it is just easier with the longer instruments. In addition, longer *f*-ratios permit the use of smaller secondary mirrors—an important consideration when trying to achieve optimum contrast. When the Newtonian's aperture increases to much over 16 inches, as is the case with so many of today's amateur telescopes, it becomes impractical to go to very high focal ratios; but it is best to keep them as high as ladder technology will allow!

6.4.6 Newtonian Diagonal Size

Image contrast is of supreme importance in high resolution observing, whether the subject is a planet or a galaxy. It is well documented that image contrast suffers in direct proportion to the amount of central obstruction in an optical system. This has been one argument against the Newtonian. In practice, however, contrast loss is inconsequential if the minor axis of the secondary mirror is held below 15% of the primary's diameter. This usually requires a low profile focuser. A rule of thumb for calculating Newtonian diagonal size appears in Figure 6.4.1. Using this formula, if one has an 8-inch *f*/8 Newtonian with the focal plane 2 inches outside of a 10-inch ID tube, he would require a diagonal with a ⁵⁺²/₈ or ⅞-inch minor axis. This will give 100% illumination over 0.25 inch at the focal plane, more than enough for planetary work. But what about deep-sky observations? The improvement in image contrast at the center of the field will more than balance the slight fall-off in magnitude at the edge of the field. Besides, who wants to look at coma anyway? Remember, most deep-sky objects display very low contrast; they will also benefit from reducing secondary size.

6.4.7 Collimation

All of the preceding discussion on optimizing telescope performance is worthless if the optical train is not properly collimated. This procedure is fairly straightforward and should be checked at the start of every observing session. Unfortunately,

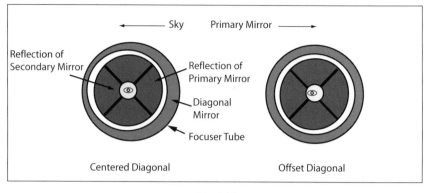

Sky   Primary Mirror

Reflection of Secondary Mirror

Reflection of Primary Mirror

Diagonal Mirror

Focuser Tube

Centered Diagonal   Offset Diagonal

Fig. 6.4.2

this is rarely the case, and poor collimation remains the prime cause of poor images (which gives the Newtonian a bad name). There is no excuse for this, since today a number of collimating tools are readily available to the amateur. One of the handiest is the Cheshire eyepiece, a device that produces a bright annulus in the shadow of the secondary (where the drawing of an eye is located in Figure 6.4.2). The Krupa collimator, described in *ATMJ* #4 (Section 4.5 on page 132), employs a similar principle and is much easier to fabricate.

In large-aperture fast Newtonians, it is common practice to offset the diagonal by slightly dropping it and moving it toward the primary so that its entire surface intercepts the steep light cone. When the offset diagonal is properly collimated, the reflection of the primary mirror appears centered in the diagonal, while the latter's shadow is decentered in the primary's reflection (Figure 6.4.2, right). Fortunately, the amount of offset is negligible in Newtonians over *f*/6 and under 16 inches aperture, so the diagonal is left centered in the tube. The mirrors are then adjusted so that all reflections appear concentric with the eyepiece drawtube when one sights through a peephole at that location.

The vast majority of material written on this subject states that this is proper collimation. However, a slightly out-of-focus star image at high power will display a decentered secondary mirror shadow! The observer will then assume that his scope has lost collimation and adjust the primary mirror tilt to center the shadow. This results in collimation at the expense of wasting much of the diagonal's surface. The correct procedure is to make the final adjustment in the tilt of the secondary, not the primary. If one employs a pinhole and makes all reflections concentric and then views through a Cheshire eyepiece, the Cheshire's bright ring will be decentered in the diagonal's shadow. This can be remedied by slightly tilting the diagonal. Tilting the diagonal centers the annulus in the secondary's shadow but causes the main mirror's reflection to be slightly off center (Figure 6.4.2, left). This is the appearance for a correctly collimated Newtonian with a centered diagonal. An excellent discussion of this approach can be found in Sam Brown's *All About Telescopes*, available from Edmund Scientific Corp.

Be sure to do a final "tweak" on a slightly de-focused star image at high power. If the secondary holder and spider are properly made, collimation should be close and slight corrections can be achieved by primary mirror adjustment. Since optics do shift, make sure to check collimation on a star at least once per night. The minor adjustments that might be required will produce vastly improved images. Remember the parabolic mirror is not designed to be used off-axis!

I hope that some of the ideas presented here will help amateurs avoid some of the pitfalls into which I have fallen over the years. Improvement in image quality can only make our hobby more enjoyable. After all, the name of the game is "Have Fun!"

6.5 Texas Star Party 1994

By Dean Ketelsen

This year's Texas Star Party broke all attendance records for the event. Coupled with the annular solar eclipse, hordes of observers made their way to the spot 8 miles from McDonald Observatory in West Texas. Unfortunately, for some Star Party veterans such as myself, they quickly filled up the allotted 900 people roster (over a month early), and returned the registration application for me, Bernie Merems, and many others. However, since I had promised to deliver some items both to and from the event, I felt obliged to show up and see what happened.

Upon arrival Wednesday afternoon (after the eclipse), all three observing fields were filled to or beyond capacity. One literally had to walk carefully to avoid tripping over scopes, even in the daytime! Vicki and I decided to park our van near the Levine's RV, which was only a stone throw from the entrance. Talks that afternoon were sparsely attended. They included optical testing (of great interest to me), and much was made of Ceravolo's interferometer and the "new" method of testing. That night saw only an hour of clear skies, but a quick trip to the upper "elite" field showed two 36-inch scopes, and the usual dozens of scopes greater than 16 inches in aperture. Peter Ceravolo's Maksutov-Newtonians also drew much attention. Peter is competing with the apochromatic manufacturers for crisp, color-free images of planets and double stars. The early fog that rolled in put an end to observing giving me an early rest after driving a good part of the day.

The next day (Thursday), observers started leaving the star party in droves. No doubt, many had come just for the eclipse and were heading home. Another interesting observation—usually you have to use off-peak hours to find an unused shower (like 6 p.m.), but there was never a wait this year, even with their record attendance!

Thursday night and Friday night have got to go down in TSP history as two of the best nights in recent memory. Usually, one gets very good transparency but mediocre seeing and lots of dew. These two nights were clear and dry with excellent seeing. I parked my old 10-inch Celestron next to Peter Ceravolo's 8.5-inch "McNewt" and John Gregory's 9-inch Mak-Cass. The old Celestron did remarkably well in the good seeing, but John's Maksutov was a clear step up, and Peter's Maksutov-Newtonian was the clear winner, revealing incredible contrast on Jupiter. The standard test object for double stars was Eta Corona Borealis—separation 0.7 arc seconds. The Celestron sort of showed a skewed—perhaps multiple—image, but Peter's scope clearly resolved it.

Other memorable sights—Roger Tanner was using his "Cookbook" CCD for the first time at TSP and getting excellent images of SL9 with his 17.5-inch. Roger is an aspiring University of Arizona student and will be moving to Tucson soon. Tom and Jeannie Clark of Florida brought out their "Yard Scope" (36-inch diameter). With it I got some great views of David Levy's latest visual comet discovery and some galaxy clusters, although a visual search for SL9 was unsuccessful. The neatest thing about TSP was exemplified with their scope. With so many large aperture telescopes on hand, you were always next in line!

TAAA (Tucson Amateur Astronomy Association) members made a sparse showing there—Glen Nishimoto, Hazel and Dick Lawler, David and Elinor Levine, Bernie Merems, and Vicki and I were the lone Tucsonians. Vicki and I left Saturday for the TAAA picnic, but I think we definitely hit the heights.

6.6   Much Better Bearings

By John Shelley

My first three viewing seasons using my home-built Dobsonian were most enjoyable. But, by the time summer came around, I started having trouble. It seemed that while the azimuth bearing was indestructible, the elevation bearings began to fail soon after the hot weather arrived.

I found that perfect tracking could be done only when the friction in azimuth was equal to that in elevation, and obtaining the required equality with a given size of trunnions meant the trial and error attachment of Teflon pads at different positions. Unfortunately, Teflon, while offering a near-perfect bearing surface, has viscosity which allowed it to flow. When this happened, my trunnions slipped past the pads, and I had aluminum against painted wood—a very poor bearing arrangement. Repairs were easy, at first. I simply added shims, and later, new pads. But the screws holding them tended to be loose and started touching the aluminum rims of the trunnions. This was a bad situation, so I started looking for a modification that would bring back the original, satiny feel without the required frequent rebuilding.

My first proposed solution was to make nested cones faced with Teflon/aluminum. The reasoning was that the angle of a cone would offer a friction surface to both the horizontal and vertical forces. A nut and bolt arrangement through the center could control friction.

Experimentation revealed that cones were not worth the trouble and made assembly difficult. I could see that bolts alone could take care of the tube weight. Also, they could be used with a set of flat Teflon/aluminum rings of large diameter

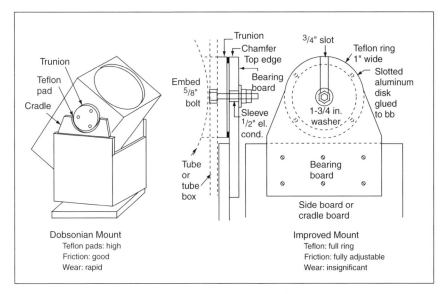

Dobsonian Mount
Teflon pads: high
Friction: good
Wear: rapid

Improved Mount
Teflon: full ring
Friction: fully adjustable
Wear: insignificant

Fig. 6.6.1 *Drawing by John Shelley.*

and double nuts for locking. (A complete Teflon ring is not needed; three or more sections of a ring will suffice, if they are evenly spaced.)

The addition of a Telrad and a 3-inch finder scope was no problem for my 13-inch telescope; I simply increased the pressure with the nuts. This face type of bearing has proven to be ideally suited for equatorial mounts also.

6.7 Telescopes for CCD Imaging, Part II
By Richard Berry

The advent of CCD imaging seems to have breathed new life into the classic reflector versus refractor debate. And, as has always been the case, no one seems to have a definitive solution because there is, in fact, no definitive solution. Reflectors, refractors, and a variety of compound optical systems each have a place in CCD imaging. In this issue, we will take a look at what the classic optical systems offer, and offer some tips on how you can decide what type of telescope is best for your type of CCD observing.

We will begin our survey with the classic achromatic refractor, which (as it happens) was the first type of telescope that I tried for CCD imaging. The setup was this: I placed an SBIG ST4 camera at the focus of my 6-inch $f/15$ refractor—the same one described in *Build Your Own Telescope*—and set out to make some images of the gibbous Moon. This lens is a truly fine performer, so I expected success. However, it turned into a very frustrating experience: the lunar craters never got sharp, no matter how I focused. It took several nights and a lot of experimenting to determine exactly what was happening.

All achromatic refractors suffer from considerable secondary chromatic aberration, but the lenses are designed to minimize the visual impact. The shortest focus of the lens occurs in yellow-green, and both blue and deep red focus farther from the lens. This works well for visual observing because the eye is sensitive to yellow-green, and the out-of-focus red and blue make only a faint purple halo around the star image. However, CCDs are most sensitive at wavelengths around 7500 ngstroms, in the near-infrared part of the spectrum; so when you focus for a CCD, you are focusing some distance beyond normal.

What you see is a sharp image formed in near-infrared light and an out-of-focus blur of red, orange, yellow, and green light. Even though blue and violet might focus with the deep red and near infrared, the defocused light from the center of the visible spectrum washes away almost all image contrast.

The importance of the chromatic blur depends on the aperture and focal length of the lens. You can, for example, get acceptable images through an achromatic finder telescope with a 50 mm f/5 lens; but somewhere not far above 3 inches aperture, classic doublets simply do not form images good enough for satisfactory CCD imaging. Star images taken at the focus of my 6-inch f/15 achromatic refractor have halos more about 30 pixels in diameter.

With apochromats, the story is more complicated. Most apos are optimized for the visible part of the spectrum with little concern for what happens in the near infrared, which is fully justified because you cannot see light in these wavelengths. However, apos vary enormously in design. Some bring the near-infrared to the same focus as visible light, and others do not. As a result, some apos form sharp images with CCDs while others do not. Given the wide variety of possible designs, the only way to know how a given apo will perform is to shoot some test images of stars and examine what you get. The most sensitive test is a through-focus star test: a lens that forms sharp star images will give identical intrafocal and extrafocal star images.

I have seen numerous CCD images made with a variety of three-element apos (made by AstroPhysics), and they seem to be exceptionally well-corrected because star images are tight and crisp. My own 4-inch f/5 Tele Vue Genesis gives crisp and acceptable images with slight haloes approximately 4 pixels in diameter. I have not seen enough images from other refractors to form any reliable conclusions and so advise testing any apo before committing serious dollars to using it for CCD imaging.

Well-corrected telephoto lenses often perform extremely well for imaging fields up to several degrees across, but you need to check out the individual lens because telephoto designs vary. To give good images, the lens must be reasonably well-corrected when it is focused for the near infrared, and the yellow-green light must still focus reasonably close to the near-infrared focus. If you shoot a star field with a telephoto and see sharp star images enveloped in soft haloes of light, you are probably looking at an out-of-focus visible-light problem. If this is indeed the case, when you place a red filter over the lens, the haloes will disappear and you will be left with crisp, sharp star images. For imaging the big H-alpha nebulosities, a red-filtered lens is great; but if you want to do galaxies, it is hardly ideal. At swap meets you can often pick up a used telephoto lens, especially one in the 135 to 250 mm range, for a pretty good price. Among brand-name lenses, I have been consistently impressed with the performance of those made by Nikon. Avoid zooms; they seldom perform as well as fixed-focus telephotos, and the mechanisms tend to "drift" when you least expect it.

After refractors come the reflectors. Reflectors are completely color-free. Furthermore, unless you plan to use a CCD that is large (i.e., more than 10 mm across the diagonal), or one that has small pixels (i.e., smaller than 10 micrometers), or have a Newtonian that is faster than $f/4$, the coma blur simply does not get large enough to be visible. However, reflectors do suffer from a variety of ills that refractors seldom exhibit. Newtonians especially are prone to field flooding because light can reach the CCD in a variety of ways. For example, light from the ground can enter the bottom of the tube around the mirror. Light may also scatter to the CCD from a short tube or enter around the base of the focuser. Such stray light can seriously degrade the performance of your CCD camera, especially in urban and suburban observing sites. Fortunately, these sources of unwanted light are easy to eliminate with: a tube-bottom baffle; a "snoot" at the front of the tube twice the diameter of the mirror; and the addition of black sealant, caulking compound or tape around the focuser. Once sealed, you should be able to shine a bright flashlight all over the outside of the telescope without increasing the background sky brightness.

In addition to admitting stray light, many reflectors introduce spikes around bright stars due to diffraction from secondary mirror supports, but this seldom poses much of a problem. Some people even find diffraction spikes aesthetically pleasing! All in all, Newtonians are great for CCD imaging.

Naturally, it makes sense to check the optical performance of a reflector by taking extrafocal and intrafocal star images. The resulting "donut-o-gram" images should be the same on either side of focus. If you see bright rims on one side and bright centers on the other side, you are looking at evidence of spherical aberration. In focus, this aberration will show up as a bright core enveloped in a soft halo. If you see signs of astigmatism or other figure irregularities with one of today's large thin mirrors, check that the mirror cell is working properly.

Next time we will cover some compound optical systems.

6.8   A Closer Look at High Magnification
By Rodger W. Gordon

Open almost any guidebook on amateur astronomy today, and invariably one finds words of warning about using high magnifications on a telescope. No doubt some of this caution is justified, particularly in the case of small 2.4-inch or 3-inch "department store" refractors, often imported from Japan or other Asian sources, which are advertised to beginners with power claims of 400x to 600x. Such an instrument usually has a mediocre lens at best, and the instrument's barrel is coupled to a mechanically unsound mount or tripod. The tyro soon finds his dreams of high power shattered by the poor optical quality and unstable mount, and the lesson learned can be expensive.

The "taboo" against high magnification, however, has crept into the more advanced amateur community to an extent where many amateurs use only low to medium magnifications and thus deny themselves a major portion of the performance capabilities of their instrument. For argument's sake, we will assume our intermediate or advanced amateur is using a telescope of very good or excellent optical quality, regardless of what its aperture or type may be. In other words, optics whose wave front errors, as measured on the P-V (peak to valley) system, do not exceed $\lambda/8$ at the image plane.

Before discussing high magnification, we must first ascertain the limits of human vision as this plays a very important role in how much magnification we require to fully exploit the resolving power of the telescope.

The unaided human eye can resolve detail which is $1'$ (minute) of arc.[6] A person with so-called 20/20 vision can resolve this separation on the common eye charts found in doctor's offices. This figure is somewhat misleading if applied to telescopic vision. The charts are of high contrast: black markings on a white background with contrast ratios of 0.90 to 0.95 (1.00 being maximum). The only features we encounter with astronomical subjects which approach these contrast levels are the umbras of sunspots, lunar terminator shadows, Cassini's division in Saturn's ring, and the shadows of satellites in transit across Jupiter when the background is a bright zone. If low contrast charts of 0.20 are substituted, the eye's resolution drops to $2'$ or $3'$ arc or worse.

Lunar and Martian features have an average contrast of 0.20 while Jupiter will range from 0.20 to 0.10 or less and Saturn from 0.15 to 0.05. The contrast on Venus seldom exceeds 0.05 and is usually less, which taxes the contrast detection abilities of both telescope and eye.

In a series of tests with students, Allyn J. Thompson found average daytime resolution to be $2\frac{1}{2}'$ to $4'$ arc. With a change in illumination providing increased contrast, a few were able to resolve $2'$ separation, but only one reached $1\frac{1}{2}'$. Thompson further pointed out that it requires a keen eye to separate the naked eye double E^1 and E^2 Lyrae, which are magnitudes +4 and +5 and separated by $3\frac{1}{2}'$.[7]

If we adopt a slightly more stringent limit of $3'$ arc as the average, this figure poses definite limits if we wish to obtain visually all that our telescopes are capable of revealing. We must magnify the smallest resolvable details sufficiently large so the eye can detect them.

Assume we now use a telescope of $4\frac{1}{2}$-inch aperture. The well-known Dawes formula R= 4.5/D states that a $4\frac{1}{2}$-inch telescope will resolve 1 arc second. Recalling that 60 arc seconds equals $1'$ (minute) of arc, we discover that to enlarge

[6] Johnson, B.K. *Optics and Optical Instruments*. Dover, 1960, pp. 43–45.

[7] Thompson, Allyn J. *Making Your Own Telescope*, Sky Publishing Co., 1947, pp. 173–175.

the smallest details to where the eye can see them at the 3′ angle, will require a magnification of 180x.[8] Even a more liberal limit of 2′ requires 120x. So, if you are average and using less than 180x on your 4½-inch scope, you are not seeing all the detail, period! The power-per-inch requirements are thus 40x or 27x based on 3′ and 2′ eye resolution.

A number of authorities (Dollfus,[9] Giffen,[10] et. al.) have shown that for minimum resolvability, the magnification of the telescope must be at least equal to the aperture of the instrument in millimeters. This works out to 25x per inch. Texereau gives a slightly higher figure of 1.25 times the aperture[11] or slightly over 31x per inch. Powers between 25x and 40x per inch are commonly employed by skilled lunar and planetary observers (usually discovered by trial and error earlier in their careers).

As the telescope aperture increases, the situation grows progressively worse. If we use the 3′ figure, a 9-inch telescope requires 360x and an 18-inch instrument, 720x, if we are to exploit their full resolving power. The difficulty here is that the atmosphere seldom allows us to use magnifications much in excess of 300x. Once the aperture exceeds 12 inches, we are almost always seeing limited rather than aperture limited—except in favored locations like the southwest or in southern Florida or Hawaii or where the trade winds dominate the weather pattern. Smaller telescopes of 2.4 inches to 7 inches in aperture are more immune to bad seeing, and it is well known that smaller telescopes are more efficient in reaching their theoretical limits on a greater number of nights. The experiences of S.W. Burnham, conceded to be the greatest double star observer of all time, clearly bear this out.[12] We then should not be surprised to find smaller telescopes capable of bearing higher powers-per-inch than their larger counterparts, and frequently the smaller instrument is optically superior to the larger one.

However, high powers must be tailored to the subject at hand. The Moon and Mars present fairly contrasty images, but Jupiter and Saturn may not. Fifty to sixty power per inch may "wash out" certain salient features on the latter objects while retaining acceptable contrast levels with the former.

We can get a better idea of this comparison if we compare Mars and Jupiter at perihelic oppositions. Both will attain about −2.7 stellar magnitude, but Mars will be 25 arc seconds diameter and Jupiter 50 arc seconds. On an area basis ($A = \pi R^2$), Jupiter will be only ¼ as bright as Mars as seen in the telescope if magnifications are identical. This allows us to use higher powers-per-inch on Mars without an unacceptable loss of brightness or contrast. Similarly, the Moon at or near first or third quarter, when it presents maximum shadow contrast, allows us to use

[8] Olcott, W.T. *Field Book of The Skies*, Appendix XVII, G.P. Putnams Sons, 1955, pp. 469.

[9] Dollfus, A. *Visual and Photographic Studies of Planets at Picdu Midi, Planets and Satellites,* ed. by G.P. Kuiper. Univ. of Chicago Press, 1961, pp. 545–546.

[10] Giffen, C.H. "Foundations of Visual Planetary Astronomy, Part I," *Journal of the Assoc. of Lunar and Planetary Observers*, Vol. 17. Nos. 3–4, 1963, pp. 69–71.

[11] Rutten, H. and M. Venrooij. *Telescope Optics, Evaluation and Design*. Richmond, VA: Willmann-Bell, Inc., 1988, pp. 217–218.

[12] Bell, L. *The Telescope*. Dover, 1981, pp. 263–264.

considerably higher powers as compared to its thin crescent or full moon presentations. Sidgewick[13] points out that magnifications of 20x per inch or lower tends to suppress find detail. With Mars, however, even twice this amount may not be enough. Color filters can be employed to reduce glare as most lunar or planetary observers are aware, and we can thus use somewhat less magnification. However, filters have advantages and disadvantages—the discussion of which is not within the scope of this paper.

There are other steps we can take to improve image brightness and contrast allowing somewhat higher powers-per-inch. When I use my 3½-inch Questar, I by-pass the built-in prism, built-in Barlow, and the included four element ocular. These conveniences are nice, but incur additional light transmission and scattering losses. Instead I employ a 9 mm Zeiss monocentric eyepiece (145x) with only 2 air/glass surfaces giving me a total of only 6 air/glass surfaces in the system. The eyepiece is attached axially for this purpose. The number of air/glass surfaces in total is less than found in many 7- or 8-element wide-angle eyepieces with 8 or 10 air/glass surfaces. The increased image brightness and contrast are quite pronounced over the view using the built-in accessories. For maximum high power resolution or for splitting tough doubles, I use either a 6 mm solid Tolles eyepiece, or a modern 6 mm Zeiss Abbe type orthoscopic (217x), or perhaps a 5 mm Zeiss orthoscopic (260x). With these oculars the power-per-inch is 41x, 62x, and 74x, respectively. On a 2.4-inch Zeiss refractor (uncoated O.G.), I often use 170x and 210x, and obtain excellent lunar images (71x and 88.5x per inch). Similar powers-per-inch were used on a 1950s 3-inch f/15 Brandon refractor or a 4¼-inch f/10 Bausch and Lomb refractor (lens made in 1944).

At this point we list a table of well-known observers of the past showing the maximum powers they employed on Mars together with my own observations. Except for my instruments, all the rest lacked modern coatings on objectives or eyepieces. It is unfortunate that many of today's amateurs are not familiar with older references, since all too often erroneous opinions are currently held that have been thoroughly discredited for decades.

In general, the power-per-inch figures are often well in excess of the usual dictum not to exceed 50x per inch. We should emphasize that these magnifications will not always be suitable for Jupiter and Saturn.

It is interesting to note the magnifications employed by double star observers. R.G. Aitken in his classic "The Binary Stars" used up to 3000x on the Lick 36-inch (about 83x per inch). Aitken also referred to S.W. Burnham's discoveries of double stars only 0.2 arc second separation with a 6-inch Alvan Clark Refractor.[14] No magnification was given, but it had to be quite high to detect the residual elongation of two stars that close together in an instrument of that size. Aitken stated that these doubles were difficult to measure in the 36-inch Lick refractor.

Although I have cautioned against using too high a power on Jupiter and Saturn, it is interesting to note that W.R. Dawes used 323x, 425x, and 460x on his

[13] Sidgewick, J.B. *Observational Astronomy for Amateurs*. Faber and Faber Ltd., 1971, p. 101.

[14] Aitken, R.G. *The Binary Stars*. Dover, 1964, pp. 29, 59.

Observer	Year	Instrument	Magnification	Per Inch
T.R. Cave[a]	1958	3¼" Brashear Refractor	250x	77x
G.V. Schiaparelli[b]	1877	8½" Merz Refractor	468x	55x
Attkins[a]	*	8½" Reflector	480x	56x
Backhouse[a]	*	4½" Refractor	370x	82x
Wood[a]	*	3¾" Refractor	200x	53x
Sill[a]	1941	7½" Refractor	400x	53x
T. Hake[a]	1941	8" Refractor	400x	50x
T. Hake[a]	1941	3¼" Refractor	184x	60x
T. Hake[a]	1941	4½" Refractor	300x	67x
H. Webb[a]	1941	5" Refractor (stopped to 4")	213x	53x
P. Lowell [c]	1894	18" Brashear refractor**	862x, 1305x	48x, 73x
R.S. Richardson[d]	1941	6" Brashear refractor	500x	83x
R. Gordon	1961	4" Unitron Refractor	300x	75x
R. Gordon	1971	3½" Questar	250x	71x

Also, W. H. Pickering (1926) recommended 55x per inch for Mars. (10)

* Before 1896

** 18" Brashear on loan from Amherst College. Later sold to University of Pennsylvania, where Walter Leight frequently used it at 962x and 1,218x on Saturn.

a. Cave, T.R. "Observations of Mars," *Telescopes, How to Make and Use Them*. T. and L. Page, editors, MacMillan, 1965, p. 133.

b. Webb, H.B. *Observations of Mars and Its Canals*. Privately published, 1941, pp. 39 and various drawings.

c. Lowell, P. *Mars*. Houghton, Mifflin and Co., 1895, p. 217.

d. Richardson, R.S. *Exploring Mars*. McGraw-Hill, 1954, p. 158.

excellent 6⅓-inch Merz Refractor between 1850 and 1854 for observations of sub-divisions in Saturn's rings. Dawes stated that 460x gave him a better view of the division in A (Encke's) than did 323x. W.S. Jacobs at Madras Observatory in 1852 used 365x on a 6.2-inch refractor for ring division observations. We might point out that Encke's division was of very low contrast (about 0.05) as were other sub-divisions that are sometimes noticed in small apertures. Unless the rings are tilted at or near maximum to Earth, many sub-divisions remain undetected and are none too easy in small apertures even when favorably presented. I have seen Encke's only a few times in 6-inch to 8-inch reflectors—though I glimpsed it in 1973 on two occasions with a 3½-inch catadioptric.

The resolving power of the human eye is fairly constant from about a 5 mm to 1 mm exit pupil. At 1 mm this would correspond to 25x per inch. Below 1 mm it falls slightly down to 0.5 mm (corresponding to 50x per inch) and deteriorates somewhat below this. It might thus be argued that there is no need to press magnification beyond 50x per inch. While it is true that powers in excess of 50x per inch will not reveal new details, they do enlarge the scale of what is visible, making it easier to see as long as the light levels/contrast levels are sufficient. In such circumstances, the higher magnification might aid in determining if that tiny object we are seeing is a crater mound or a crater pit. It is also fun to push optics to 100x per inch or more to examine the perfection of the stellar diffraction pattern. The aesthetic satisfaction of owning an instrument capable of rendering such high powers without image breakdown is a source of pride to the lucky owner, and he can feel justly proud of their equipment.

There are other reasons for using high powers. Astigmatism is less bothersome with smaller exit pupils. A person suffering from severe astigmatism, who needs to wear glasses when observing at low power, can frequently remove them for high magnifications. There is ample evidence that myopia and presbyopia do not seriously interfere with telescopic vision. It is said that W.R. Dawes of "Dawes Limit" fame had myopia of such a severe nature, that he once passed his wife on the street without recognizing her! (The reaction of Mrs. Dawes is not recorded.)

Telescopes work best for contrast and definition when we use as few air/glass surfaces as possible. This means eliminating prism diagonals, mirror diagonals, Barlow lenses (except when additional eye relief is necessary), and eyepieces with many elements and air/glass surfaces. Each additional element in the optical train slightly degrades the image, and these losses will mount up noticeably regardless of whether the optics have no coatings, standard coatings, or multi-coatings. My best oculars have only two or four air/glass surfaces. Even today the now discontinued (and much lamented) Zeiss monocentric is still the sine qua non of lunar and planetary oculars, but top quality four element—four air/glass surface oculars such as the Plössl or Abbe orthoscopic are excellent substitutes when maximum contrast and definition are required. A really good eyepiece can increase contrast by 20%, 30%, or more over an indifferent one. A 20% loss of contrast is endured when we go from a perfect telescope with zero obstruction to a perfect telescope with a 30% obstruction. (based on the redistribution of light in the disk/ring system), so it behooves us not to increase losses further by using the wrong eyepieces; i.e., those of lesser quality.[15]

I am not adverse to low power viewing. In fact, my repertoire of low power/wide angle eyepieces is probably greater than the total number of eyepieces most have in their collections. Few have ever seen, for example, the entire Moon at 166x with room to spare as I have on my 4¼-inch *f*/10 refractor using a Jena 12.5 mm super-wide-angle 90° eyepiece (plus 2x Barlow lens); and the aesthetic views of Milky Way star clusters and star fields with a 40 mm 65° Erfle, or a 28 mm military Zeiss 80° eyepiece are most gratifying in one's less serious moments. However, the low end of the power spectrum is not an end unto itself, and the observer who uses a 16- or 18-inch reflector at perhaps 80x or 90x on Jupiter or Saturn and extols these "high" power views is obviously missing much of what his instrument is capable of showing.

---

[15] Dunham, Jr., T.H. "The Influence of Central Stops on Performance of a Telescope." Univ. of Rochester. (Unfortunately, I only have a copy of this article; and the publication date and publisher are not included, but a similar article may be found by the author in *J. Opt. Soc. America*, 1951, pp. 41,290.)

Fig. 6.8.1 *Rodger W. Gordon has been an amateur astronomer for over 42 years. His articles have appeared in* Review of Popular Astronomy, Star and Sky, Sky and Telescope, The Astronomical League Reflector, Zeiss Historica *and the* LVAAS Observer *since 1962. He divides his time between astronomy, birding, long walks, and writing long letters. Rodger is shown with two of his favorite binoculars: a 6 x 42 U.S. Navy Sard (1943) with a 12° field of view and a military Zeiss BLC 7 x 50 Kriegsmarine glass with triple-layer multi-coatings, also from 1943.*

The usual rule of thumb (often found by trial and error) is 30x to 40x per inch for planetary observing, and a somewhat higher 50x per inch for double stars. This writer will not dispute those figures; they are an excellent general guide. However, at the same time, there are instruments and objects that will bear more magnification when seeing permits; and it would be foolish not to exploit these opportunities when they occur. So, do not be afraid to push the power. You may be quite surprised at the results.

6.9 From the Bench Binocular Collimation, Part II

By William J. Cook, with photos by Cory Suddarth

In *ATM Journal* #5 (see Section 5.8 on page 191) we introduced the first in a series of three articles on binocular collimation and explained the difference between collimation and "conditional alignment." In Part II, we will describe some of the more common alignment conventions and explain some of the pro's and con's associated with them.

6.9.1 Eccentric Ring Alignment

Alignment on higher quality binoculars is usually performed by adjusting "eccen-

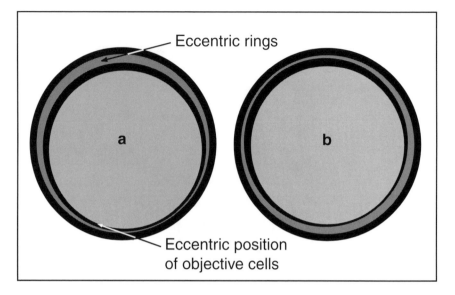

Fig. 6.9.1 *Figures* **a** *and* **b** *illustrate the two extreme positions encountered in eccentric ring adjustment. The lens in cell* **a** *will have the greatest amount of lateral motion (perhaps as much as 3⁄16-inch) while any decentering of the lens in cell* **b** *will be the result of the two lens elements differing slightly in concentricity or wedge in one or both elements.*

tric" rings that are part of each objective cell (see Figure 6.9.1). These rings work as interlocking cams that, when turned, move the mechanical axes of the lens as needed to perform alignment. When the thickest parts of the cams are adjacent (Figure 6.9.1a) turning the outer ring will produce the greatest amount of motion; and when the thickest part of one cam is adjacent to the thinnest part of the other cam, lateral motion will be kept to a minimum. (Figure 6.9.1b)

One might ask why there needs to be allowances made for any type of lateral motion at all. The reason is that some degree of wedge comes in just about every lens element. Thus, even though the lens elements might be "centered" to each other, once cemented into one unit they will invariably exhibit some degree of eccentricity. On occasion this can be very helpful, especially when the eccentricity in the lens cells falls just short of allowing alignment to be attained.

On the highest quality binoculars, the small amount of movement afforded by eccentric ring alignment is more than adequate. The prism cluster or prism shelves are uniform in manufacture and in the way they hold the uniformly manufactured prisms. Oddly enough, this is the same type of alignment convention used on the vast majority of inexpensive binoculars as well. Unfortunately, at this end of the spectrum, where quality control is not on the cutting edge of anyone's concern, there is often so much lateral motion that one or both of the eccentric rings must be shimmed after alignment is achieved just to keep them there. While shimming has long been an accepted technique in many types of optical repair, it is the bane of most conscientious technicians.

6.9.2   Push-Pull Alignment

Another convention for performing alignment is based on setscrews that pass through in the main body housing and push on the center of each prism forcing it (as needed) to ride up on a flat steel spring. (Figure 6.9.2) This convention has been utilized for years, and is one you are apt to find used in conjunction with eccentric ring objective housings.

Although the texturing of most vinyl or leather covered instruments hides the locations of these setscrews, an experienced technician will know where to look in order to keep damage to the covering to a minimum. Figure 6.9.3 shows a rubber-clad binocular with the rubber peeled back in such a way as to show the locations of the collimating setscrews.

While this type of alignment arrangement is widespread and will most likely allow alignment to be rapidly obtained, it is the least likely to insure that alignment will be preserved for a long period of time. It is much easier to see cause and effect with this type of convention. However, the screws are often too short or too flimsy. In addition, the springs are often too weak to maintain alignment.

Figure 6.9.4 shows a similar push-pull arrangement. However, in this case there are small flaps over the screw holes so that alignment may be performed and the flap re-glued with no damage to aesthetics.

Figure 6.9.5 shows still another push-pull arrangement. This time, however, the screws are used to move the prism cluster and not just the individual prisms. This type of assembly is more prone to allowing the instrument to be knocked out of alignment. However, the ease with which a reasonably secure alignment can be obtained makes it very desirable. This particular prism is from a waterproof Swift Sea Hawk.

On this instrument and its close cousin, the Bushnell Navigator, you will find that the objective lenses have no allowances for lateral motion. However, alignment may—in many cases—be performed without going inside the binocular at all. Once the offending side has been determined, the technician will remove the rubber covering from the back-plate (a job that makes skinning a catfish look easy), remove the three screws that seal the holes through which a small jeweler's screwdriver must be placed, insert the screwdriver and start the alignment process. Another pleasant feature concerning this type of alignment convention is that, with a bit of effort, alignment may be achieved while you are looking through the instrument.

6.9.3   Three-Point Objective Alignment

Probably the most important binoculars to the amateur astronomer are the 20 x 80s and 11 x 80s. Unfortunately, many of these instruments have the weakest of all alignment conventions. Most have three small setscrews equi-spaced around the outside of the objective housing with which alignment is supposed to be achieved. After the objective retainer (which secures the lens cell and beauty ring in place)

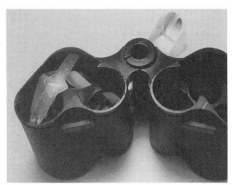

Fig. 6.9.2 *This shows one of two steel straps that push against the prisms in each telescope of some binoculars.*

Fig. 6.9.3 *This shows the position of the collimating setscrews in the most common arrangement. The screw on the left adjusts the prism nearest the observer, and the screw on the right adjusts the prism nearest the objective.*

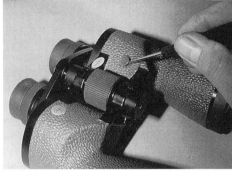

Fig. 6.9.4 *This push-pull arrangement is facilitated via small flaps covering the adjustment screws that can be re-guled.*

Fig. 6.9.5 *This shows the screws used to collimate the Swift Sea Hawk binocular. Three screws have hefty springs (circled, upper right) and three smaller screws (in the box, near the center), press against the prism housing changing the angle of the prism just as one would find in a telescope mirror cell.*

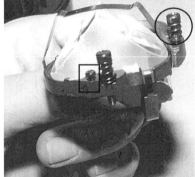

Fig. 6.9.6 *This illustrates the prism assembly (or cluster) from Figure 6.9.5 after it has been installed into the binocular. Here, technician Cory Suddarth tightens the screw holding the coil spring so that there will be adequate tension to maintain collimation once it has been achieved.*

has been removed, the screws will be exposed and ready for tweaking. It will not take a great deal of observation to determine that there is an awful lot of glass depending on those set screws (which are often not much longer than they are wide) to hold, and manufacturers seem to place a great deal of faith in the ability of the retainer ring to apply enough thrust pressure to hold the lenses in place. Too often they do not.

Because of the number of these units that come into our shop each year, I called on some of the importers to suggest that they allow us to retrofit these instruments in such a way as to allow for a more permanent means of alignment. However, the feeling seemed to be that (1) most people are not critical observers and would not care for, or even recognize an improved level of performance; (2) the additional cost would be prohibitive; and (3) the market (especially the American market) would not be pleased with anything which would detract from the aesthetics of the binocular.

Anyone who has had problems keeping their 80 mm binoculars collimated and is willing to "ugly them up" a little for the sake of performance may want to make the modifications themselves.

To do this, simply machine new beauty rings (with a wall thickness of ⅛ to ³⁄₁₆ inch), which will fit very snugly over the objective housings. Then, with the objective lenses out of the instrument, drill and tap three holes (³⁄₅₆ will do nicely) through both the new beauty ring and the objective housing wall and install your new, more "beefy" setscrews.

Be advised that this method will not work on any of the popular Fujinon 70 mm binoculars. These instruments will handle more than an average amount of abuse before requiring collimation. When it does become necessary, it is performed by the eccentric ring method.

It is important to remember that to this point we have been discussing alignment only and not collimation. For those astronomically and optically inclined, alignment may be all that is required. After all, if the instrument is not going to be passed around from person to person (each with a different IPD—interpupillary distance), the need for stringent collimation is diminished.

In Part III, we will discuss the collimation process and the standards involved in making sure that you are getting the most from your binocular.

Issue 7

7.1 Too Many Diffraction Rings
By Richard A. Buchroeder

Refractors have a reputation for excellence, based in part on the observation that they seem to show lots of diffraction rings. W.H. Pickering wrote in 1930, reproduced in *ATM*-2 "Reflectors vs. Refractors", p. 613, 1957 edition, that he saw as many as a dozen rings on a bright star like Arcturus with his 12-inch refractor, while he saw hardly any at all with his reflectors. He went on to discuss what he thought distinguished the refractor from the reflector.

Now, in recent years I have made a hobby of collecting examples of very fine optics and comparing the star images produced by each. One curiosity was that apochromats always seemed to show fewer rings than achromats! I also came to suspect that if you saw more than three diffraction rings, the rest were not a symptom of excellence, but a fault with your optics.

Therefore, I telephoned a few people who owned fine telescopes and asked for their observations on the subject. John Pons of North Hollywood, California, owns a 10" Zeiss and has access to others up to 12", all old lenses with uncoated surfaces. He noted that he too could see as many as a dozen rings when the seeing settled down. Bob Ariail of Columbia, South Carolina, made observations with an uncoated 6" Alvan Clark (1887), an Astrophysics 7" EDT, and a 4" Takahashi fluorite. He complained of poor seeing conditions, but estimated that he saw more rings with the Clark than with the other (coated) optics. Julie Burger of Camano Island, Washington, observed with her 4" Zeiss APQ cemented triplet APO, and reported seeing no more than three rings. Jack Mutch of Medford, New Jersey, observed with an old 4" Zeiss E-type achromat, an 80 mm AS uncoated doublet, and a 4" APQ and reported that he saw about the same number of rings with all of them. However, he reported comparatively poor seeing.

Jack says that he always wears contact lenses, and sees artifacts at night looking at streetlights, and suggests that there may be an eye-telescope interaction as well. Bob Royce of Northford, Connecticut, observed with 6" oiled doublets and reports seeing only two or three diffraction rings.

Well, this whole issue came up because I had acquired a superb 110 mm $f/14.3$ Zeiss E-type objective, circa 1935, from Kevin Kuhne to replace the abominable objective of a 4-inch Unitron that I had previously acquired. I noticed when I first tested the lens on a star that it was about $\frac{1}{4}$ wave over-corrected, as evidenced by hard Fresnel rings outside of focus and softer ones on the inner side of focus. I had Kevin re-space the doublet with a custom-made 3.8 mm spacer and used it again. The spherical aberration was eliminated, and the color correction shifted from F–B correction to modern F – C. In other words, the performance was modernized and made nearly perfect. However, when I observed things like Antares, I was annoyed that as many as 7 rings were visible on bright stars. Although resolution and contrast were nice, it was apparent that the green companion of Antares would likely be obscured by the rings.

I mentioned to Rodger W. Gordon, noted observer and telescope critic, that there seemed to be more than a reasonable number of rings around bright stars; he suggested there must be something wrong with the lens! I had it tested by Bob Goff with his mercury-lighted Foucault Tester and Phil Lam with his Shack Laser Interferometer. The lens was superbly smooth and accurate.

Next, I speculated that secondary color must be the problem, with the extra rings being blended Fresnel patterns of out-of-focus wavelengths.

A green filter still showed lots of rings, and it finally dawned on me that the uncoated lens surfaces were producing ghost images that, being coherent with respect to each other, could interfere and produce additional rings—as well as modify the nominal diffraction pattern!

Therefore, I asked my friends to examine stars in their telescopes, and the evidence suggested that stray light was the cause of the extra rings.

To verify this, I first asked Phil Lam to apply a partially-silvered film to the two inner surfaces of the Zeiss doublet on the premise that if AR (anti-reflection sputtering) coatings would eliminate rings, then partial silvering (enhancing the reflected light) would produce more. Well, the image of the moon was a lovely sepia color, but Vega was too faint to draw any conclusions about interference rings!

The ultimate test was to get the lens completely AR coated. Greg Lowe of Photon Sciences, Tucson, Arizona, used the modern coating technique called "sputtering" to produce an ultra-low AR coating on all four surfaces. Sputtering is a low temperature procedure that minimizes risk. I also asked that no aggressive cleaning operations be performed, as I did not want to risk a "figure" change on the excellent Zeiss optics.

Greg was not entirely happy with the results. Because the glass is probably 60 years old, the surface has some cosmetic problems due to age, which resulted in small blemishes in the coated surface. However, I was delighted with the results, which in my opinion were stunningly successful. At some angles, the glass surfaces simply seem to disappear!

I then put the lens back on the Unitron tube, installed it on my 3" Unitron Mount (I prefer the 3" mount to the 4", which is too heavy for easy handling), and did some observing.

Vega was just west of the zenith and perfect for testing, as it was earlier when it was used to test the uncoated lens.

We often have excellent seeing in Tucson; much of the time a 6" telescope can reach diffraction-limit, and it is almost always possible to do so with a 4" objective. What I found was that Vega had only two or three rings visible by direct vision. With averted vision, fleeting additional rings were visible.

I repeated the experiment on several other stars on that and additional nights; the results were consistent. The AR coatings had dispelled 3 or 4 of the 7 rings that I could regularly see with the lens before it was AR coated.

Intuitively, it would seem that the inner two radii, which produce the most intense computed ghost image, were responsible for the large number of interference rings around bright stars. However, since I did not test the lens with any but completely coated surfaces, I cannot be sure. It is perhaps no coincidence that prior to the widespread use of AR coatings, many manufacturers did AR coat just the inner two radii. This has, on any basis you want to rate it, a considerable advantage. First, it reduced the number of ghost combinations from seven down to just one. Second, the old coatings were too soft for handling, so those surfaces were protected. Third, it increased transmission by 7%. Finally, it would not surprise me if some of the old-time designers realized that spurious rings were being generated by failing to coat the objective optics. The ghost would be harmless if it merely caused a veiling glare; but by forming rings, it concentrates light in a way that harms double star and related visual tasks.

7.2 Stellafane 1994

By Diane Lucas

I had not been to Stellafane since 1991. The improvements that had been made were really worthwhile and added to everyone's enjoyment of the weekend. Most of the activities took place at Stellafane East, the new property purchased and developed by the Springfield ATMs. The light rain and drizzle had not dampened spirits appreciably when I arrived Friday night in time to start greeting old friends and making new ones, and to attend Friday night talks. The talks touched on annular eclipse reports, a table-top grinding machine for up to 24-inchs, daytime observing, an observatory for a mirrors up to 32-inch telescope, etc. The cold front blowing in convinced many people that it was time to close up early.

Saturday was cool and clear! A great day for Stellafane. A new part of the activities was a session on telescope making, which was so popular Friday that it was repeated Saturday. Most of my Saturday was spent looking at and talking about telescopes and other astronomical activities. I never did get to the Saturday afternoon talks, but for once I got to see almost enough telescopes. There were two 32-inch telescopes on the hill. Mario Motta's won a recognition award, but John Vogt from Huntington, New York had an $f/4$ instrument with a very unusual mirror made from fused quartz tubes and plates. Unfortunately one of the ribs cracked when, contrary to instructions, the coating company supported it on an edge while coating. Forrest Hamilton from Randlestown, Maryland, had a very easy-to-adjust-and-use 6-inch Newtonian binocular with both intraocular and relative tilt for

Fig. 7.2.1 *One of the best traveled and most intriguing telescopes at Stellafane was that owned by Clyde Bone of San Angelo, Texas. Clyde received a Merit Award in the Mechanical Design category for his 20-inch Mersenne-Naysmith telescope.*

I arrived at Riverside a day early this year and had the opportunity to talk to Clyde before the rush set in. The conversation was enjoyable, but each time I see the telescope at another event, and see how much time Clyde has to offer to our hobby, I have to force myself to stop thinking about how long I have to wait for retirement

easy alignment. Jon Kern from New Orleans, Louisiana, had a tricolor, three-objective, radial grating telescope for super accurate color eclipse corona photography. Cap Hossfield was there with another demonstration of a possible gravity-wave detector. Then there were all the ones that I never had a chance to see.

7.2.1 Evening Program

The highlight of the evening program was the telescope awards. I had cleverly managed to get pictures of most of these and talk to their makers during the morning and afternoon.

The usual special awards were presented. John Martin of the Springfield Amateurs received a special recognition award for his service to Stellafane. Youngest and oldest also got awards with Allen Mackintosh (who now lives in Cornwall, England) receiving a prize for the oldest attendee (85) and sharing one for traveling the furthest.

A special and impressive memorial for Walter Scott Houston was presented. His wife and two daughters were present; this concluded with the audience lighting candles for a moment of silence.

I did not stick around for the evening address but wandered off to look through telescopes, and not just at them. The original Stellafane area around "The Heavens Declare the Glory of God" pink clubhouse was reserved for telescope optics judging and for deep-sky observing with no car traffic or white flashlights allowed.

It was a great Stellafane!

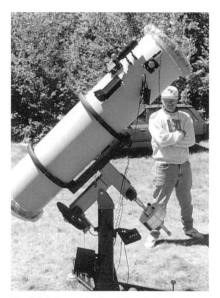

Fig. 7.2.2 *Paul Pompa of New Hartford, Connecticut, with his 20 x 80 binoculars mounted atop his parallel motion binocular mount.*

Fig. 7.2.3 *Ron Arbey's 12.5-inch f/4.8 Newtonian took first place for craftsmanship at last year's Stellafane.*

Fig. 7.2.5 *Jon Kern of New Orleans, Louisiana discusses his tri-color, radial grating, solar camera pictured in Figure 7.2.1.*

Fig. 7.2.4 *(left) Steve Watkins of Houston, Texas, received a Merit Award in the Mechanical Design category for his 10-inch split-ring equatorial.*

Fig. 7.2.6 *Peter Hirtle of Seattle, Washington, prepares to demonstrate the operation of his award winning telescope.*

7.2.2 The Background on Peter Hirtle's Tri-Schiefspiegler
by Peter Hirtle

My Tri-Schiefspiegler did not start out as the instrument I ended up with. I fabricated the main part of the mount about 10 years ago when I took a welding class at the local community college. In its original concept, it was to have been for an 8-inch *f*/5 photographic telescope. That project got stalled for reasons I have now forgotten, and the partially machined fabrication sat in a closet for about four years waiting for a new inspiration. I had made a 4.25-inch *f*/27 Schiefspiegler many years ago and have always been pleased with its performance; but like anyone who still has warm blood in their veins, I wanted something bigger, faster, and more compact.

My first stab at satisfying that "want" was a 6-inch *f*/15 Yolo. It turned out well and has given me some of my best observing; but because it is fairly bulky and finicky about being transported, it was not the answer.

While thumbing through back issues of *Sky & Telescope,* I came across Anton Kutter's articles on his Tri-Schiefspiegler (January and February, 1975). I had rejected that design in the past because of the long back focal length which requires a very tall focuser and the necessity of making a test plate for the convex secondary. This time, it occurred to me that the previous objectionable back focal length might make it through the abandoned mount sitting in the closet. Sure enough, with some modifications, it did. I still objected to making a test plate that would be used only once and then collect dust on a shelf. That was solved by enticing other members of my astronomy club to make Tri-schiefspieglers. They still use the test plate regularly.

One of the modifications I made was to scale the design from $f/14.7$ to $f/15$. There were a few reasons for that. One was to increase the back focal length slightly so the focus would reach the eyepiece. Another was the fact that increasing the focal ratio decreases the sensitivity of any design. It was more appealing to me to say $f/15$ instead of $f/14.7$ when asked "what kinda 'scope izzat?"

Making the optics was straightforward. I used the Foucault test on the primary mirror. For testing the tertiary, I made a Foucault tester with a halogen flashlight bulb and used a 7 x 50 binocular behind the knife edge so I could see the mirror, which was more than 1,000 inches away. The secondary was tested using the test plate under a mercury light source. Rough testing was done with the light by itself, but for final testing I used a collimating lens over the mirror and test plate. With my eye and the light close to the focus of that combination, I could see the fringes more accurately and with better contrast. Slow strokes, lots of pressing and fairly hard laps help produce the smooth surfaces that are so important in this and any other high performance telescope.

I was careful to build the tube assembly as closely as possible to the dimensions given by the ray-trace. Maybe it was just dumb luck, but collimation has proven to be simple. Centering the reflection of each mirror in the next; then tilting the primary, while observing a star, to tune out the last bit of astigmatism was all that was required.

The performance was similar to a good refractor. Planets and faint fuzzies were seen with excellent contrast and definition. Anyone who says you need a short focal ratio for deep-sky observing does not know the facts. The tube assembly was designed and built so that an absolute minimum of stray light gets to the focal plane. At RTMC, one experienced observer commented that he was disoriented because the field was so dark that it was difficult to see the field stop of the eyepiece!

The mount is the most distinctive part of this telescope. Every part of it is built with rigidity in mind, because it does not matter how good the image is if it will not stay motionless. The polar shaft is 3 inches in diameter and both axes ride on pre-loaded ball bearings. The slow motions have friction clutches to allow hand slewing, and the controls are located within easy reach from the observing position. The drive is a stepper motor which turns the worm for a 360-tooth worm wheel. The final reduction is a 6.75-inch friction disk driven by the worm wheel's 6 mm shaft. The hour circle is connected to the drive; so once set, it reads correctly as long as the drive is running. It and the declination circle are located next to the eyepiece and are easy to see. The focuser is a modified Crayford design. Not counting the stepper motor, I used a total of 16 ball bearings in the mount.

I do not claim to have any original ideas for this instrument; I just used a bunch of old ones in a new combination. In many respects, with its solid steady mount, excellent image quality, fixed eyepiece position, and conveniently located controls, this is the ultimate telescope—although I will be the first to point out its faults. With its four reflecting surfaces, only about 60% of the light makes it to the focal plane. Enhanced coatings would improve that, but I have unanswered questions about their longevity and effect on scattered light. At a little over 100 lbs, it does not lend itself very well to the guerrilla astronomy we do here in the Pacific Northwest— we need to get set up, get the observing done and get back inside between rain showers. Also, I am accustomed to using a finder telescope that points in the direction that the telescope is aimed and have never liked right angle finders. The finder in this does not have just one, but two right angles. Setting circles are the solution, but accurate polar alignment takes time. I have always liked the more intimate experience of star-hopping.

One of the most common questions I am asked is, "How long did it take to make?" Although I do not believe anyone should keep track of the time they spend at a hobby, I am certain that the hour count would be well into four figures.

7.2.3   Merit Awards Stellafane 1994

Craftsmanship

- Matt Marulla, Nashua, New Hampshire, 12.5-inch f/5 split-ring Newtonian. Matt is a software engineer and has installed a motorized eyepiece focuser and motorized primary-mirror collimation device.
- Mike Jean, Athol, Massachusetts, 8-inch f/6 Dobsonian.
- Peter Hirtle, Seattle, Washington, 6-inch f/15 Springfield Tri-Schiefspiegler.
- Jerry Wolczanski, Warrenton, Virginia, 5-inch f/7 Springfield Dobsonian (fixed eyepiece position).

Mechanical Design

- Peter Hirtle, Seattle, Washington, 6-inch f/15 Springfield Tri-Schiefspiegler.
- Clyde Bone, San Angelo, Texas, 20-inch f/5 Mersenne Naysmith convertible Newtonian. Commercial primary furnishes converging light to an f/5 convex paraboloid secondary, which reflects parallel light via a conventional diagonal just in front of the primary out of the altitude axis to another diagonal and then to a Genesis refractor, which forms the final image. This makes a very convenient sit down position for observing. The four reflections plus the refractor do significantly decrease the available light; the diagonal in front of the primary can easily replace the convex diagonal to create a conventional Newtonian when maximum light is required.

- Mario Motta, Lynfield, Massachusetts, 32-inch equatorial Newtonian. This telescope is designed for a permanent observatory. It made a one-time trip to Stellafane for an exhibition in a U-haul truck. The weight of the primary was significantly reduced by sand blasting cavities between supporting ribs.
- Steve Watkins, Houston, Texas, 10-inch f/8 split-ring equatorial.

Antique Instrument Restoration

- J. E. Woodward, South Salem, New Hampshire, has restored an antique brass refractor with case. The maker is unknown, but the telescope came from Scotland.

Special Award

- Martin Hamer, Wilton, Connecticut, for a portable observatory trailer.

Junior (under 16)

- Jesse Flaherty, Hingham, Massachusetts, 6-inch f/10 Dobsonian.

Optical Performance (less than 12-inch)

- Steve Watkins, Houston, Texas, 10-inch f/8 split-ring equatorial Newtonian.
- Tom Calderwood, Waltham, Massachusetts, 6-inch f/5 Newtonian category.

Multi-Surface and Catadoptrics

- Peter Hirtle, Seattle, Washington, 6-inch f/15 Springfield-Tri-Schiefspiegler.

7.3 The Perfect Telescope Is
By Ed Villareal

The perfect telescope is a single objective with a binocular viewer.

This was not always apparent to me, and in the past, I had often considered building a 6" Newtonian binocular (two complete tube assemblies). Fortunately, before I ever got around to gathering up the project supplies—matched objectives, tubes, spiders, etc.—I had the opportunity to use my buddy's (Bob Lindsey) binocular viewer. After using the viewer on both his 6" refractor and his wife Karen's 10" SCT, it occurred to me that the current thinking on building binocular telescopes is flawed.

Q. Why do we build 'em?

A. Because we know that the image processing that occurs in your brain, when it is fed two images, produces superior image quality in the form of contrast improvement and critical detail detection compared to the image you perceive when you only supply your brain with a single eye image.

Q. Is that so?

A. O.K., you caught me; we all know that the real reason we build binocular telescopes is so that we can see the 3-D effect.

Q. Well, then shouldn't we build two telescopes in order to see that 3-D effect just like 7 x 50 binoculars?

A. No, that is the flaw in the recent binocular trend.

Regular terrestrial binoculars produce the effect of depth perception because the left and right eyes receive two slightly angularly offset images which your brain translates into a 3-D image. The slightly differing angles of view are because the objects we normally daytime view are physically close to us in distance, and the spread between your eyes (interpupillary distance, or a slightly wider distance between binocular objectives) is sufficient to allow real triangulation and relative distance evaluation to occur with brain processing. With viewing of celestial objects, everything is, for all practical purposes, at infinity. The view from one to the other is identical. The view from here is the same as the view from 50 yards away. Whether the views supplied to both eyes come from one telescope or two does not matter. The 3-D effect is totally a construct in your brain because the two views are identical—still, it is a nifty and enjoyable effect. Ever notice the 3-D effect at the terminator on the Moon at high magnification using one eye? I once saw a 3-D view of M42 with a 20-inch telescope and a single eye, but it takes a lot more light and contrast to produce that effect with one eye than two. The 3-D effect happens at fairly small apertures using both eyes. I've seen it well on planets and star clusters with a 6-inch scope equipped with a bino viewer (light equivalent of about two 4-inchers).

To build a single 6-inch tube assembly from commercial components, at 1994 prices, costs approximately $300. About $400 makes an 8-inch, and let's say $700 for a 10-inch.

Commercial binocular viewers are a price item at about $500 for an excellent example. This turns out to be a bargain.

From the above it appears that the binocular has an advantage at two 6-inch = $600 compared to one 8-inch plus viewer = $900. That is too simplified a view (just ask anyone who has ever built a binocular telescope). I have certainly glossed over the engineering required to mount the two tubes and a mechanism to adjust the i.p.d. and focus, etc. It is more fair to consider the bino viewer as a one-time expense addition to your eyepiece collection. After all, it is usable with future telescopes. In that case, the comparison is $600 to $400 for two 6-inch against one 8-inch. That is too easy to decide. Now you can use standard telescope mounts and have no collimating between tubes to worry about. And don't forget that the 8-inch has better resolving power than the 6-inch.

If you currently own a telescope, you have most of a binocular telescope already!

I am a definite technical rookie on binocular viewers. Perhaps someone out there with extensive practical use experience would be kind enough to write up a guide on what are the important considerations in selecting a binocular viewer? I'd be grateful.

7.4 Astrofest 1993

By Diane Lucas

This was probably the largest Astrofest yet with almost 700 preregistered. In order to keep the rental costs down for the very convenient 4H camp where the event was held, only the Astrofest staff were allowed on the grounds before 12 noon on Friday. However, the campgrounds began to fill immediately upon opening, and by dinner time, there was a huge crowd of amateur astronomers and telescope makers ready to participate in this year's events with some of the flea market sellers opening up almost as soon they parked their vehicles. The Northwest Suburban Astronomers were busy selling "astrodogs" for those who did not want to leave the camp. There were clear skies for a while but some rain that evening.

Saturday was the big day with one of the largest flea markets ever, keeping everyone busy walking around looking for bargains. Talks were going on all day in the dining hall, except for breaks for lunch and dinner. I listened to a couple of these: John Dykla described how global changes in the atmosphere could be monitored by recording the brightness of earthshine of the Moon; Tom Clark showed a great group of slides of the night sky taken in Australia near Ayers Rock; and A. J. Sehgal described some basic methods of image processing and the convolution-deconvolution method used in his relatively expensive "Hidden Image" software. I missed the afternoon talks, which included Richard Berry's discussion on CCD universe exploration.

One of the noticeable changes from last year's Astrofest was the larger number of telescopes equipped and set up for CCD or video image recording. These advances in amateur astronomy require more money or ingenuity and better optics to obtain really good images, plus a computer to assist in image collection and processing to achieve optimum results. Still, they permit amateurs an opportunity not only to obtain really nifty pictures of deep-sky objects but also to allow them an opportunity to do valuable scientific research. Amateur astronomy has certainly changed since the 1950s, when I began my hobby of star gazing. There is still a great deal that can be done by amateurs using simple scopes (even those without a drive), the super-automated image collection telescopes, and all the variations in between.

There were a lot of great telescopes on the field this year. I was especially impressed by Dale Asbyry's achievements in making a scope right down to all the optics and silvering the mirrors. Frederick Schebor's extremely compact package holding a 5-inch $f/5$ refractor was also impressive.

During the Friday night observing session, I really enjoyed the images in Al

Fig. 7.4.1 *Richard Walker, of the Genessee Astronomical Society, has kindly provided assistance in gathering information on Astrofest.*

Fig. 7.4.2 *Dale Asbyry of Delavan, Wisconsin is a telescope maker in the Bob Cox and Russell Porter sense of the words. He received his award for "Outstanding First Scope." This classic 8-inch f/7 was made, start to finish, by Dale himself. This included the primary, secondary, Ramsden eyepiece, finder, and the silvering of both mirrors.*

Fig. 7.4.3 *With his first attempt at telescope making, Joe Del Santo of Hanover Park, Illinois, earned a Merit Award for "Excellent First Scope." The instrument is a 12.5-inch f/5 Dobsonian with a rocker-box as proposed by Richard Berry in* Build Your Own Telescope.

Fig. 7.4.4 *(Right)*

Fig. 7.4.5 *(Left)*

Fig. 7.4.6 *(Right)*

Fig. 7.4.8

Fig. 7.4.7 *Steven Overholt*

Fig. 7.4.9 *Walter Piorkowski of South Beloit, Illinois, took home a Merit Award for "Excellent New Idea for Innovative Use of a Microscope Stage for Guidescope Adjustment." This 10-inch f/3.7 was somewhat dwarfed by the guidescope. The adjustment device for polar axis alignment was also well constructed and usable with a fully loaded mount.*

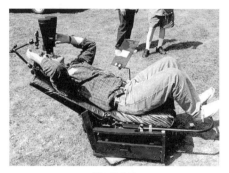

Fig. 7.4.10

Fig. **7.4.11** *This mahogany telescope is named "Hercules" and is the property of Mr. Jeff Morgan of Chicago. More than that I do not know. Perhaps Mr. Morgan will write to give us additional information*

Fig. **7.4.12** *Frederick Schebor shows off his 5-inch f/5 "VPR" (very portable richfield). This telescope, built around a 5-inch Jaegers objective, includes several self-contained eyepieces and diagonal; the Telrad finder was shrunk to fit the available space.*

Fig. **7.4.13** *Here Frederick. Schebor is about to install the 5-inch objective and make the telescope operational. For those yet to be attacked by aperture fever, a 5-inch refractor is a very respectable instrument for wide-angle viewing, and having an instrument such as this come in such a convenient package will insure that it will be used often.*

Woods' (St. Louis, Missouri) Stevick-Paul telescope. This year the collimation and alignment were perfected, and the images were pinpoints right across the field. This is a truly great design, and the instrument provided the best images I have seen in an off-axis, multiple-mirror telescope.

Having to leave early Saturday evening, I missed the observing that evening as well as the presentation of the awards for the telescopes and accessory category.

Fig. 7.4.14 *Ron Hawk of Flint, Illinois, received his Merit Award for the "Practical Accessories and Refinements" to this very heavily modified 13.1-inch Dobsonian (above left and right). Accessories include an easy to adjust counterweight system (note the metal strip above the ALT bearing) and a quick-change filter slide, which was mounted inside the tube.*

Fig. 7.4.15 *(Right) Howard N. Klauser of Addison, Illinois, received a Merit Award for "Excellent Craftsmanship" in making his 8-inch f/10 "Daley" solar reflector.*

Fig. 7.4.16 *(Left) Mr. Piorkowski's guidescope is mounted to his primary telescope with non-adjustable rings. The use of the micrometer stage allows him to compensate for flexure of the two instruments. It also allows him to move around the field of the guidescope to have some control over the object he chooses to guide on.*

Fig. 7.4.17 *David DeRemo of Cedar Springs, Michigan, received a Merit Award for "Good Craftsmanship for a 22-inch f/5 Dobsonian (left) and Companion Podium" (below). The podium was a well-made observing tool with back-illuminated negatives as star charts and drawers to hold extra eyepieces, books, and observing aids.*

Fig. 7.4.18 *Here Scott Jamieson of Waukesha, Wisconsin, presents his 4-inch f/15 refractor. Scott received a Merit Award for a "Very Stable German Equatorial Mount Using Low-Tech Bearings." The bearings for this mount were Teflon and Formica, as one might expect to see on a Dobsonian, but they were adapted in such a way as to create an inexpensive, smoothly moving (but stable) mount.*

Fig. 7.4.19 *Well-known amateur Steven Overholt of Santa Margarita, California, makes the final adjustments on his 30-inch f/3.75. Steve received the award for "Largest Scope on Site Transported in the Smallest Car." This instrument has some interesting features. First of all, the eyepiece is positioned just past the end of the truss, thus allowing for a shorter than normal instrument. Also, Steve has incorporated some interesting baffles around the secondary mirror and between the secondary and eyepiece.*

Fig. 7.4.20 *A close-up of Steve Watkins' mirror housing and drive mechanism.*

Fig. 7.4.21 *Steve Watkins of Houston, Texas, received a Merit Award in the Mechanical Design category for his 10-inch f/8 split-ring equatorial.*

Fig. 7.4.22 *(above) Matt Marulla of Nashua, New Hampshire, stands beside his 12.5-inch f/5 split-ring Newtonian. The instrument has a motorized eyepiece focuser and motorized primary mirror collimation (below).*

Fig. 7.4.23 *Forrest Hamilton's, easy to adjust and use 6-inch Newtonian binocular has intraocular and relative tilt adjustment for easy alignment.*

7.4.1 Merit Award Recipients

- Frederick S. Schobor, Michigan, 5-inch $f/5$ "VPR" (very portable rich-field) for a collapsible refractor and accessories.
- Howard N. Klauser, Addision, Illinois, Excellent Craftsmanship for an 8-inch $f/10$ "Daley" solar reflector.
- Ron Hawk, Flint, Illinois, Practical Accessories and Refinements on a heavily modified 13.1-inch Dobsonian.
- Joe Del Santo, Hanover Park, Illinois, Excellent First Scope for a basic 12.5-inch $f/5$ Dobsonian.
- Paul Nelson, Oliva, Minnesota, and Steve Messner, Northfield, Minnesota, Excellent Design and Craftsmanship for a remote controlled CCD Imaging Telescope.
- Dale Asbyry, Delavan, Wisconsin, Outstanding First Scope for an 8-inch $f/7$ Classic Design Newtonian.
- Walter Piorkowski, South Beloit, Illinois, Excellent New Idea for innovative use of microscope stage for guide scope.
- John Phelps, Orlando Park, Illinois, Frugal Design for an inexpensive tracking platform.
- David DeRemo, Cedar Springs, Michigan, Good Craftsmanship for 22-inch $f/5$ Dobsonian and companion podium.
- Scott Jamieson, Waukesha, Wisconsin, Very Stable German Equatorial Mount for a 4-inch $f/15$ refractor using simple low-tech bearings.
- Steven Overholt, Santa Margarita, California, Largest Scope on site transported in the smallest car.

7.5 Astrofest 1993—A Second View

By Richard Walker

It seems that as Astrofest for one year is ending, on the trip back home, you start to immediately plan for the next year. Last year was no different, and it was great to get back to Kankakee, Illinois for this year's Astrofest.

Astrofest gets bigger every year. The unofficial attendance announced at the Saturday evening awards and door prize ceremony was over 800.

Registration started early, somewhere around 4:00. From that time on, a new but common sight all over the grounds was the "Star" T-shirts being worn by amateur astronomers who had volunteered time to help The Chicago Astronomical Society with the ever growing task of achieving a successful Astrofest. It worked. This year's Astrofest was great.

The skies were clear (but bright) until around midnight on Friday when the

Fig. 7.5.1 *Snapshots from Astrofest 1993*

clouds rolled in. This worked out great for those of us who had been up early to get on the road for a long drive. We did not have to think up some pitiful excuse to go to bed before dawn. The 4 or 5 hours of sleep was barely adequate.

Saturday dawned bright and early. No one could sleep in late on Saturday morning at Astrofest; the flea market started as soon as there was enough light to set things up. In fact, I saw one hardy soul setting things up before it was light enough to see your feet on the ground in front of you!

Once again, it was possible to buy everything from telescope parts to complete telescopes of every description. Also, this year saw a much larger selection of books and magazines. I picked up a copy of one of Tom Clancy's latest novels—I am sure that somehow it is astronomy related. The flea market this year was huge, stretching from the hall well past the first set of cabins. It was a good mix of commercial vendors and amateurs, like you and me, trying to sell some surplus items—usually to then turn around and buy something else, of course! A few of the vendors were set up in the telescope field.

The morning talks started at 10:00 with a series of short presentations on a variety of topics, including video and CCD imaging. The talks are a great way to learn what others are doing and a great place to pick up tips for your own projects.

The astrophoto judging started around 10:00. The photos were judged by the attendees; a ballot was part of the registration packet. Everyone was given the opportunity to see and pick their favorites from both the deep-sky and solar system categories, and even though there were not too many entries, they were great, and choosing just two was not easy.

Telescope judging started after lunch at 1:00. Everywhere you looked around the Astrofest field, you could see the registration forms that had been attached to telescopes waving in the light wind. The panel of telescope judges started the not-so-easy task of looking at all of the telescopes entered in the contest. This involves walking through the entire field, looking at all of the scopes, taking pictures, and in most cases, talking to the owner/builders.

Large telescopes continue to expand in popularity, with 20- and 25-inch scopes not that unusual anymore. The largest scope this year was Steven Overholt's 30-inch, which he brought to Astrofest in a Ford Festiva!

The results of all of this judging were announced at the Saturday evening awards and door prize ceremony, which took place after dinner. Once again, the air inside the hall was thick with anticipation (and heat and carbon dioxide—the only thing missing was oxygen), and many of us stood outside straining to hear the announced winners.

After the ceremonies, we all went back to our telescopes to find the skies clearing for another night of observing. The skies were clear but bright, so most of us were sticking to the brighter objects, including Saturn. As usual, there was a lot more talking than observing going on. It is probably one of the main reasons that so many people go to Astrofest every year: to see old friends and to make new ones. There was also a lot of CCD and photographic imaging going on in the telescope field and in the "electric area". Of course, we all looked through a variety of telescopes and at a good many objects; the old rule of never looking through a telescope larger than you can afford or can haul around was being violated everywhere.

Sunday found everyone wiping the dew off of telescopes and tents, packing up, saying good-bye, and getting ready for the long drive home. See you all at Astrofest 16, next year.

7.6   The Joys of Low-Power Viewing
By Al Nagler

The majesty of the universe has fascinated me since I was a teenager in the '50s. While I have always enjoyed all aspects of amateur astronomy from telescope making to astrophotography to viewing planetary details to planetary nebulae, I am, if anything, partial to spectacular views of Milky Way star fields. I regard this kind of viewing as a constantly refreshing aesthetic experience. To me, "celestial", or heavenly, connotes wide angle star-field vistas. Whether you own a large or small scope, refractor, SCT or Dobsonian, or just binoculars, the Universe is full of delights in low power viewing.

Let's look at what low power means, its relationship to field of view, the meaning of "RFT", and what specific low power views (in my experience) are particularly exciting. While any telescope can be optimized for some degree of low power viewing, you might refresh yourself with some exploring with good astronomical binoculars (I own 7 x 50, 10 x 50 and 14 x 70). I heartily recommend Phil Harrington's book *Touring the Universe Through Binoculars* as a reference tool that is equally valid for telescopes, since most binocular "spectaculars" are even more so with a telescope.

7.6.1   Field Considerations

Low power is a pretty vague term, and perhaps misleading, since a "low power" view in an eyepiece with a 40° apparent field is not the same as a view with an 80° apparent field at the same magnification. What we are looking for here is the widest effective true field possible with any given instrument. As an aesthetic pursuit, I think we want to "frame" any object in its context, much as we would frame a portrait. Seeing the Pleiades just barely getting into the picture is no match for seeing the jeweled pattern set against a background spray of tiny diamond points. Having a telescope with a potential field of many degrees allows you the luxury of framing the scene with your palette of eyepieces.

How large a field can you get with a telescope? Let's assume your scope can use 2-inch eyepieces. Since the field area of a 2-inch eyepiece can be 3 times the area of a 1.25-inch, it is a great idea, for example, to get a 2-inch diagonal fitted to an SCT to appreciate its field potential.

The field of any telescope is determined by the eyepiece field stop and the focal length of the telescope. The 47 mm barrel inside diameter of a 55 mm,

2-inch Plössl eyepiece is the largest you can get. (Longer focal length eyepieces only serve to reduce the apparent field and are useless in my opinion.) Divide 47 mm by the focal length, and you have the approximate "tangent" of the true field. Therefore, a 500 mm f.l. telescope could give you 5.4°, a 1000 mm f.l. = 2.7°, 2000 mm f.l. = 1.35°, etc.

7.6.2   f/# Considerations

For visual observations $f/\#$ is basically irrelevant. The $f/\#$, or photographic speed, is important for photography; but a fast $f/\#$ telescope objective will produce the same brightness as a slow telescope with the same aperture and magnification. No binocular manufacturer ever specifies the $f/\#$ of the objective, just aperture and magnification, which gives the exit pupil spec (aperture/magnification). However, for visual observing, the fast $f/\#$ implies a short f.l., which does give the most field. A 100 mm aperture telescope at $f/5$ = 500 mm f.l. (5.4°), while a 100 mm $f/10$ = 1000 mm f.l. (2.7°).

Visually, therefore, we can say that a "faster" scope has more potential field, but no more image brightness than a slow scope of the same aperture!

7.6.3   Exit Pupil Considerations

The exit pupil is the image of the objective (or mirror) that is formed by the eyepiece. It is where you position your eye to see the full field. There is a myth that a 7 mm exit pupil (which matches the maximum pupil for those lucky enough to be youthful) is magically best for low power viewing. Not! You may require a larger pupil (lower power, larger field) just to fit the desired object into the eyepiece. Example: if you wanted to see the entire Andromeda galaxy with its companions (a 4° field), you could do so with a 100 mm aperture, 500 mm f.l. telescope and a 55 mm Plössl eyepiece (total field 5.4°). Note that the magnification is only 9x (500 mm f.l./55 mm f.l.) and the exit pupil is 11 mm (100 mm ap./9x or 55-mm eyepiece f.l./$f5$). The view is great, but the numbers sound horrible: your 7-mm eye pupil is "sampling" an 11 mm exit pupil. Are you, therefore, losing brightness? Absolutely not! Since your eye pupil is fully filled, you have maximum brightness for extended objects, even though you are only using $^7/_{11}$, or 64 mm, effective aperture. In effect, you traded scope aperture for maximum field with no loss of brightness or resolution at that magnification! However, if you increased the magnification, let's say by using a higher power wider field eyepiece, you would see more detail over a similarly sized field. In fact, as you continued to increase magnification so the exit pupil gets smaller and smaller, the subject will indeed be reduced in brightness, but the view will generally improve. Why? Because the sky background dims at the same rate while contrast is maintained between the background and subject. The larger size of the subject with maintained contrast gives the best view! You are limited in only two ways as you increase magnification: (1) You may run out of field, even with wide-angle eyepieces. (2) If the power is high enough to render the background so dark that the eyepiece field stop is almost invisible, more power will reduce contrast.

Stated as a conclusion in my article on magnification in the May 1991 issue of *Sky and Telescope*: *For best low-power viewing, use the highest power that properly frames the subject.* You will see the most detail and best contrast with the least contribution of eye defects.

7.6.4   Low Power Limits

While the example for a refractor shows that there is no arbitrary limit to how low you can go, that is not true for reflectors or SCTs that have central obscurations. If you had a 40% obscuration (SCTs range from 30% to 45%), an 11 mm pupil would produce a 4.4 mm diameter black spot in your pupil. Not too pretty for viewing comfort or light loss! Using the 55 mm Plössl, an 8-inch f/6.3 SCT with a 45% obscuration produces an 8.7 mm exit pupil with a 3.9 mm black spot right in the best part of your eyeball. So ironically, a fast SCT is not ideal for low power wide angle viewing. Choose an f/10 model with an f/6.3 focal reducer for the least obscuration. An even better choice would be an f/4.5 Dobsonian or Newtonian with the secondary obscuration limited to about 20%. Even though the pupil is 12 mm, the central obscuration is only 2.4 mm. Increase the power a bit to get a 7 to 10 mm pupil, and you have a spectacular wide field with no obscuration problems at all.

Remember though to consider a coma corrector such as the Paracorr, which increases the diffraction limited area of a parabola 36x by eliminating the coma that robs resolution outside the very center of the field. To me, no view is spectacular if the stars look like blobs.

7.6.5   "RFT"

The Rich Field Telescope is simply one that has sufficient field and aperture to provide exciting Milky Way and other rich field views.

I remember an article in the classic book *Amateur Telescope Making*[1] that excited me as a teenager about RFTs. (Can you imagine any teenager today getting

---

[1] Walkden, S. L. "The Richest-Field Telescope—a Plea for Low Magnification." *Amateur Telescope Making*—Book 2. A. Ingalls (ed.), Scientific American, Inc. 1968, pp. 623–647. Of special interest in this article are a letter and photo from Clyde Tombaugh in January 1935 describing the 5-inch f/4 reflector he built which gave a 2° field at 20x. Here's a partial quote: "The sights through this instrument are truly marvelous, especially at Flagstaff when the night sky is very transparent. The 'double dark hole' or dark nebulae in Sagittarius stands out beautifully, as it does on a moderate exposure photograph. Doubtless, the strong effect is partly due to the fact that the rich star-cloud wisp in its vicinity is resolved into stars in my 5-inch, and the absence of stars renders the dark hole conspicuous. The one faint star in this hole is easy to see. The Lagoon nebula in Sagittarius shows up well too. The dark lanes cutting through it are plainly seen. The Trifid nebula, on the other hand, is but a ghost. Since the Trifid shows up strongly on photographs, it is reasonable to infer that it is bluish, and the Lagoon nebula more yellowish. The real Sagittarius cloud, of course, is somewhat disappointing because it is too far away to be resolved.

Perhaps the most beautiful and richest star-fields to be found for an instrument of this size is in the Cygnus region. Much of this region is resolved into stars, and I have found some spots to run as high as 600 or 700 stars per field of view. It is a superb sight! Almost any place in this region will have 250 stars per field—the field of view being 2° in diameter. One of the astounding things is the way this type of instrument shows up the North America nebula in Cygnus. The object just about fits the field of view. The 'Florida' and 'Mexico' contrast well against the dark 'Gulf of Mexico.' The dark hole representing the 'Hudson Bay' shows up conspicuously. I have easily traced the Andromeda nebula for a full degree on either side of the nucleus. The galactic star clusters are beauties, but the globular star clusters are not so good—simply an amorphous nebulosity because their stars are too faint to be resolved."

excited about RFT instead of TV?) By analyzing how many stars of differing magnitudes are visible, it concluded (roughly) that 3-inch to 6-inch scopes could see more stars and more total star brightness in an average field than larger or smaller scopes. Of course, this was during an era when no sharp wide-angle eyepieces were available, and a 6-inch scope was considered large. However, anyone who has seen the Omega Centauri globular in a large telescope knows that many objects are stirring and inspiring beyond belief compared to viewing them in a small RFT.

As for which RFTs are best, of course I am slightly partial to fast APO refractors, particularly if they have flat-fields. Larger aperture $f/4$ to $f/5$ Dobsonians with small diagonals, 2-inch focusers, and a coma corrector are every bit as good if the mirror is well-made. In this case, do not confuse such RFTs, which can also give superb high power views, with $f/5$ doublets (comet seekers), which because of color limitations, are not very good at higher magnifications.

7.6.6   What to View

We all have our favorites. Here are some of mine in no particular order:

1. M24. This region is filled with stars of all brightnesses and a lovely embedded cluster. Choose a magnification that allows it to fill the field so that the sky background is darkened to enhance contrast. At a low enough power, M17 (the Omega) can be seen in the same field.

2. The Sagittarius star cloud. Under the dark skies at the Texas Star Party, the dark nebulae are just beautiful.

3. The Scutum star cloud with the magnificent M11 (Wild Duck) cluster at its side.

4. The Double-Cluster in Perseus is always a crowd pleaser, but with low enough power, you can see Stock 2, a large open cluster in the same field. I was pleased to open a few incredulous eyes to this field at a Japanese star party I attended a few years ago. Similarly, one of my Bermudan friends recently challenged me to show him something in my 4-inch $f/5.4$ refractor that he could not see in his 4-inch $f/10$. This view proved the point. The double cluster is also a wicked test for optical quality of eyepieces as well as telescopes, with brilliant stars at the edge of the field and little in the center.

5. Anywhere in the Milky Way, and especially in Cygnus.

6. Large open clusters. Aside from the usual Pleiades or Beehive, try the Hyades and the Coma Berenices region.

7. A special favorite of mine is the region around NGC 6231 in Scorpius. At very low power or with the naked eye, it looks like a comet, but at higher powers, it appears like a spray of diamonds as I have seen it follow the arc above the treetops at the Riverside convention.

8. Galaxies such as Andromeda, M33, and the M81–M82 pairing.

9. M8 and M20 in the same field.

10. The North American Nebula (you must use a nebula filter).

11. All three sections of the Veil in one view.

12. Go south. Far south. The southern sky is like a new universe to northerners. And what a feast. Eta Carinae dwarfs the Orion Nebula. The Magellanic Clouds, Coal Sack, Jewel Box (I'm nostalgic already), and of course, the entire Milky Way so bright it casts shadows on the Australian outback.

Well, I'm tired of writing. I'm going outside to view. Care to join me?

7.7 Letter to the Editor[2]

Dear Bill:

I would like to comment on Al Nagler's article "The Joys of Low Power Observing." As the owner of a 4-inch *f*/5.5 Tele Vue Renaissance, several other RFTs, and a number of wide angle and super wide-angle binoculars, I've done actual star counts per field and used Webb's *Atlas of the Stars* for checking—an uncluttered atlas that goes below 9th magnitude.

In his article, Nagler uses the example of an eyepiece/telescope combination he says will give the widest possible field with that eyepiece and that focal length.

Although a 55 mm eyepiece of 50° apparent field used on a 4-inch *f*/5.4 telescope will give 9.8x and a true field of 5.1° (Nagler uses 500 mm focal length and a 5.4° field of view at 9x for his example), thus giving the widest field possible with a 2-inch eyepiece of that focal length, it does not show the maximum number of stars per field. This is due to considerable reduction of the effective aperture. Suppose we substitute a 40 mm eyepiece with a 65° or 70° field of view? Such eyepieces are readily available. Assuming the 70° F. O. V., our 40 mm eyepiece used on a 540 mm focal length 4-inch gives 13.5x and a 5.2° field of view. In the Nagler example of 500 mm, it would give 12.5x and a 5.6° F.O.V. In using the 40 mm eyepiece, we obtain a 7.5 mm to 8.1 mm exit pupil (depending on which focal length we use), which allows us to employ almost all of the 102 mm aperture of the instrument.

The higher magnification darkens the sky background, and the field will now be much richer, since more of the faint background stars will be seen. A 7 mm or 8 mm exit pupil is the maximum younger people enjoy. However, suppose we are in middle life? As we age our maximum entrance pupil shrinks in size, but we still wish to use as much as possible of the effective aperture of our telescope. Let's substitute a well-known 32 mm, 2-inch eyepiece of 80° apparent field of view. Using a 540 mm focal length, as in our example, gives us 16.8x and a 4¾° true field. The exit pupil will be 102/16.8, or about 6.1 mm—which is about what a person in their late 40s or 50s will have for eye pupil size. Sky Publications sells an inex-

---

[2] Comments from Rodger Gordon concerning "The Joys of Low Power Observing" by Al Nagler.

	Model	Config.	Field of View
1	Edmund *Satellite* scope (late 1950s-early 1960s)	5-5x	12°
2	Unitron *Satellite* scope (1960s)	6x	12°
3	USN 6 x 42 (fairly easy to find)	6 x 42	12°
4	Jaegers *Panorama*	7 x 35	12.5°
5	WWII German "flak glass" (some coated)	10 x 80	6.5 7. or 8°
6	Zeiss *Deltar* 1937-1942 (some coated)	8 x 40	11.2°
7	Various 7 x 50 binoculars manufactured 1960s-1198s	7 x 50	10° x 11°
8	Bushnell *Rangemaster* 1951-1982	7 x 35	10° x 11°
9	Fujinon binocular, current	16 x 70	4°
10	High quality 20 x 80 binoculars, current	20 x 80	3.5°
11	Zeiss *Jenoptem*, some other 10 x 50s	10 x 50	6.5° to 7.5°

pensive device to check maximum eye pupil size. By selecting the right eyepiece for our pupil size, we can obtain the maximum number of stars per field without reducing the effective optical aperture.

Let us go a step further and use the expensive Leica 30 mm, 90° eyepiece. Now we get 18x, a 5° field of view, and an exit pupil of 5.7 mm—again in the range of what many older people have. In all of these cases, we are obtaining the entire effective aperture of the scope, or nearly so, and showing the maximum number of stars possible. In some cases a minimum of 1,500 stars per field can be expected in the densest regions of the Milky Way (Cygnus, etc.), and the count is likely to well exceed that in favored locations with dark skies!

Wide-angle binoculars also allow large numbers of stars to be seen. Some older (but rare) large binoculars will show even more stars, some 3,000-5,000 per field! Low-power, 6x and 7x, wide-angle binoculars will also show huge numbers of stars per field. Unfortunately, there are very few modern low-power 7x or 6x binoculars with 11° or 12° and larger fields, and the one or two that are available are of very poor quality or have very little eye relief. So we have to look to yester-year in our search.

The table above contains a number of instruments that can still be found that will provide outstanding wide-angle viewing of stellar fields.

The most stunning views of the Milky Way I have ever seen were with a homemade 32 x 125 binocular. The objectives were 5-inch *f*/5 Jaegers, two very large 90° prisms were used, with the erecting system and eyepiece assemblies from a Jaegers 7 x 35 Panorama binocular adapted to the system. The 90° apparent field eyepieces allowed a 2.8° true field. No less than 2,500 stars were jam-packed into that field!

I prefer wide-sky RFT-type views—such as Al Nagler advocates—to Deep Sky views. The latter yields us fields of view from less than 1° out to perhaps 1.5°. Due to the peculiar aspects of star distribution beyond 12th magnitude, the number of stars per field actually diminishes here, although fainter stars are seen with the larger aperture. No large "light buckets" with their narrow fields can come close to showing the number of stars per field as many low-power wide-angle instruments.

For those who have never experienced wide-angle viewing as described here, you are missing some of the most spectacular sights the heavens can produce!

Rodger W. Gordon

7.8 Telescopes For CCD Imaging, Part III
by Richard Berry

CCDs are revolutionizing amateur astronomy because, with homebuilt telescopes (and homebuilt CCD cameras, too!), they allow us to reach sky objects long considered beyond the realm of the amateur. However, CCDs place unprecedented optical and mechanical demands on telescopes. Reflectors must be light tight for CCD imaging, refractors must offer superior color correction, and all types of telescopes must form clean, tight images to produce satisfactory images with the new generation of CCD chips with very small pixels.

However, long before new CCD users run into optical difficulties, they discover that they must deal with the tracking demands of CCDs. Both commercial and homebuilt mounts and clock drives have, for many years, been designed primarily to satisfy the needs of visual observers. If a drive could hold a celestial object in 2 or 3 minutes of arc in the center of a high-power field of view for half an hour, the drive was considered entirely adequate. Astrophotography was possible with drives of this quality with careful, attentive, and continuous guiding; and it is a tribute to their skill and patience that an entire generation of astrophotographers did as well as they did.

With CCDs, the tracking times are shorter—typically just 2 to 4 minutes—but the standards of tracking accuracy are also higher. The first experience many new CCD users have is getting trailed and smeared images whenever exposures run longer than 15 to 20 seconds. Getting round, sharp star images with a 10-micron CCD chip on a telescope having a focal length of 40 inches requires tracking to somewhat better than 2 seconds of arc during the whole integration time. The drive errors in homebuilt telescopes—telescopes well suited for visual observing—are often 10 times, and sometimes as much as 100 times, larger than required for sharp, round images. The performance of commercial drive units, especially those on Halley-era telescopes, is in the same ballpark. Yet better tracking—and great CCD images—is often possible with a clock-drive tune-up.

The first step in improving any drive system is to characterize and quantify the drive errors. Fortunately, CCDs make this process quite easy. What you must do is to set up the telescope, polar align as accurately as possible, and then take a series of short exposures extending over several turns of the slowest element in the drive train. With worm-gear drives, if the worm turns once every 4 minutes, then you should record images for 15 to 20 minutes. Rotate the camera so that RA lies along the left-right axis of the image and declination is up-down. Select a field near the equator and on the meridian, and adjust the drive rate so the telescope follows a star without drifting ahead or behind significantly.

With the Cookbook camera, you can do this automatically in multiple exposure mode. Program the camera to take 10-second integrations at 15-second intervals, which will give you around 60 images. Measure the x,y position of a star in each image, and then plot the position against time in each coordinate. You should get a graph showing the motion of the star image.

Next, quantify the plot. If you are taking images with a TC-211 CCD (such as the Cookbook 211, Lynx-PC, ST-4, and Electrim ED-1000 have), each pixel is 16 micrometers wide. For example, given that the telescope has a focal length of 1,200 millimeters, then one pixel corresponds to $^{0.016}/_{1200}$ = 13.3 microradians or 2.7 seconds of arc. If the RA graph shows a clear sinusoidal variation of 11 pixels peak-to-peak, then the drive has a periodic error of 29 arcseconds, which is fairly typical of a good-quality drive made for visual observing. (Of course, if the error curve shows nothing but very small errors, do not change anything. Count yourself lucky, and get on with your imaging!)

In CCD imaging, however, if a star drifts by more than half a pixel during the integration time, star images may appear elongated. In this example, the drive error is roughly twenty times larger than desirable. What to do next?

The first step is to examine the whole drive system looking for loose, misaligned, or worn components. To function well, drives must be tight and square—after all, on an 8-inch diameter worm wheel, a one-pixel drive error corresponds to 0.00004 inches of loose or misaligned metal. Check that the bearings are tight. Many of the older commercial German equatorials used Teflon or Nylon bushings on the polar shaft, and these materials wear down. They can be replaced with new bushings. Is the worm square on the worm wheel or tilted? (Tilts cause periodic error.) Disengage the drive motor and turn the worm shaft by hand. If you detect a significant variation in friction as you turn the shaft, you are feeling the root cause of the periodic error. A bit of dry molybdenum disulfide or graphite brushed on the worm and wheel can make a big difference.

Once you have eliminated any obvious mechanical faults in the drive system, set up the telescope and repeat the test. You will probably see some improvement and perhaps a lot of improvement. Quite often the component parts of the drive are more precise than their assembly; after they have been carefully reassembled, the same parts give considerably better performance. In commercial units, there is a reasonable chance that you are a more skilled mechanic than the original assembler. A missing or misadjusted tension screw, for example, might pass unnoticed during years of visual observing only to show up during CCD imaging, so check out every detail of the drive and set it right.

After the second test, reassess. If you can live with the residual drive errors, forget about further testing and get on with some real imaging. If the drive errors have persisted, you need to isolate the cause of the problem and cure it. At the level of accuracy required for good CCD tracking, every drive is unique; you are on your own. However, I'll tell you how I tweaked up my Byers 812 so that I can run my standard 4-minute integration with a Cookbook 211 or Cookbook 245 and get nice round star images almost every time.

I had used my Byers 812 (manufactured to carry 8- to 12-inch telescopes) with a 6-inch $f/5$ Newtonian telescope for a lot of conventional guided film-type astrophotography. (By the way, it always helps to use what might normally be considered an "oversize" mounting. The drive hardly feels the weight of the telescope.) The drive would glitch occasionally, but I just guided it out. When I started shooting CCD images, the mounting had sat in an observatory little used for several years, and the mechanical parts had a few isolated patches of corrosion but nothing big. After getting the drive rate zeroed in, I found I could shoot perfect 5-minute images about two-thirds of the time, but the remaining third of the time the images were badly smeared. I ran a 20-minute sequence of images that showed very little periodic error with odd "spikes" in the tracking graph.

The spikes meant that the drive appeared to slow almost to a halt for about ten seconds and then bounce back. The excursions did not seem to show a regular pattern. After studying the drive in action, I finally realized what was going on. The Byers 812 had a 30° sector gear that was fairly thin perpendicular to its plane of rotation. As the worm turned, therefore, the gear bent slightly, but so long as the coefficient of friction between the worm and wheel remained constant, the wheel turned at a uniform rate.

However, when a patch of dirt or corrosion came between the metal surfaces, the section flexed, and rotation of the polar axis would stop until the patch moved past the point of contact. Stars would trail on the CCD. Then the wheel would spring back and catch up to the moving sky. I put a small amount of engine assembly grease with molybdenum disulfide on the gear, and it eliminated the glitch. More recently, I obtained a new sector from Ed Byers and will install it before putting the drive back into service.

7.9   From the Bench
Binocular Collimation, Part III
By William J. Cook

In the first installment of this article, we discussed the difference between collimation and the ever-present "conditional alignment". In Part II, we described a number of the most common conventions used in the collimation process.

In this issue we wish to explain just what is required to achieve true collimation.

For many of us who have worked with binoculars every day for years, explaining the details of the process is not at all easy. We do not think about it; we just do it. Each time we sit down at the collimator, our biological computer automatically routes our thought process to the directory which was triggered by the brand name, model number, or buzzwords we see before us. Then, coming into the picture is the current condition of the instrument along with its age and the potential difficulty we might face in trying to obtain or manufacture parts.

For this reason, we will simply present the collimation instructions found in the Navy's rate training manual *Opticalman 3 & 2* (for third and second petty officers in the optical repair specialty).

The following information has been presented directly from *OM 3 & 2,* and no attempt has been made to alter the text to address the dozens of design and collimation conventions offered by the hundreds of products currently on the market.

7.9.1 Collimation

Prism cluster assembly is a very important part of binocular collimation. If this is not done properly, you may not be able to collimate your binoculars without disassembling one or both clusters to find the problem.

The assembled binocular (except for objective seals, lock-rings, and objective caps) is mounted on an Mk 5 collimator, as shown in Figure 7.9.1. Notice that the binocular is mounted upside down, and the right hinge lug is clamped in the fixture so the left body is free to move.

Using an auxiliary telescope with rhomboid attachment (Figure 7.9.2), focus both eyepieces on the collimator target and set the focusing rings to zero diopters (within ¼ diopter of each other). At this time, the eyepieces should be of equal height within ¹⁄₁₆ inch. (Check with a straightedge.) Also check for lean by comparing the magnified collimator target, seen through the eyepieces, with the smaller collimator target picked up by the rhomboid attachment. Correct any problems noted before proceeding.

To align both lines of sight with the hinge axis, use the tall-of-arc method, explained next.

1. Swing the left body all the way up to show approximately 58 mm on the IPD scale. Using the auxiliary telescope, adjust the screws on the collimator fixture to align the normal and magnified target image. See Figure 7.9.3a.

2. Now swing the left body down to obtain the widest separation between eyepieces (74 mm). What you see through the auxiliary telescope should be similar to Figure 7.9.3b. The smaller target (A) is a stationary reference picked up by the rhomboid. The magnified target (B), seen through the left barrel, shows how far the line of sight deviates from parallelism with the hinge. *Note:* To make collimation easier, the objective eccentric rings should be set for minimum displacement when assembled.

3. To adjust the line of sight, first, construct an imaginary equilateral triangle shown by points (A), (B), and (C) in Figure 7.9.3b. Point C must always be figured in a clockwise direction from point B, regardless of where the magnified crossline ends up after swinging the left body. *Note:* The small crossline may appear to move.

Fig. 7.9.1 *Binocular mounted on an Mk 5 collimator U.S. Navy illustration.*

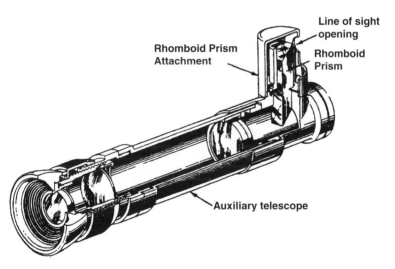

Fig. 7.9.2 *Auxiliary telescope Mk 1 with rhomboid prism attachment-cutaway. U.S. Navy illustration.*

4. With the left telescope still at 74 mm IPD, manipulate the objective eccentrics to place the magnified target image in the area of imaginary point (C). Try to make this adjustment without disturbing the binocular on the collimator fixture. *Note:* First, rotate the entire objective assembly (inner and outer eccentric). If this does not move the magnified target to point (C), throw some eccentricity into the objective and rotate the complete assembly again.

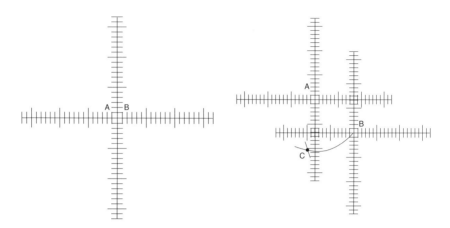

Fig. 7.9.3 *a. Collimator crosslines superimposed at 58 mm interpupillary distance. b. Preliminary step in binocular tail-of-arc collimation.*

5. After you are satisfied that the crossline is near point (C), swing the left body back to 54 mm IPD and adjust the collimator fixture to superimpose the two crosslines (Figure 7.9.3a). Repeat steps 2 through 5 until there is no displacement of the magnified crossline when you swing the left barrel from 54 mm to 74 mm.

    If you find there is not enough eccentricity in the left objective to satisfy collimation requirements, you will have to shift prisms. After shifting prisms, check zero diopters, equal eyepiece height, and lean again before re-collimating.

6. Once the left body is collimated, swing that body down to approximately 64 mm, realign the two crosslines by adjusting the collimator fixture, and adjust the right objective to superimpose the two crosslines seen through the auxiliary telescope. Recheck the left and right bodies to be sure you have perfectly superimposed the crosslines. *Note:* if you cannot superimpose the line of sight with the right objective eccentrics, shift prisms and proceed as indicated in step 5 above.

7. Tighten the eccentric lock screws, replace gaskets and rings, replace and lock the objective lockring, and replace the objective caps. *Note:* Tightening the eccentric lock screws and the objective lock rings may throw collimation off. Recheck and adjust as necessary.

Collimation tolerance for both lines of sight in a binocular are specified as (a) 2' step (vertical displacement), (b) 4' divergence (outward separation), and (c) 2' convergence. These tolerances represent government performance standards for binoculars. You will not find an optical shop supervisor who will accept these sloppy tolerances. If your collimation is not perfect, it is not good enough.

Complete overhaul information for Navy hand-held binoculars is found in

NAVSHIPS 250-624-2.

At this point, we must warn you about differences in imported binoculars. U.S. military binoculars are the best in the world, and they cost about $350 each. If you pay $35 to $120 for an import, the cost is comparable to quality. Optically, imported prismatic binoculars are excellent, but collimation is difficult due to loose tolerances, no eccentrics on some models, and sloppy prism mounting. In fact, prisms are glued in some cheaper models. Power is also excessive, sometimes as high as 20x in a hand-held instrument.

Another cheap shortcut found in many imports is the use of a center focusing arrangement, rather than individual focusing eyepieces. In this system, a large focus knob mounted on top of the hinge drives a multiple lead shaft up or down inside the hinge. Attached to the top of the shaft by another hinge are the two eyepieces which slide up or down on tubes screwed into the top cover plates. To account for differences in diopter setting between the left and right eye, the right eyepiece is focused in the normal manner.

The major shortcoming of center focus binoculars is that close fitting of sliding components is not possible if the focusing mechanism is going to work at all. The loose fitting eyepieces, combined with high magnification and sloppy prism mounting, will be the ultimate test of your skill and patience if you ever attempt to repair such binoculars.

Reflecting on the above information, one might conclude that the information is aimed at only one variety of binocular. If this is your conclusion . . . you're right. Even so, there is a point to be made. The point is that collimating binoculars (even to equal or surpass the most recent military specifications is not beyond the ability of those who possess the right amount of fingers and toes and who have the patience to think things out.)

Furthermore, the brevity of the explanation also serves to illustrate that there is much more to deal with when considering design conventions than in the actual collimation process.

7.9.2   The Basics:

1. True the prisms; i.e., be sure they are at 90° to each other.
2. Install the objective lenses with their eccentrics neutralized. (see Figure 6.9.1b).
3. Align the "swinging barrel" to the axle.
4. Align the "stationary barrel" to the first.

If these steps are accomplished, the binocular will be properly aligned for observers from six years of age to 6 ft 6 in height.

7.9.3   Helpful Hints:

1. Many times collimation on Zeiss style binoculars (or binoculars with a two piece body) has been lost due an impact which caused the objective

housing of one of the telescopes to be jarred out of alignment. Quite often this can be remedied by simply unscrewing the offending piece, cleaning the threads, and reinstalling it. This may be difficult because the threads have been stripped. However, if the housing is seen to be undoubtedly askew, you are going to have to perform this operation anyway.

2. With most collimation errors, it will not be possible to tell which telescope is causing the problem.

    Looking backwards through a Porro prism binocular, one can see a series of rings made up of stops at the prisms and in the eyelens. If a collimation problem is severe, one may look backwards through the instrument to see which side has a series of apertures that are not concentric.

    Doing this may negate the need to align two telescopes instead of one.

3. While the information presented here might inspire some of our more energetic optical tinkerers to build a collimator and auxiliary telescope, the fact remains that most people are not going to have these handy little tools lying around the house. Those individuals resort to conditional alignment by performing the following procedure.

First, determine which of the telescopes is out of alignment (if possible). Then, determine which method of alignment is necessary for the instrument in question. Next, find a small, high-contrast target—such as a street lamp a mile or more away. Then with the binocular correctly focused, move it slowly away from your face and try to keep the target centered in the ever-shrinking exit pupil.

As you move the bino away from your face (to perhaps 8 to 10 inches), you will see the image appear to separate. (In the worst of cases, the alignment will be so far afield that one image will leave the field of view with the binocular no more than 2 or 3 inches from your face.)

Now, alter a single adjusting screw (or eccentric ring) by about a quarter turn and have another look. If the images do not separate as badly as the instrument is moved away from your face, you are on the right track.

Repeat this process until you can hold the bino a foot or more away from your face and still have the target visible in both eyepieces. When this has been accomplished, you should have no further problems with double images. Of course, without the use of an auxiliary telescope and collimator, the test results will be totally subjective. However, just as a proponent of the Ronchi test might say in discussing the value of that test versus the Foucault test: "When one finds himself in water over his head, he doesn't need to know the depth of the water. He needs to know how to swim."

Finally, when this form of conditional alignment is attempted, it is imperative that you stare at the entire field of view and not at the object. The brain **wants** to see one image and will try to bring the two images together whenever possible. This can give the viewer a false sense of accomplishment that can be swiftly washed away when the brain stops trying to do you a favor and sends you either a double image or a headache.

Also, if you have followed the various steps noted above, you should not be overly critical from one observing session to another. *If you look for a mis-alignment . . . you'll find it.* Some amateur astronomers and bird watchers profess that one can determine whether or not an instrument is collimated (or conditionally aligned) by whether or not you can see an elongated field of view (thus indicating that the two fields are slightly overlapping.) This is true, but can also lead you to make some serious errors in judgment.

If, while looking through a perfectly collimated instrument, you look for the field of view to be elongated, you will be bringing your attention inside the instrument instead of being on a distant object, and as a result, you will cross your eyes, thus putting yourself out of collimation.

Good Luck!

7.10 Table Mountain Star Party 1994

By William J. Cook

For several years, Table Mountain near Ellensburg, Washington, has been "the" place to go for individuals and small groups looking to escape the light pollution of urban areas and to lessen the atmospheric blanket by a mile or so. At an altitude of over 6,300 ft, and under relatively dark skies, one can enjoy observing from high rolling meadows punctuated by stands of 100 foot high evergreen trees.

As an added attraction, those visiting Table Mountain for the first time are undoubtedly pleased with the quality of the roads leading to the summit. Unlike some other popular observing sites across the country—for which a half-track is the preferred means of travel and a 5-minute rain can lead to spending an extra day at the site—the long and winding road leading to the top of Table Mountain is nicely paved for all except the last mile or so. Even then, the remaining road is hard packed dirt and gravel.

In recent years, this location has gone beyond offering a user-friendly getaway for the amateur astronomers of Washington state and has become the site of the Table Mountain Star Party. This gathering, which started as a cooperative effort between the Seattle Astronomical Society and the Spokane Astronomical Society, has been growing by leaps and bounds, with the Rose City Astronomers becoming major players and with participants and vendors coming hundreds of miles to take part. This year, organizers were anticipating about 500 attendees, and total attendance was about 20 persons shy of that figure.

Enjoyment of Table Mountain activities is often left in part to the equivalent of meteorological Russian Roulette because the weather up there can be quite unpredictable. Even though the event is held in the middle of summer, one might encounter a bit of snow. Last year, I left Seattle with a couple of boxes of ATM Journal T-shirts and hopes that I might be able to sell them to ATMA milling about the summit. I was still hopeful as I turned off the freeway just west of El-

Fig. 7.10.1

Fig. 7.10.2 *Kreig McBride's 8-inch refractor was a big hit with this year's Table Mountain crowd. The scope, riding atop an all-wooden tripod fashioned after the description in Richard Berry's* Build Your Own Telescope, *was especially popular as word reached us that Comet Shoemaker-Levy had begun its attack on Jupiter. Kreig is the founder and organizer of the optics/telescope-making workshop held each fall in Bellingham, Washington.*

Fig. 7.10.3 *Standing as another tribute to the wonders of Boeing Surplus, Paul Ham's 4.25-inch Dobsonian features a polished (rotating) titanium tube and single curved-vane secondary holder. With its long (f/10) focal ratio and tiny secondary mirror, Paul's first attempt at telescope making has netted him an instrument of apochromatic proportions. Details on the construction of this instrument will be featured in* ATM Journal #8.

lensburg. However, shortly after beginning my negotiations of the hair-pin turns, my hopes began to grow quite thin, and as I pulled into the observing site, I was greeted by a man in a parka and a watch cap, with little gray clouds being formed by his breath.

This year, with the weather cooperating nicely, I parked in a location near the registration tent (so that I would be highly visible while trying to drum up ATMA memberships) and headed off for the telescope fields. About the only restrictions one will find on where to place his telescope are created by the trees and one's desire for privacy; yet there is one field which one might look at as the main attraction. Here, on approximately 2 acres of meadow, you will find the most unusual telescopes being presented by some of the most seasoned telescope makers. There are a number of smaller "fields" each separated by fences, roads, tents or vehicles. And while amid these areas one can find any number of high quality and unique instruments, they seem to be the domain for the more timid telescope enthusiasts—those who perhaps do not want their commercially made 60 mm refractors stomped in the darkness by those on their way to one of the big Dobs, or who just came to the event to enjoy the heavens without getting caught up in all the telescope hoopla.

While the event was billed as a "star party," it was plain to see that the location was gaining popularity with the telescope-making crowd as well. Until Steve Swayze arrived from Portland with his 40-inch $f/5$ Dobsonian, the main telescope field was dominated by Kreig McBride's 8-inch refractor. This beast, with its bright yellow and green paint job (which looked like something one could do with a John Deere tractor and a magic wand) never had a shortage of those waiting to enjoy its images. As you might expect, this interest was heightened as word arrived via the radio that comet Shoemaker-Levy had begun its attack on Jupiter.

After getting a look at all the instruments in the main field, my "glass thumb" began to twitch, and I realized that I had obviously missed something important. With that thought haunting me, I set out to prove my instincts wrong or come up with something new and nifty to stare at.

About 40 yards distant, I came across a telescope that captured my interest and later the interest of the telescope judges as well. The instrument was an all aluminum 6-inch $f/5$ Newtonian by Canadian Gary Wolanski and sported many interesting features, which are described in Section 7.12 on page 280.

Another interesting telescope was presented by former Seattle Astronomical Society President, Paul Ham. This instrument, (a 4.25-inch $f/10$ "baby Dob"), which was the first Paul had ever made, featured a simple but elegant design. Dobsonian in origin, the telescope featured a highly polished (rotating) titanium tube and a single vane curved secondary holder to reduce the diffraction spikes cause by conventional 4-vane spiders.

In *ATM Journal*'s first call for papers, I made the mistake of saying that without innovations there was not much we could say about Dobsonian telescopes that had not already been said—a comment which apparently offended a couple of folks who apparently overlooked the word "innovations" and did not realize

Fig. 7.10.4 *Bruce Kelly's 17.5-inch Dobsonian featured electronic focusing and collimation.*

Fig. 7.10.5 *(Right) Ken Ward of Snohomish, Washington received a Merit Award for his 17.5-inch Dobsonian. Ken makes virtually all of the mechanical parts himself and finishes most of his instruments with high-gloss, auto-body finish. This instrument can be broken down into pieces small enough to fit neatly into the trunk of a mid-size car.*

that my telescope of choice is a 10-inch $f/6$ Dob. Bruce Kelly, a member of the Seattle Astronomical Society, seemed to understand quite well; and this year presented a 17.5-inch Dobsonian with some major innovations, not the least of which were remote controlled collimation and electronic focuser.

Ken Ward of Snohomish, Washington, received a Merit Award in the "Professional" category for his 17.5-inch Dobsonian in turquoise and blue. I first noticed Ken's work last year in the form of a beautifully executed 10-inch $f/6$ Newtonian. With experience in auto-finishing, Ken is producing some exceptionally good-looking telescopes.

Outside of the opportunity to look at and through so many interesting telescopes and to have a bit of a getaway with my two sons (Bill and Sean), the high point of my stay on Table Mountain was the opportunity to talk with Richard Berry who had been invited to speak on CCD technology. We had tried to get together at Riverside; however, he was in such demand that I was rarely in shouting range. With things being considerably more low-key at Table Mountain, we had enough time to discuss each other's work at length and get better acquainted.

I was very impressed with the number of telescopes entered in this year's

Merit Award competition and asked a few telescope makers to offer manuscripts for publication in the *Journal*. So far, we have received four manuscripts. These include an article on Howard Banich's "Springtonian" and Paul Ham's award-winning Dobsonian. I look forward to receiving other manuscripts and to Table Mountain '95.

7.11 Merit Awards Table Mountain Star Party 1994

Dobsonian Telescope

- Bruce Kelly (17.5-inch purple and blue Dobsonian w/electronic collimation).
- Karl Schroeder (10-inch Dobsonian system, including telescope, observer's chair, cartop telescope carrying tubes, and chart stand).
- Steve Swayze (40-inch Dobsonian—need we say more?).

Workmanship - Amateur

- Robert Wilson (17.5-inch blue honeycomb fiberglass Dobsonian—150# total weight—and eyepiece and filter case).
- Howard Banich (motor-driven equatorial platform for 20-inch Obsession).

Workmanship - Professional

- Ken Ward (17.5-inch turquoise and silver fiberglass Dobsonian).
- Charles Jones (12.5-inch machined photographic German equatorial mount).

Optical Excellence - Amateur

- Hulan Fleming (4.25-inch $f/23$ Schiefspiegler).
- Vladimir Steblina (8-inch $f/7$ mirror).
- Howard Banich (8-inch $f/4$ mirror).

Best First Scope

- Paul Ham (4.25-inch baby Dob).

Best Use of Common Materials

- Sam Johnson (4.25-inch Kutter Schiefspiegler on bowling ball mount).

Innovative Design

- Howard Banich (8-inch $f/4$ "Springsonian"—A Springfield Dobsonian combination mount).

Judges' Choice

- Gary Wolanski (Beautifully machined 6-inch Newtonian equatorial mount).

Many thanks to Jim Girard for providing the above list of Merit Award Winners.

7.12 King of The Hill

By Gary Wolanski

The plan to build a light and more portable mount came soon after my first successes in astrophotography. I wanted it to maintain the same rigidity of my heavy-duty mount but in a lighter package. It also had to be very user friendly; no nuts, bolts, or wrenches that can get lost in the dark.

In January of 1993 work began on the equatorial head. It was entirely designed around a 6 x 30 finder scope that was to slide into the polar shaft when the project was finished. This immediately set up the starting dimensions for the project. The head, with the exception of the thread-on declination shaft and tapered bearings, was totally made of aluminum. The right ascension and declination drive use gear reduction DC motors that are voltage regulated. A small hand controller is used for the dual axis correction. The six-penlight battery pack lasts approximately 70 hours. The total weight with battery pack is 14 lbs., which is not much considering it can easily support three times its own weight. The tripod has three 3¼-inch diameter legs that slide into a triangular base. A threaded shaft through a wedge plate forces the legs into tension, while securing the head which rotates 360° for polar alignment. It is also made completely of aluminum and weighs a mere 12 lbs.

The scope that rides on this mount is a 6-inch $f/5$ Newtonian. It has a telescoping primary mirror, used when experimenting with different optical devices; a three vane spider, which lowers diffraction spikes in astrophotos; and a quick-release spring-loaded rotating tube assembly. The scope does not present a challenge to the mount, but this makes taking photos a pleasure. I worked approximately 10 hours a week, and in March of 1993 the project was completed. The only problem I encountered was the threading of the right ascension gear; but thanks to a friend suggesting I use a spiral tap, the problem was solved.

I guess I always wanted to go against some of the old standards, such as solid steel shafts, 60 Hz drive motors, castings, non-spring loaded worm drives, etc., just to see whether it would work. It seems my suspicions have now been verified; the mount performed beautifully. The total cost for the materials was $260.00, which sounds good, so let's not mention the labor. My next project is to build an autoguider using the old quad photoelectric diode design.

Fig. 7.12.1

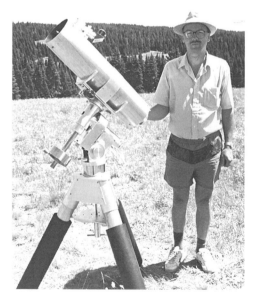

Fig. 7.12.2 *No, Gary didn't think up "king of the hill." He didn't have to. When the dust had settled and the judges had to decide which instrument would receive the "Judges Choice" award (based on which instrument on the mountain each would like to take home with him), Gary's 6-inch f/5 was the instant and unanimous winner.*

Fig. 7.12.3 *Gary's mirror mount allows the mirror to have a great deal of movement along the optical axis. This allows the instrument to accommodate any number of accessories without having to alter the focuser profile a great deal.*

Fig. 7.12.4 *(Left) Here we see still more detail in Gary Wolanski's craftsmanship. He machined virtually every piece of this instrument, including the gears. (Below) The off-axis guider.*

Fig. 7.12.5 *This photo can attest to the rigidity of the mount and the smoothness of the instrument's operation.*

Issue 8

8.1 The Shelley Collimating Aid
By John Shelley

After transporting a large Newtonian to a viewing site and setting up in near darkness, it is wise to check its collimation. The easier this job is made, the better; especially when you cannot count on others for help.

The collimating aid described below is very convenient because of its self-contained light source. It is composed of a common sight tube with the addition of an illuminated target. This target is a tiny LED (light emitting diode) mounted on the side of the tube, along with a battery and switch. The LED is seen only after its image makes the rounds through the optical path. Then it is viewed through the un-silvered mirror that centers it on the axis of the tube.

First, the device is used to frame the diagonal by sliding it in and out of the focusing mount until the diagonal exactly fits the inside edge of the collimator's tube end. Most deviations from proper diagonal alignment will be obvious immediately.

The scope should have a prominent center mark on the primary mirror and an offset center mark on the diagonal. With these present, a succession of images should be seen through the peep-hole as follows: Starting with the mark on the diagonal at the center of the field of view, one will next see the mark on the primary followed by an outline of the secondary and the inner end of the collimation aid. Lighting the LED will present a tiny dot of light in the otherwise dark tube, and this should be brought into coincidence with the primary's center mark by adjusting the primary. When all images are brought into alignment, star checks may be performed.

8.1.1 Construction

Refer to Figure 8.1.1 on page 284 for details on the parts. A chrome-plated, brass tailpipe made for a bathroom sink is cut off 7 inches from the small end. Next a peep-hole end cap is made from plastic stock, preferably opaque. This part can be greatly refined and fabrication simplified if an arbor is made upon which to turn it. The plastic can be drilled using it as a guide; and then, with the two firmly screwed together, the plastic can be turned to 1¼-inch O.D. with a lathe or an electric drill. Once all prominent corners are sawed off, a file can be used to bring it down to proper size and to shape the forward half to fit firmly into the tube.

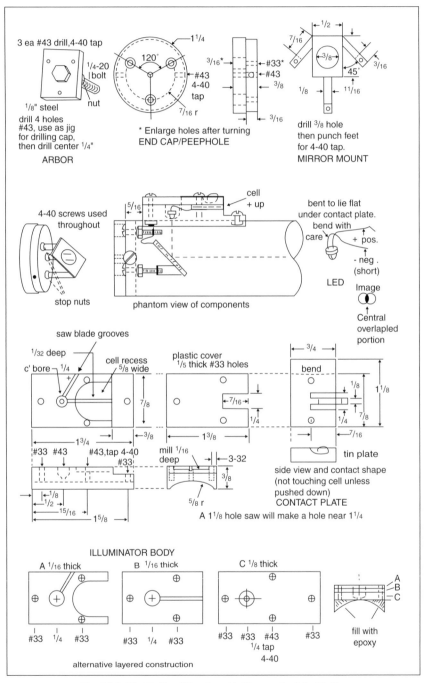

Fig. 8.1.1 *The Shelly collimation aid. Diagram by John Shelley.*

Similar plastic stock is used to make the illuminator; the contact parts are made from thin sheet metal. The original device was made in a well-equipped workshop; but in retrospect, an alternative plan of construction using layers of parts is offered for those less well equipped.

The final part is a three-legged support for mounting and adjusting the mirror. Metal from a tin can may be used because of its availability, and it is easily cut and worked.

When the components are finished, the main tube is marked and drilled for the mounting holes and the LED aperture. The mirror is a section of microscope slide glued to the support with three dots of contact cement. This assembly is mounted on the three adjusting screws that are retained in the cap with stop-nuts.

After all parts are attached, the mirror is adjusted to align the LED image with the optical axis. A temporary peephole disk is needed at the inner end to guide the eye during this operation.

8.1.2   Parts

> lithium cell: Radio Shack 23-167
> LED: Radio Shack assortment 276-1622

8.2   Winter Star Party 1995

By Diane Lucas

Attending the Winter Star Party in the Florida Keys is a rare treat for someone who has only lived in upstate New York and northern Ohio. Just the thought of lots of sun and warm weather, to say nothing of the chance to observe below the 25th latitude, is incentive enough to leave Ohio in February. Driving the 1300 miles to the Keys is a great introduction since extra jackets and sweaters can be peeled off on the way, until the final arrival in shirtsleeves with open car window.

It is impossible to spend enough time looking at all the nifty telescopes during the daytime, observe all night, and meet and talk to all those great observing and telescope-making folk. So the following is a sampling of what I saw in the line of telescopes from Thursday afternoon to Sunday morning.

Since I am currently making the lens for a similar telescope, I headed instantly for one that really does not look like a telescope at all. Gayle Riggsbee's hydrogen alpha solar telescope looks more like a prop for the weary amateur. Gayle is from Atlanta, Georgia, and has spent a lot of time perfecting this telescope. It uses a 70 mm $f/30$ lens made from a red energy-rejecting filter to minimize the light falling on the Daystar hydrogen alpha filter. The image is directed to the eye through a diagonal, a polarizing filter, and an eyepiece. This eyepiece is at a very convenient position for sit-down observing. The solar image can be fine-positioned using the four red buttons on top of the telescope tube near the eyepiece, and the solar image can be fine-tuned using the hydrogen filter tilt control and polarizing filter rotation.

Fig. 8.2.1 *Bob Ward's C-14-locked, loaded, and ready for an evening of CCD work.*

Fig. 8.2.2 *Close-up of the head assembly for the telescope in Figure 8.2.1.*

Fig. 8.2.3 *Roland Christen's FastMax 235 Maksutov-Newtonian.*

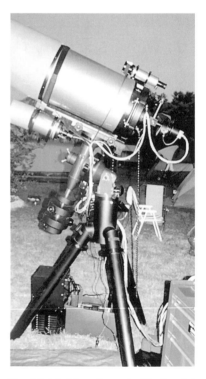

Fig. 8.2.4 *I did not learn the identity of the owner of this instrument but was quite impressed by the machine work and the rigidity of the mount.*

Fig. 8.2.6 *This large aperture Dobsonian rests atop a rock-er-"box" executed in tubing. While perhaps not as elegant as some of its highly polished and inlaid wooden cousins, any observer who has spent most of an evening jury-rigging bearing surfaces because his mount was not sealed quite as well against moisture as he had thought since the last time he was caught in a drizzle, can appreciate the real beauty here.*

Fig. 8.2.5 *A 12½-inch Newtonian by Charles Riddel of Light Speed Telescopes.*

One of the vendors was Jerry Armstrong, who was selling some great paintings of stellar objects. He also had a computer there and was showing CCD images made using a 16-inch *f*/6 in Winston, Georgia. I brought a Mars painting home with me.

Bob Ward, who is the chairman of the WSP, had a nice setup for doing CCD photography using his 14-inch Celestron. Many of the parts, tools, and equipment are carried to the observing site using a large red Sears tool cabinet which also acts as an observing stand.

There was a large crowd of nighttime observers using CCD cameras including several of the Cookbook cameras made by their owners. Richard Berry gave an afternoon talk concerning construction details of this camera and showed a shot of Mars made with Don Parker's 16-inch telescope and Richard's Cookbook Camera. The Mars shot just whetted our appetite for the next talk on Mars, given by Don Parker. ALPO and Don are truly doing professional work during Mars' closest approaches to earth.

With the need for solid vibration-free mounts to permit CCD imaging, some mounts are truly impressive. I took a couple of shots of a nifty aluminum pier and equatorial mount supporting a pair of refractors that I would like to have in my backyard.

If you had deep pockets and wanted to buy your telescope, Roland Christen was there with an Astrophysics Newtonian-Maksutov that was beautifully made and ideal for wide-field observing or photography. It was a great looking, and easy to use and maintain telescope.

If you would rather have a Newtonian that is engineered with much attention to detail and performance, Charles Riddel of Light Speed Telescopes, Inc. brought a beautiful 12.5-inch high precision telescope.

In the middle of all the techno-whizzy telescopes, there were still many that have no motors. Many of these were the big Dobs, made or bought by avid deep-sky observers. One of my photos was a big Dob with a unique welded aluminum construction for the box. Tom Clark was there with a new 160 lb, 24-inch in which a great deal of care had been taken to reduce the weight, yet retain the ease of operation. He had also modified the mirror portion of the 36-inch Yard Scope.

Another big telescope with an equatorial mounting was Cliff Gosney's 18-inch f/4.5 Newtonian "Knobsonian" telescope, which had a truly alarming number of round black knobs used for assembly and adjustment. A clean looking design, I regret that I was unable to make it back and have a look through it at night.

8.3   A Short History of Telescope Testing
By Peter Abrahams

Telescope making techniques have evolved since Galileo, but the problems that he faced remain to challenge telescope makers of today. The literature on historical telescopes includes a meager supply of information on how they were made. This article is the first of a series on this subject, which will include both the techniques used by early makers and the testing of antique instruments using modern optical techniques.

At the 1994 Convention of the Antique Telescope Society, a brief presentation addressed some of the historical techniques used to test early telescope optics. This amplification is far from definitive but will highlight some of the landmarks in the history of optical testing. In *The History of the Telescope*, Henry King notes that the telescope makers of the seventeenth century observed a page of printed type set at a distance. This technique is still useful for a quick indoor test for coma in a close-focusing instrument. In the 1600s, the periodical *Philosophical Transactions* was favored for this test because many other books were so poorly printed, that the aberrations seen might be from the print and not from the lens. Star tests were also used, but their full meaning was not clear. Even the eminent Johannes Hevelius thought that the spurious disks produced by stars in his 150 ft telescope were measurable stellar diameters, and he used a micrometer to quantify his instrumental error. A future installment of this series could discuss the many varieties of false test results, including simple mismeasurement, erroneous interpretation, misapplied test procedures, inadequate test equipment, and the fa-

Fig. 8.3.1 *William Lassell's 48-inch on the island of Malta. (Woodcut from Royal Astronomical Society, circa 1867.)*

mous 1990 "failure to test as an assembled instrument."

John Mudge was a mid-eighteenth century English ATM who might have been the first to use zone tests on mirrors. He observed a distant object using high power, masking the edge and then the center of his speculum. If the object stayed in focus at all positions of the mask, the mirror was judged properly corrected. Also in England during the 1700s, the Reverend John Edwards tested his Gregorians by observing a ½-inch black ring drawn on a card from a distance of around 200 feet. By racking in and out of focus and comparing the defocused images on either side, the primary could be judged over-corrected or under-corrected. If the blurring was equal on both sides, the mirror was considered parabolic. Most early reflectors were $f/10$-$f/12$ or more and did not require parabolizing; the exceptions being mainly tabletop telescopes, the beautiful (and scarce) handheld Gregorians, and enormous instruments such as those by Herschel or Rosse.

John Hadley was a more professional English telescope maker of the 1700s, who tested mirrors with a lighted pin hole at the focus and an adjacent eyepiece to inspect the image. If the out of focus image was any shape other than a circle, then the various diameters of the speculum (at one o'clock, five o'clock, etc.) were judged to have different curvatures. Comparing the two images on either side of focus shows the degree of parabolic correction. Henry King quotes Hadley at length from a book written by Robert Smith and published in 1738 entitled *A Compleat System of Opticks*. This comprehensive 500 page compilation includes much material on telescope making.

The Earl of Rosse was a notorious tinkerer and gadgeteer who used Mudge's zone test on his enormous mirrors. Even before the overall figure is true, determining the focus of areas of the mirror will show where further work is needed. One of his final tests was to view a watch dial from 50 feet. Rosse also made speculum flats and tested them by observing with a telescope the reflection of a distant object in the flat. Considering how difficult it is to make an accurate flat, the inaccuracy of this test was probably appropriate. Another chapter in the false test saga could be: Inadequate Tests Applied to Inaccurate Optics for Results that Go Beyond False.

In contrast to Rosse's achievements is the inspired career of Leon Foucault, his French contemporary. One of Foucault's many accomplishments was the publication, in 1859, of his development of the older pinhole technique into the Foucault test. Viewing the reflected pinhole across a straight edge causes a pattern of shadows that clearly shows the figure of the mirror. This simple improvement granted an enormous increase in the perfection of telescope optics. Foucault went on to analyze the patterns made by mirrors as they are polished from spherical to parabolic form. He also experimented with moving the light towards the mirror and the straight edge away from it, and with the various ellipsoidal mirror forms that can be made in this way. In his fifty-year lifetime, Foucault also devised the famous Foucault pendulum that was the first experimental demonstration of the rotation of the earth. He was also responsible for an accurate determination of the velocity of light, measured the distance to the Sun with improved precision, and pioneered the silver on glass telescope mirror.

In 1864, Henry Draper published an authoritative guide to the Foucault test in his book, *On the Construction of a Silvered Glass Telescope.* This was very popular throughout the remainder of the nineteenth century and in 1904 was reprinted with additional material by George Ritchey, a true master of the art of optical testing. Ritchey also relied on the Foucault test and developed an elaborate mask with small arcs cut out of it at different radii. He made many hyperboloidal secondaries for Cassegrain configurations and tested these against the intended primary mirror and a full size flat in an autocollimation test. One of Ritchey's colleagues at Yerkes was Frank Wadsworth, who in 1902 published a pioneering article on the caustic test. The shorter focal ratios of modern telescopes use deeper parabolas that are very difficult to test with a Foucault tester. The caustic test uses a mask with holes cut in it to reduce the area under test and is capable of great accuracy.

The testing of refractor objectives is not as well documented in the historical literature. Many makers had a pragmatic, hands-on approach that was not easy to describe, and many were very reticent concerning their work. The highly secretive telescope makers of yore, laboring in their closed-door shops, allowing employees to work only in selected stages of the process, and keeping no written records of their techniques, could make a fascinating chapter in this history—with an appendix on the cloak and dagger techniques of their spying competitors. Lens testing was very complicated because the low quality of glass available to craftsmen meant that the final lens curvature was very different than the mathematically correct figure. Testing was followed by local correction of the lens and more testing.

Fig. 8.3.2 *Lord Rosse's "Giant of Birr Castle." This telescope had a four-ton metal mirror—the largest ever cast.*

Alvan Clark and Sons used a pinhole test and many hours of labor to test and correct their objectives. With the advent of astronomical photography, lenses had to be tested at the blue wavelengths that film was most sensitive to; and the human eye is least sensitive to this color range. The Clarks used a spectroscope for testing photographic lenses, and a corrected lens showed that blue light was brought to the same focus as the other colors. John Byrne, an apprentice of Henry Fitz (a self-taught apprentice due to Fitz's secret ways), suggested in a circa 1880 catalog that astronomers test their own refractors using a star test. If a star is red on one side and green on the other, the lenses are not on center with each other. An objective with low chromatic aberration will be fringed with reddish blue inside focus and yellow green outside focus. Spherical aberration can be star tested by examining the (unnamed Airy) disks for even roundness inside and outside of focus. Collimation of lens elements can be checked by removing the eyepiece and viewing through the objective at a candle held in front of the tailpiece. The four small reflected images of the flame should be centered in the actual flame. Finally, John Brashear was unique among telescope makers in publishing details of his testing procedures, necessitating a separate article on his techniques.

There are many interesting topics in modern telescope testing. The Ronchi test was published in 1922 and uses a grating to form an image of the mirror's surface as a series of lines. The Hartmann test was the standard for large observatory instruments earlier in the 20th century. A mask with holes cut in it is placed over the mirror, and photos are taken from prime focus. By shifting the mask and taking pictures in front of focus and behind focus, a very accurate picture of the mirror's surface can be deduced. The 200-inch at Mount Palomar went through years of testing as engineers first hoped it would sag under its own weight into a proper figure, then refigured it, and finally made a corrector lens for it. The Hubble Space Telescope is a paradigm of how to test and how not to test. The most accurate objective does not a telescope make. Since historical aspects of telescope making are the theme of this article, it should be noted that one thing HST did right was in realizing the historical importance of the project and including, from the beginning, an organization devoted to recording and documenting the program. In 1982, the Space Telescope History Project was founded; and in 1989, Robert Smith's book, *The Space Telescope*, was its first product.

The testing of antique optics using modem techniques is a subject that could fill a book. Polarized light was used by nineteenth century makers to check their optical glass and the finished lens. Inhomogenous glass would usually reveal streaks or splotches where the molten glass was not thoroughly mixed. An antique lens can be placed between crossed polarizers and will frequently reveal imperfections. Autocollimation tests are performed on assembled telescopes and also reveal inhomogenous glass; by placing a flat in front of the objective and a Foucault test straight edge at the focus, inhomogenous glass will show areas of light and dark corresponding to high and low density of the glass in the lens. This test is recommended by several optical engineers as the most sensitive test of glass quality in antique refractors. The Foucault test can be used with filtered, monochromatic light to show aberrations at each wavelength. Spherical aberration, in particular, varies with wavelength—a phenomenon known as spherochromatism.

Finally, interferometers are of great use in analyzing the surface quality of old lenses and mirrors. Testing reflectors can reveal all types of aberrations. Since most interferometers use monochromatic laser light, chromatic aberration in a lens will not be shown unless several different colored lasers are used. Coma, spherical aberration, and astigmatism can all be seen as variations in the parallel lines seen on an interferogram. Many antique lenses have been tested in this way, including several of Galileo's lenses, as reported in the July 9, 1992 issue of *Nature*. Two eyepieces and two objectives were examined. One eyepiece had a flat side polished to a fraction of a wavelength of light, a virtuoso feat that was not required to maintain image quality. One objective was considered nearly diffraction-limited. One eyepiece showed a ring-shaped pattern that might indicate that it was made on a lathe, and it should be noted that here we have optical testing used as a tool for historical investigation of early technology. The chromatic aberration that was not revealed by this test is the imperfection that the lenses suffer from, and that could not be reduced until the doublet lens was developed. At single wavelengths, his telescopes are of very high quality.

There are many other fascinating aspects of the history of optical testing. The Reverend W. R. Dawes derived his Dawes limit for the theoretical resolution of a point source and gave Alvan Clark an important career boost by publicizing the quality of his telescopes. George Airy realized the importance of the rings of light that surround a star when viewed through a telescope, and the Airy disks are the first thing that a user looks for when evaluating an instrument. Fraunhofer used monochromatic sodium light in optical testing, an important advance in the field. However, if the literature on historical telescope making is sparse, the available references on the testing procedures of early makers is extremely so. Henry King included a number of details on the subject in his history, and the absence of a bibliography for this article is due to the very few references used. This conclusion is to be considered a plea for copied reference material on the subject of telescope making techniques of old, and modern testing of early instruments, to be written and credited in an ongoing series of articles.

8.4 The Complete Dobsonian

By Tom Osypowski

Ten years ago I built a 16-inch *f*/5 Dobsonian and have spent the time since then enjoying the views this scope provided me and my observing comrades. Early on, while using this 16-inch, I became convinced that some kind of drive was needed to realize the full potential of the telescope. So began what was to become a decade devoted to the development and advancement of equatorial platforms. The long-time relationship with my 16-inch telescope, on its series of ever-improving platforms, convinced me that this was the way to go for big aperture telescopes.

It was no surprise that when the time came for me to plan a new telescope project, I again chose a Dob on an equatorial platform.

After carefully comparing prices and weights of mirrors in the 20- to 25-inch class, I decided a 22-inch was a good compromise, giving me almost twice the light gathering power of my 16-inch for an affordable price. Also, it was a mirror I could lift as a unit in its cell. As in my previous telescope, I planned a snap-off cell for the 22-inch. Because of the terrain I observe from and store the telescope in, wheelbarrow handles or casters are not practical. I need to be able to lift and carry all parts of the scope.

Okay, 22 inches was the choice. But how long—or more precisely, how short—and still get great images? An *f*/3 would be super—no ladder would be needed. On the other hand, the images would not be very good! An *f*/5 would give great views, but I would need an 8-foot ladder. No thanks! Somewhere around *f*/4—3 steps up a 6-foot ladder—would be the highest I would have to go. That sounded right. The final choice was *f*/4.2, which provided a 92-inch focal length. I have had no regrets. The mirror (made by Steve Dodds at Nova Optical) is truly excellent. Planetary viewing at high power is outstanding. Low power fields, using a Paracorr and 35 mm Panoptic eyepiece, are sharp almost to the edge of a ¾° field. As a bonus, the Paracorr (acting like a mild transfer lens), allows fine prime focus photography using a fairly small (4-inch) diagonal. All of these observing options were achieved by a 22-inch that was a full 8 inches shorter than the common 20-inch *f*/5s!

I knew that a lot of people would be looking through this telescope—what with my astronomy classes, school star parties, and private sessions with friends—so it had to be convenient for even a novice to use. Making it short was the first step.

There was also "the half-step syndrome" to deal with. All big Dob users know what this is about. One step on a ladder is just a little low, the next up is a tad too high; so you spend a lot of time either viewing on your tiptoes or in an awkward squat. Neither position is comfortable, and neither position allows full viewing potential. With my old 16-inch I dealt with this problem by making a set of blocks of various thicknesses that could be attached to the ladder steps. This system worked, giving me comfortable access to the eyepiece, but I did spend a lot of time taking blocks on and off. Others have dealt with the "half-step syndrome" by actually installing half-steps on their ladders. This also works. However, both removable blocks and half-steps do pose some danger. People are not used to navigating modified ladders, especially in the dark. I did not want to invite accidents; however, I still wanted everyone to observe with maximum comfort. The solution was a rotating upper cage assembly. This allows the eyepiece to be placed perfectly for each viewer.

The mechanics of this assembly are very simple. The cage simply rotates on a separate ring to which the strut tubes are attached. The lower ring of the cage is confined to a rotational movement by eight aluminum clips attached to the fixed ring. Thin Teflon between these rings and on the bearing surfaces of the clips make for smooth rotation. Is there loss of alignment when moving the cage? Not if the alignment is good to begin with—as it should be in order to get maximum performance out of an $f/4.2$ system.

I had some problems at first until I realized that my primary was ½-inch off-center in its cell. In this case, the rotating cage alerted me to that fact and helped me achieve better alignment. Remember, also, the cage was usually rotated in small angles, say 5° or 10° at a time.

After building in viewer comfort with the rotating cage, my next consideration was the focuser. I wanted this piece of equipment to be particularly accessible and user friendly. Decades of experience with both making and using focusers have taught me a few things. First, the Crayford style wins the award for ease of construction and rigidity. Properly constructed, it can handle the heaviest Barlow/eyepiece combination out there with zero image shift. Second, most commercial focusers, regardless of price, are not easy to use because the adjustment knobs are too small, too awkward to get at, and too hard to find in the dark.

So with these considerations in mind, I dug into my metal scrap bin and came up with the ¼-inch thick aluminum plate, tube, and angle needed for a sturdy Crayford focuser. The tube glides on four small ball bearings attached to a piece of right angle stock. The "pinion" rides in Delrin bushings mounted on a solid aluminum block with a cutout to fit around the focuser tube. This block carrying the pinion is pinned at one end to the base plate. The other end is pushed against the focuser tube with an eccentric cam to supply tension to the system. The tube has a flat milled along its length—the only machining needed for the whole project—

Fig. 8.4.1 *Tom Osypowski and his 22-inch f/4.2 Dobsonian on a dual-axis equatorial platform at RTMC 1994.*

Fig. 8.4.2 *A two-speed Crayford focuser with large wooden knobs (surfaced with non-slip tape) makes focusing easy.*

Fig. 8.4.3 *Close-up of the rocker-box for Tom's 22-inch Dobsonian. Note the detachable mirror cell. The handle on the left is pulled to "rewind" after a one-hour tracking run. The cast aluminum side bearings are from Dave Kriege at Obsession Telescopes.*

for the pinion to bear on. So far, a pretty straightforward Crayford focuser.

The knobs are the distinguishing feature. First, I extended the pinion a couple of inches out to one side and put a 2¼-inch diameter x 1-inch thickness hardwood disk at the end of it. This worked quite well. Such a knob was easy to find in the dark, and its large size made for easy adjustment. Still, this was an *f*/4.2 optical system. I wanted more focusing control, so a second knob was added below the

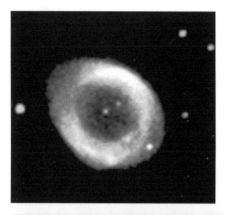

Fig. 8.4.4 *This ST-6 image of the Ring Nebula (M57) was taken by John Sefick. It was an unguided 20-second exposure from a 25-inch Obsession resting on the platform described in the article.*

Fig. 8.4.5 *This photo of M42 is from a 10-minute exposure at the prime focus of a 16-inch f/5 Newtonian.*

first with a belt and pulley arrangement. The 2:1 step down on the pulleys gave twice the focusing sensitivity. This did it. I have never used a better focuser. The large knobs and their placement out and away from the focuser body make them convenient to find and use in the dark. In addition, rubberized friction tape around the perimeter of the knobs gives a soft, no-slip feel.

The final and most crucial touch to the 22-inch was putting it on an equatorial platform. A big Dobsonian reaches its full potential with tracking. No doubt about it, without tracking, high-power observing can be a real chore. At 400 to 600x or more, you are always watching a moving target, always anxious about needing to re-center the object. With tracking you're *observing* the object, not chasing it, as it floats stationary in the "sweet" spot of the field. There's no hurry, no anxiety. You view longer; you can wait for the seeing to clear—wait for that planet to sharpen or for that spiral structure to pop out. You have time to change the eyepiece, put in a filter, or get the note pad again, without having to worry about the image leaving the field of view.

Group viewing is also enhanced with a tracking Dob. Here's an example: A couple of weeks ago our club had a big star party at our deep-sky site. It was a good night so I cranked up the power to 560x on the 22-inch for the central star in the Ring Nebula. A large group of people lined up behind me to get a look at the star, some for the first time ever. I was comfortably seated in my chair the whole time while everyone got to see the Ring Nebula as it should be seen—centered in a high-power field of a large telescope under good skies. The rest of the story is this: There was a 25-inch *f*/5 Dob set up right next to me. A good telescope by a well-known manufacturer. No one saw the central star in that scope. The owner could not use the kind of power needed for the job; and even with lower powers, he spent most of the night scrambling up and down his 8-ft ladder re-centering things.

For those so inclined, tracking ability also opens the big Dobsonians to the world of color astrophotography and CCD imaging. The latest equatorial platforms have all the controls needed for prime focus photography and even auto-guiding capability with no field rotation. All this in a squat, triangular platform that sits low to the ground—adding just a few inches of height to the telescope— and is light enough to be carried with one hand like a suitcase. With a big Dobsonian on an equatorial platform, you get to have your cake and eat it too. Here is this easy-to-point, rock steady, alt-azimuth mount carrying your telescope and an equatorial drive to give it tracking. The best of both worlds! I can be reached through my web site: www.equatorialplatforms.com.

8.5   The Jet Engine Telescope
Building a 4.25-inch *f*/10 Titanium Tube Dobsonian
By Paul Ham

Four years ago I revived a long-dormant interest in astronomy by purchasing my first telescope: a 10.1-inch Coulter Odyssey. The Coulter performed adequately on the planets, the Moon, and double stars, but only when stopped down with a cardboard mask. Since most of my observing involves solar system objects under light-polluted Seattle skies, it was soon apparent I needed a slower *f*-ratio telescope with a smaller aperture. I was also eager to try my hand at telescope making.

I decided to build a 4.25-inch *f*/10 Newtonian on a Dobsonian mount. Such a telescope would offer superb portability and would stand apart from the usual 6-inch *f*/8 first scope. In addition, a 4.25-inch *f*/10 mirror would be relatively easy to grind, polish, and figure to a high optical standard. Lacking any previous experience in telescope making, I read several books on the subject, made line drawings, and then went to work on my first telescope.

Fig. 8.5.1 *Paul Ham's Jet Engine Telescope won the Best First Telescope Award at the 1994 Table Mountain Star Party.*

8.5.1 Design Objectives:

1. A portable telescope well-suited to high resolution lunar and planetary observing.

2. An optical configuration optimal for medium to high magnification, with low diffraction and high contrast.

3. A fully rotatable tube, allowing comfortable viewing positions.

4. Innovative use of simple, readily available materials, minimizing reliance on commercial components.

8.5.2 The Fuselage

For the telescope tube, I decided to use an exotic, space-age material—titanium. Fortunately, surplus titanium tubing is available locally from Boeing Surplus Sales for only $8.00 per lb. Boeing uses this material as ventilation conduit on their jetliners. The tubing I chose had an outside diameter of 5 inches and a wall thickness of 0.03 inch. With considerable difficulty, I managed to saw off a 48-inch length, which cost only $28.59.

The finished tube length was to be 44 inches—I used the remaining 4 inches to construct the curved-vane secondary support. Aluminum would have worked well enough, but I wanted to do something different. Titanium is probably the ideal tubing material. It has the following desirable qualities:

1. Extremely high strength-to-weight ratio. The tube will not sag under the weight of components.

2. Flexure memory. It springs back to its original shape after being squeezed.

3. It takes a high-polish finish. Aggressive buffing yields a beautiful, chrome-like finish impervious to rust and tarnish. No painting or protective sealants are required.

4. Rapid, efficient heat dissipation.

5. Resistance to dew formation on tube surface.

Titanium presents only one drawback: it is difficult to work with. Cutting, drilling, and polishing offer a considerable challenge. I ruined a hole saw while cutting an opening for the focuser. Trimming the ends of the tube required a radial-arm band saw equipped with a tempered blade.

To polish the tube I used a one-horsepower bench grinder with a felt buffing wheel. Seven hours of effort brought the metal to an almost mirror-like finish—unpolished titanium has a dull gray appearance. It was necessary to draw the tube slowly, at an angle, and with considerable pressure across the wheel. I intermittently applied a soft-metal buffing compound to the spinning felt.

I now had a showpiece titanium tube, but it suffered from a problem common to all metal tubing: vibration. I rectified the situation by lining the interior with black flocking paper. This completely deadened all vibration. Flocking paper has excellent light-absorption qualities as well—better than flat black paint. It is also much easier and neater to work with than paint. Since metal tubes radiate about half their heat inward, flocking paper offers yet another advantage. It insulates the tube interior thus shunting heat away from the optical path.

Next, I mounted counterweights totaling more than 5 lbs on either side of the tube's rear end. I used barbell weights obtained from a sporting goods store. By remarkable and most fortunate coincidence, round lead fishing sinkers fit perfectly in the center holes. This combination provided nearly optimum counter-balancing. It shifted the balance point far to the rear, allowing for a very low and highly portable Dobsonian mount.

8.5.3   The Diagonal Holder

I used the extra 4 inches of titanium to make a 2-inch-wide, one-piece curved vane. The titanium's 0.03-inch edge-on profile keeps diffraction and light loss to a minimum. The vane is attached to the tube by titanium brackets which I fashioned with files, a hacksaw, and two vise-grip pliers. According to Peter Francis in his book *Newtonian Notes*, a semi-circular curved vane yields low diffraction and no diffraction spikes, but provides limited stability. The properties of titanium, however, dramatically enhance stability in such a vane.

I made the diagonal holder from common fasteners and a wooden dowel at a cost of about $12.00 (Figure 8.5.2). I used silicone rubber to fasten the secondary mirror to the diagonal dowel. A 3½-inch length of 10-24 threaded rod attached the diagonal holder to the vane. I split the rod down the middle using a hacksaw. To

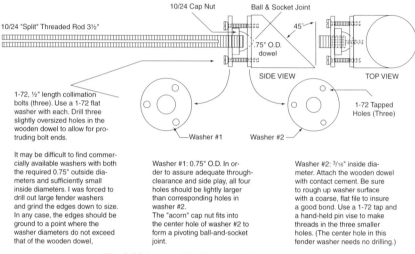

10/24 Cap Nut

Ball & Socket Joint

10/24 "Split" Threaded Rod 3½"

45°

.75" O.D. dowel

SIDE VIEW

TOP VIEW

1-72, ½" length collimation bolts (three). Use a 1-72 flat washer with each. Drill three slightly oversized holes in the wooden dowel to allow for pro-truding bolt ends.

Washer #1

Washer #2

1-72 Tapped Holes (Three)

It may be difficult to find commer-cially available washers with both the required 0.75" outside dia-meters and sufficiently small inside diameters. I was forced to drill out large fender washers and grind the edges down to size. In any case, the edges should be ground to a point where the washer diameters do not exceed that of the wooden dowel,

Washer #1: 0.75" O.D. In or-der to assure adequate through-clearance and side play, all four holes should be lightly larger than corresponding holes in washer #2. The "acorn" cap nut fits into the center hole of washer #2 to form a pivoting ball-and-socket joint.

Washer #2: 3/16" inside dia-meter. Attach the wooden dowel with contact cement. Be sure to rough up washer surface with a coarse, flat file to insure a good bond. Use a 1-72 tap and a hand-held pin vise to make threads in the three smaller holes. (The center hole in this fender washer needs no drilling.)

Fig. 8.5.2 *Diagonal holder assembly (scale =1:1).*

make this cut, I pressed the rod into a shallow groove made with a file in a piece of scrap plywood. Bracing the plywood between my feet and using my free hand to steady it, I hacksawed through the wood and down the middle of the rod (Figure 8.5.3). (One must be sure to stop about ⅜-inch before the end of the rod.) I used a small point file to ream out and smooth the cut slightly so that the "split" rod would neatly fit over the vane. Two 10-24 hex nuts (and washers) placed fore and aft on the split rod secure the diagonal assembly to the vane. They also allow the assembly to be positioned properly within the tube. I coated the wooden diagonal holder with polyurethane and painted the entire assembly ultra flat black.

8.5.4   Diagonal Mirror and Focuser

Keeping diagonal size and focuser height to a minimum allows for optimum con-trast. I selected a diagonal mirror with a minor axis of only ¾ inch—the same diameter as a penny. Such a diagonal imposes a mere 17.6% obstruction on the primary mirror—an impressive figure considering the average Schmidt-Casseg-rain has a central obstruction of about 35%.

Wanting a Crayford-type focuser but lacking the means to make one, I or-dered a NGF Mini-4 focuser from Jim's Mobile, Inc. The focuser came with a nearly flat base—an awkward fit on a 5-inch diameter tube. Armed only with a rasp file, I proceeded to shape the base to fit the curvature of the tube (Figure 8.5.5).

The resulting racked-in focuser height was 1.3 inches, and the total drawtube travel a mere 1.1 inches. This kept the distance between focal plane and diagonal to a minimum. In order to keep it from protruding into the light path, I sawed a full inch off the drawtube. To further enhance contrast, I located the diagonal/focuser

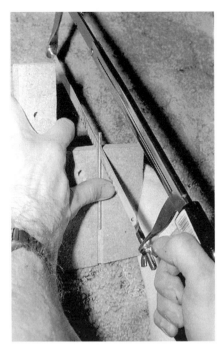

Fig. 8.5.3 *Splitting threaded rod, (in the words of Mr. Spock), "using stone knives and bear skins."*

Fig. 8.5.4 *The diagonal holder and support assembly. Notice how the split rod fits around the vane. Titanium brackets and simple fasteners attach the vane to the tube.*

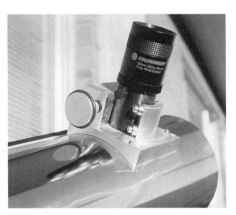

Fig. 8.5.5 *Rasping down the base of the focuser took five hours and created a wealth of calluses and blisters!*

5 inches from the front of the tube. Providing an extra overhang of at least one tube diameter not only shields the diagonal from stray light but also prevents it from dewing up.

8.5.5 Woodworking

I used AC grade oak plywood throughout and applied three coats of outdoor grade polyurethane sealant to all wooden parts. The panels were joined with Elmer's professional woodworker's glue. I made all straight and radial cuts with an electric jigsaw, and a willing student at a local vocational college used a Forstner bit drill press to cut circular holes in the mirror cell back-plate.

8.5.6 Primary Mirror Cell

The mirror cell is a simple compression spring design. Three $\frac{1}{4}$-20 hex-head collimation bolts fit tightly through $\frac{1}{4}$-inch diameter holes in the back-plate. They protrude through the mirror cell and thread into T-nuts on the mirror side of the cell. The bolts run through compression springs sandwiched between the back-plate and the mirror cell. Quarter-inch washers, fore and aft, act as "mending plates," preventing the springs from chewing through the wood.

I used three blobs of silicone rubber to fasten the mirror to the cell. (See Richard Berry's *Build Your Own Telescope* for specific details. In fact, I highly recommend this book as a general reference.) The entire back-plate/mirror-cell assembly attaches to the tube by means of a wooden ring. The latter fits tightly over the rear end of the tube (Figure 8.5.6). Three $1\frac{1}{2}$-inch diameter holes in the back-plate provide ventilation, allowing for thermal equalization. These holes are arranged symmetrically near the center of the plate to prevent light leakage around the mirror and into the tube.

8.5.7 Rotating Tube Cradle

Both the cradle assembly and the mirror cell are variations on designs appearing in *Build Your Own Telescope*. I used a jigsaw with a rip-fence compass attachment to fashion the circular tube rings. The cradle ring/tube bearing surfaces consist of strips of ABS plastic on the rings and 0.010-inch Teflon strips on the tube. Pairs of 0.010-inch thick Teflon gaskets allow rotation between anchor rings and stationary rings.

Anchor rings on the front and rear of the assembly are secured by friction to the tube. I hand-worked these with rasp and rat-tail files until they fit tightly over the tube. Precision eliminated the need for fasteners. Due to the small tube diameter, I had to extend the stationary inner rings outward so the altitude bearings would reach the Dob box side panels. These panels are 10 inches apart—the minimum practical distance. Two $\frac{3}{8}$-inch wooden dowels extend between the inner rings to increase the rigidity of the stationary cradle assembly (Figure 8.5.7).

I used $\frac{3}{4}$-inch plywood to build the Dobsonian mount. All other flat wood parts are made from $\frac{1}{2}$-inch plywood. The 1-inch-thick altitude bearings are con-

Fig. 8.5.6 *Primary mirror cell assembly. The assembly is easily removed for maintenance. The wooden "docking ring" fits flush with, and tightly over, the tube. A 0.010-inch-thick Teflon gasket prevents the coated wooden parts from sticking to each other. Notice how the round lead fishing sinkers fit perfectly into the center holes of the barbell weights.*

Fig. 8.5.7 *This shot shows the wooden rocker-box and system which allow the tube to rotate freely without being allowed to move fore and aft.*

structed from the circular remnants of tube ring construction. Each bearing consists of two ½-inch-thick disks glued together. The bearing surfaces are ABS plastic. ABS is much easier to work with than Formica, works just as well, and is less expensive. Try it!

8.5.8 Performance

After seven month's work, my telescope was ready for its first flight . . . I mean, "first light." Construction had set me back a total of approximately $250, and I had completed the project just one week prior to Comet Shoemaker-Levy 9's much-awaited encounter with Jupiter. A diffraction-limited mirror with an insignificant secondary obstruction yielded refractor-like images.

At a public star party, it was generally agreed that my little scope provided more sharply resolved impact site images than much larger Dobsonians and SCTs. I was able to resolve all four craterlets on the floor of the lunar crater Plato, stars appeared as tiny pinpoints, and Airy disks showed no diffraction spikes against a very dark background. Seeing permitting, this telescope will provide useful images at magnifications in excess of 80x per inch of aperture.

A polished titanium telescope has turned out to be something of an unexpected novelty. Metal tubing, in general, does not seem to attract much attention in ATM circles; I cannot recall the last time I saw a metal-tube home-built telescope. Yet last July at the Table Mountain Star Party, the "Jet Engine Telescope" was given the "Best First Scope" award.

Making a first scope requires considerable time and patience. If you are mechanically inclined, interested in self-learning, and able to improvise, you do not have to be a highly skilled ATM to build a high-performance first telescope.

8.6 Telescopes for CCD Imaging, Part IV
Imaging on a Mountaintop
By Richard Berry

Table Mountain must not be much of a mountain because you can drive right to the top. Located in Washington state about 15 miles north of Ellensburg, this broad rise to an elevation of 6,500 feet features piney woods, lovely open meadows, and nice dark skies—the kind of skies that the average east-coast observer would kill for. Westerners, of course, have their own favorite mountains with even better skies, well away from Ellensburg and several other towns 50 or 60 miles away. Nevertheless, Table Mountain is a good site, and it is the place where about 600 observers gather every year for the Table Mountain Star Party.

The trouble with Table Mountain is that it has no electricity. When I decided to go to the Table Mountain Star Party this year, I started to ponder how I would shoot CCD images from the mountaintop. Those of you who know me know that I shoot my images with a homebuilt Cookbook CCD camera, but what I have to say should apply to pretty much every type of CCD camera; only the details will vary. Those of you who know me also realize that 99.44% of the CCD imaging that I have done has been from civilized locations with electric power; and when I imaged at home, I did so from the small observatory pictured in *The CCD Camera Cookbook*, or nowadays from the shop building beside my house. In other words, I had never been a roving imager running on 12-volt batteries before.

One option was to take some eyepieces with my telescope and stick to visual observing. However, a couple weeks before the event, people started to call and ask if I would be there with the Cookbook camera. Without really thinking, I said, "Certainly!" I was trapped not only by the technical challenge but also by the expectations of people whom I did not even know!

Now the Cookbook camera was never designed for use on mountaintops. Veikko and John had electricity in their houses, as did I. Because it makes so much sense to operate computers and highly sensitive digital cameras in a backyard observatory environment, we figured most people would observe from the back yard. We did not think about mountaintop observing until we encountered Californians. Only then did we seriously start thinking about portable observing with the Cookbook camera.

California amateur astronomers are different. Most people in the United States do most of their observing in their own backyard or within a few blocks or a few miles of home. Californians hardly ever observe at home or anywhere near it. They hop in a car and drive 100 or 150 miles to observe. This is the natural consequence of California's big, well-lit cities full of people an hour's drive from mountains and deserts with dark skies.

What is true of the Cookbook camera is true of most other CCD cameras: to shoot images, you need power for the CCD camera, power for the computer, and power for the telescope. The Cookbook camera normally uses 120-volt AC power, the computer I use needs 120-volt AC power, and my telescope normally uses 120-volt AC power. However, the Cookbook power supply sold by Coherent Systems will run from 12 volts DC and the antique Orion VFO for my Byers 812 will run from 12 volts DC, so (in theory) I was home free on two out of three. That left (in theory) only the computer, a 386/25 MHz in a mini-tower and a 14-inch monitor, to power from batteries.[1]

Here's how the options stacked up: For about $100 I could buy a heavy-duty deep-cycle marine battery that would run the power supply and the VFO. Cables would run about $20 more. The total drain on the battery would be about 7 amperes, so a fully charged battery would easily last the night. For $200, I could get an inverter to make 120 volts AC for the computer and monitor, but the draw would be over 10 amps, suggesting strongly a second battery for another $100. Or, for about $1,500 I could get a laptop computer that would draw only an ampere, and one battery would do the whole job.

At this point, I will admit to feeling somewhat defeated. People who buy their CCD cameras have already laid out several thousand dollars in cold hard cash, so spending another pile of cash to go portable is no big deal. As a Cookbook camera builder, my CCD imaging has been inexpensive, and I wanted to keep it that way. The laptop option was attractive at the "nice new toy" level, but I really have no use for a laptop. For years I have used an NEC 8201 (Kyocera's other version of the Radio Shack's Model 100) for writing on the go, and a fancier laptop would not capture keystrokes any better. The bottom line was that portable observing costs a minimum of $400, and as much more as you want to spend. No wonder folks get sticker shock when they think about CCDs!

---

[1] Nowadays, of course, I use a laptop computer that runs nicely from batteries.

Then, I got lucky. Mel Bartels and I had done several joint observing sessions in my yard with his 20-inch *f*/5 computer-controlled Dobsonian. The results were promising, so we were thinking of teaming up at Table Mountain to shoot more images. Well, it turned out that Mel had a 650-watt generator that he was willing to bring.

To make a long story short, Mel's generator worked out wonderfully on Table Mountain. The generator is a Honda model 650 rated for 450 watts continuous and 650 watts peak. We put it 25 feet away from the telescope and ran its output through a line conditioner. On three quarts of unleaded gas it would run 3 hours, and the best thing was that this generator was *quiet*. It purred along for hours while we observed and nobody minded its sound. Sadly, it was cold on Table Mountain this year (about 35° F with a 20 knot wind); we were all miserably cold, but Saturday night was beautifully clear once we got going. A dozen other CCD enthusiasts and I shot several hundred CCD images.

The most important result of the imaging for me was making direct comparisons between images taken from the pretty good skies over my house (the Milky Way is clearly visible, but somewhat washed out) to the very good skies over Table Mountain (the Milky Way gleams). My conclusion is that good skies make for better CCD images. Although CCDs are relatively insensitive to the amount of sky background, when the sky is black, the images are better. If you can, do your CCD imaging under good skies.

There is no moral to this story. However, I have learned that it is entirely practical to rent, borrow, beg, or buy a generator to go CCD imaging in the boonies, and really great (quiet, easy to start) generators exist. Powering the CCD camera from batteries is expensive but not unduly so. The killer comes when you try to power a 120-volt computer and monitor from batteries or when you buy a low-power laptop you have no other use for. (If you already have a laptop computer, then lucky you!)

One month later, at the Oregon Star Party in the Ochoco Mountains 50 miles northeast of Bend, Oregon, I used Mel's generator again for two nights; this time taking over 50 megabytes of wide-angle color images of the Milky Way using a 5.5-millimeter aperture *f*/3.2 lens. It was great fun, and the generator purred happily providing power for CCD imaging.

8.7 Tips for the Big Dobsonian Builder
By Bill Russell

Undertaking the building of such a large telescope—the heart of which is a 160-pound, 24-inch full-thickness mirror—forced me to come face to face with some interesting design problems.

How could I achieve a silky smooth motion at both axes of rotation without inducing unwanted vibration? My goal was to build an instrument that would allow magnifications up to 517x (using a 4.8 mm Nagler) without tracking difficulties. I eventually solved this problem by using industrial bearings. Traditional Ebony Star laminate with Teflon is utilized to prevent backlash, but the resistance to movement proved to be too great. To reduce resistance, industrial bearings are incorporated at both axes. This combination delivers just the right feel.

For the declination axis, pillow block bearings with heavy (adjustable) truck valve lifter springs are used. The weight is transmitted off the Teflon and redirected to the ground. This configuration allows me to fine-tune the resistance. Slightly turning the hold-down bolt creates a positive feedback loop between the observer and telescope. If it gets windy, I can stiffen the resistance.

The azimuth axis required a different approach. The trick was to get most of the weight to the center on a race bearing. By stacking spacers, one can shim the center so most of the weight is carried at the center without lifting off the outer bearing surfaces.

The second set of bearings (I used 16) are placed farther out from the center where they provide smoothness and eliminate any existent wobble. Finally, the Teflon is placed so that it lightly touches the opposing laminate surfaces. This feature provides the slight resistance needed for control. Through trial and error, I was able to achieve the feel of a very lightweight instrument in a scope that weighs approximately 700 lbs.

All shafts are custom-turned, 1¼-inch stainless steel with ¼-inch inserts to receive 8200-byte encoders. This enables sky vector computing and use of a Celestron Astro Master to locate objects quickly.

The scope is mounted on 14 piano-type casters. It can be easily rolled and winched on and off a pickup truck via a specially constructed ramp.

To prevent movement when in use, I fashioned four damping supports, each with a ½-inch threaded rod. Turning a knob adjusts the damping pad. This keeps the scope from rolling and minimizes vibration.

Another simple enhancement was to provide baffles. On a calm night the baffles significantly increase contrast. But if it gets windy, they can be incrementally removed from closed tube to open truss in three minutes!

The last major design problem was how to keep a full-thickness, 24-inch mirror cool. Over time I developed a strategy on how to keep the mirror cool in the daytime so thermal equilibrium is reached more quickly at night. I currently employ thermal wraps, an active Peltier cooling system, and a thermal container to keep down the mirror temperature.

I strongly recommend exhaust fans to cool the mirror and ventilate tube currents. I used two CFM fans. Through the use of rheostat, the fan's rpm can be controlled (the fans create no noticeable vibration). Each fan circulates in opposite directions (one intake, the other exhaust). It is my opinion that utilizing fans actually improves image steadiness by expelling tube currents.

To test my cooling strategy, I bonded a thermal sensor to the back of the mirror and a second sensor to the outside surface of the mirror cell. Using a Fluke K/J model 52 thermometer, I got a continuous readout of the mirror temperature, the

Fig. 8.7.1 *Bill Russell's 24-inch Dobsonian.*

Fig. 8.7.2 *Azimuth axis fabricated with sixteen bearings and Teflon/Ebony Star laminate.*

Fig. 8.7.3 *A 32-inch gear will drive the azimuth axis when the software is ready. The author hopes this will allow him to use his CCD camera to its full potential.*

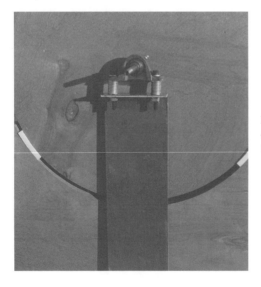

Fig. 8.7.4 *The declination axis is fabricated using a pillow-block bearing, bolts, washers and truck valve lifter springs*

Fig. 8.7.5 *Dual fans (one intake-one exhaust) operate at variable speeds and a 3-way switch allows the fans to be operated independently. One battery powers the fans, another the onboard computer.*

Fig. 8.7.6 *The outer thermal blanket protects the mirror from heat and dust.*

Fig. 8.7.7 *Bill Russell's 24-inch Dobsonian is mounted, as of 2002, in a roll-off roof observatory.*

temperature of the ambient air, and most important, the difference between the two. It has been my experience that in order to get the smallest Airy disk and better star images, the temperature differential should not exceed 4° Fahrenheit—of course, zero would be optimal.

However, I discovered that these measures can still be defeated if the temperature drops drastically at nightfall—at Riverside, for example. I hope to eventually win this battle by storing and transporting my telescope in an insulated trailer rather than in the exposed bed of my pickup truck.

Currently under construction is a computer-assisted drive for the telescope. A series of clutches will allow the user to move the telescope in the normal Dobsonian mode. Then, once the user stops moving the scope, the drive will engage, and the telescope will begin to track. This will allow me to finally get my money's worth from my CCD!

Throughout the various stages of construction of this telescope, I have learned a simple fact: By building your own scope, you can ensure that only the best components will be used and that no compromises will be made on quality or design. With few exceptions, most commercial companies cannot afford to invest similar attention in their products and have any hope of reaching the amateur market. Since the builder provides the loving labor and has no need to generate profits, a dream scope can be created, a scope like no other!

8.8 The Springsonian Telescope
By Howard Banich

I built the first version of my 8-inch $f/4$ Newtonian in high school and, in true ATM fashion, tried a new mount configuration every few years looking for the ideal design. Five attempts, and a couple of decades later, I think I am finally getting there. So I am not in a hurry, but I really did not give this much thought until it became obvious that setting up and observing with my 20-inch scope was easier and more pleasant. Not only that, my back would get just as sore with the 8-inch because it was always so awkward and inconvenient to set up and use. I needed to come up with something quick and easy.

Fortunately, an idea that had been brewing for a long time finally came together. More specifically, I wanted the ability to easily roll the scope outside, put in an eyepiece, sit comfortably upright, and observe—with the eyepiece at the same height regardless of where the scope was pointed. A few weekends of surprisingly straightforward fun in the garage brought forth what is essentially a Dobsonian on wheels with the focuser poking through the center of one of the altitude bearings.

The name Springsonian was promptly coined by my friends in the Rose City Astronomers (RCA) and seems appropriate—it acknowledges features that make both the Springfield and Dobsonian so easy to use, and I hope it is suggestive of the Springsonian's unique function. Although the eyepiece does not stay in the same place, it does stay at the same height.

The first thing I considered was the height of the eyepiece. It should not be higher than your eye when comfortably seated in your observing chair. In my case this is about 52 inches. I purposely chose an upright posture (no slouching!) to insure I would not guarantee myself a stiff back and neck. Seated eye level determines the overall configuration of the Springsonian—and means any instrument with a focal length much over this measurement is not suitable for this mount.

Fig. 8.8.2 *Pointed near horizon.*

Fig. 8.8.1 *Note how focuser stays at essentially the same height while the scope is pointed near the zenith (above) and at the horizon (Figure 8.8.2). The aluminum bar at the top of the rocker box is used as a handle when wheeling the scope around, a stop to prevent the tube from being pointed below horizontal, and a stiffener for the top of the rocker box. The sturdy looking observing chair has an adjustable seat.*

Fig. 8.8.3 *(Below) Tube assembly with the framework/cradle. The cradle is hinged open showing how the tube assembly is clamped in place (note the small pads of Sorbathane that grip the tube) and showing the large cutout the focuser is centered in. The sliding counterweight (the wing nut adjusts the tension) can be removed for transportation. This is also a fine view of the blistered Kydex skin of the formerly smooth tube.*

Fig. 8.8.4 *Close-up shot of the wheels. These pressurized wheels were borrowed from my handtruck to make use of their shock-absorbing capabilities. Although hard to tell from this photo, they are about ½ inch above the ground when the tripod is upright. They become load bearing only when the tripod legs are tilted.*

Fig. 8.8.5 *The locking bolt is shown lying next to the hole into which it would be inserted to lock the azimuth axis. It is attached to the tripod with a piece of cord so it does not get lost. It is also the same size as the center pivot bolt of the rocker box in case I lose the pivot bolt (which I actually managed to do once). When not in use, the locking bolt is stored in a slot immediately below where it is tied down.*

Fig. 8.8.6 *The Springsonian won a Merit Award at the 2002 RTMC Astronomy Expo.*

An adjustable observing chair is still a good idea, however. I have found most people have a different seated height than I do, and an adjustable chair allows them to fully enjoy looking through the Springsonian.

8.8.1   A Few Notes on the Design

The framework/cradle holding the tube assembly is hinged and is held closed by two clasps. The pads in contact with the tube are Sorbathane (recycled from an old pair of athletic shoes), a shock absorbing material with an easily gripped texture. This insures the tube will be firmly clamped in place and dampens vibration as well. This also allowed me the luxury of not having to modify my precious tube assembly. As it turned out my choice for the outer layer for the Sonotube, Kydex, was a poor one and has since taken on a wrinkled, blistered look after being in the sun for a few days. I will fix this anytime now.

The large altitude bearings are meant to minimize the height of the rocker box and provide a large cutout surrounding the focuser. The cutout area is to allow easy access to the focusing knobs with gloved hands. Note that I have positioned the focus knobs in the center of the bearing rather than the drawtube—but then I have big hands. Although this pushes the drawtube forward of the exact center of rotation about an inch, it is hardly noticeable while observing. The asymmetrical positioning of the bearings on the cradle keeps as much weight forward of the center of rotation as possible.

The counterweight is an old ankle weight, now encased in a small box. Designed to be adjustable, mostly because I had only a rough idea where the balance point would be, it has the added virtue of being removable for transportation. Anyone with moderate mathematical skills could figure out the balance point in advance.

The cradle extension board (what the counterweight slides on) was put on the opposite side of the tube from the focuser to act as a light shield. At least it works that way from my backyard, as my neighbor's light is in a position where this generally works well. Two shorter extensions, one on each side of the cradle, would work if you are surrounded by lights and would require less weight for balancing the tube/cradle assembly. I elected to put the counterweight on the side rather than the top or bottom of the tube to keep the balance point constant in all positions.

The height of the rocker box, even though shortened by the large altitude bearings, is much taller than a normal Dobsonian rocker box. Additional reinforcement is needed to make it rigid enough for high-power observing. I added the large aluminum angle inside the box and two small buttresses on the outside, but there are other possibilities (a welded metal framework encased in plywood would probably work well). I found it important to keep my seated position at the eyepiece in mind while designing the rocker. It is easy to over build it for stability and make the eyepiece difficult to approach.

The tripod makes up the difference needed to bring the focuser to the 52 inches I need. A 10-inch f/5 or a 12.5-inch f/4 would not need a tripod, and may be the largest practical sizes for this design. Regardless of the tube assembly, the

Springsonian is larger and heavier than a comparable Dobsonian mount.

I can not emphasize enough how essential wheels are to the overall success of this design. They allow for quick, no-lifting set up and truly enhance overall ease of use. The wheels are load bearing only when the scope is tilted and do not detract from the tripods inherent stability. Since quick, no lifting, and easy are all things I like in a scope, wheels will be a big part of all my future projects.

A bolt is inserted in a hole drilled though the bottom of the rocker box and ground board (top of the tripod in this case) to lock the azimuth axis when the scope is wheeled around. Teflon and Ebony Star are used for both altitude and azimuth bearings.

An unexpected bonus from having the focuser at the center of rotation of the tube/cradle assembly is that image shake from focusing is greatly reduced. Additionally, the weight of the eyepiece cannot unbalance the scope. A 35 mm Panoptic with a Parracor balances the same as an empty focuser. The comfortable, upright seated position makes it ideal for comfortable viewing, especially at high power, and has finally given me the satisfaction of easily and painlessly using my old 8-inch. Honestly, the Springsonian turned out better than I thought, and I have thoroughly enjoyed the entire process of designing, building, and observing with it. Now, if only I can figure out how to set up and use my 20-inch as easily. . .

Issue 9

9.1 Stellafane 1995

By Diane Lucas

The weather for Stellafane did not look very encouraging on my trip from Ohio to Vermont. I kept chasing and being overtaken by rainstorms. This year I finally obtained a reservation at the Hartness House and believe it or not, ended up sleeping in the Russell Porter room.

Upon arrival Friday afternoon, it rained like it was never going to stop. Fortunately, Burt Willard showed up to open the museum. I had not seen this display for several years. The museum occupies three rooms at the end of the long tunnel from the Hartness House to the Hartness Turret Telescope; it has many more exhibits of Russell Porter's telescope making equipment and instruments now than when it first opened.

One excellent thing about Stellafane is that, when it stops raining, the extra water drains away quickly. Saturday morning the Stellafane crew was out with a pickup truck shoveling gravel onto some of the slippery uphill "roads" on the grounds. Saturday the humidity was ferocious and the temperature hot and hotter as the day progressed. A new area for the flea market on higher, drier ground was used this year and was soon overflowing with lots of vendors. I managed to escape with no more new old-stuff than I could carry in one load. Then I went up the hill to look at telescopes and talk to their owners.

The telescope judging should have been a little easier for the judges this year since there were only 17 entries. First place in Excellence in Mechanical Design went to Peter Wraight of Ridgefield, Connecticut, for his 6-inch $f/5$ reflecting binocular. He has invented a very different way to assemble the mechanical parts. The two telescope tubes are mounted parallel to each other, one on top of the other and with the mirror positions offset by the diameter of one of the tubes. The top scope has its eyepiece out the side in a conventional Newtonian position. The lower telescope has a larger diagonal, and its light path goes through the upper scope to an eyepiece mounted parallel to the upper telescope's eyepiece. This arrangement minimizes the number of reflections as compared to other binocular reflectors. There are push buttons on the side of the binocular—easily accessible while looking in the eyepieces—to control the motorized pan and tilt of one of the mirrors. The intraocular adjustment is just as easy.

Second place in Mechanical Design went to Joe Castoro of Coram, New

York, with his 25-inch Newtonian, and third place to Roy Diffrient of Monkton, Maryland, with an 18-inch Newtonian. Eric Allen of Edmond, Oklahoma took home the fourth place with an 8-inch Newtonian.

In the Excellence in Craftsmanship judging, Roy Diffrient received a second award with first place. Joe Castoro took home a second, second place award. Mike Ward of Ballston Spa, New York, was awarded third place for his 12½-inch Newtonian which breaks down into a very compact package, while fourth was presented to Richard Loiselle of Loeminster, Massachusetts, with a 14.5-inch Newtonian. Loiselle used only hand tools to construct his rotating tube telescope and claims to have spent only $478 on this project.

Two craftsmen from Drummondville, Quebec, Real Manseau and Francois St. Martin, were given a Special Award for their display of replicas of Newton's Reflector, Gallileo's Refractor, and Cross and Back Staffs. These instruments were very artistically and accurately done.

An Innovative Component Award was given to Jim Vail of Idaho Falls, Idaho, who has made another significant improvement on a clock-driven camera mount. The original barn door camera mount with a motor-driven threaded rod is not suited for long focus lenses for long exposures since it accumulates several arc seconds of error in 15 minutes. Dave Trott has improved this device by adding a second arm.[1] One version of this double-arm barn door drive only has a few seconds of error after 60 minutes, but this device was not stable in all positions. Vail has modified this design and made it more compact and rugged so that it can support a small telescope. He also has invented an alignment device for easy setup of the mount.

Stellafane optical judging is done in several categories which include both large and small aperture Newtonians. Steve Eldridge of Londonderry, Massachusetts, won an award for Optical Excellence for Large Aperture Newtonians with his 16-inch f/6.4.

Two awards for Optical Excellence for Small Aperture Newtonians were presented with the first going to Eric Allen of Edmond, Oklahoma, with an 8-inch f/5.4, and the second to Bruce McDonald of Pasadena, Maryland, for his 8-inch f/7.91.

Other instruments included a telescope brought by Pierre Lemay of Blainville, Quebec. This reminded me of two telescopes brought to Stellafane years ago by Norman James whose spheres floated in liquid in contrast to Lemay's design. Lemay's telescope mirror, mounted in a fiberglass sphere, was rotated by two motor-driven rollers. A ball provided the third support to the sphere, and small adjustments to its position relative to the two rollers provide declination adjustment. The eyepiece and diagonal support were easily removed and packed inside the sphere to provide easy transportation of the optics.

Cap Hossfield, who was the AAVSO guiding light of solar observations for years, has since turned to other activities. Several years ago he brought a device

[1] *Sky & Telescope*, Gleanings for ATM's, 2/88 and 4/89

Fig. 9.1.1 *Flat polishing machine built by Russell Porter. The machine is described in the March 1947 edition of* Scientific America.

Fig. 9.1.2 *Peter Wraight and his reflecting binocular.*

Fig. 9.1.3 *Close-up of eyepiece assembly on the Wraight binocular. Note the interesting finder to the left of the eyepieces.*

Fig. 9.1.4 *The rocker-box for Roy Diffrient's 18-inch f/4.5 houses batteries and switches for lights, a fan, and a dew removal system.*

Fig. 9.1.5 *(Left) 12.5-inch Dobsonian ready to use and...*

Fig. 9.1.6 *(Right) ...ready to travel.*

Fig. 9.1.8 *This model of Newton's reflector earned a Special Award for its creators, Real Manseau and Francois St. Martin.*

Fig. 9.1.7 *Richard Loiselle and the 14.5-inch Dobsonian he made exclusively with hand tools.*

Fig. 9.1.10 *Jim Vail's clock drive has collapsed the awkward appearing barn door drives into a compact sandwich of moving parts with excellent accuracy.*

Fig. 9.1.9 *Jim Vail with his clock driven mount.*

Fig. 9.1.12 *Cap Hossfield tuning his gamma ray burst detector.*

Fig. 9.1.11 *Pierre Lemay's "ball" tele-scope.*

Fig. 9.1.13 *The base of Pierre Lemay's ball telescope. The ball rides on the bearing on the left and is driven by two rotating cylinders. The knob to the left of the bearing is used to adjust the declination of the base.*

to the top of Breezy Hill that he was hoping would serve to detect gravity waves. He is still working on that project but has more recently turned to an attempt to detect gamma ray bursts; he brought two loop antennas, one horizontal and one vertical, collecting the signal from one of the East Coast VLF Navy radio stations at 21.4 kHz. He is collecting the phase difference between the horizontal and vertical signal and examining it for gamma ray effects. So far he has made several improvements in sensitivity of this device but does not have any positive results yet.

During the day, the newly dedicated McGregor observatory with the 13-inch $f/10$ Schupmann telescope was open for examination. Four of the main "movers" on this project were honored at the evening talks. Philip Rounseville contributed at least 600 hours of glass pushing, Charles Thayer was the head carpenter, Scott Milligan was the optical designer, and John Martin was the overall design manager. These amateurs and all the others who have helped deserve an immense

Fig. 9.2.1 *Left to Right: Charles Thayer (carpenter), John Martin (observatory designer), Scott Milligan (optical designer), and Philip Rounseville (optician).*

Fig. 9.2.2 *Stellafane's new McGregor observatory houses a 13-inch Schupmann telescope— the largest instrument of its kind in the world.*

amount of credit for completing this 8-year project. Milligan and Martin both presented talks during the afternoon on fabricating the telescope and the observatory.

9.2 Dedication of The McGregor Observatory
By Maryann Arrien

On July 15, 1995, the Springfield Telescope Makers welcomed the family of Douglas McGregor to a rocky hilltop known as Schupmann Hill in order to dedicate a new observatory to their departed comrade. There, adorned with a wrap-around gold ribbon, stood a white building with foot thick reinforced concrete walls. This

Fig. 9.2.3

Fig. 9.2.4 *Doug McGregor*

was no ordinary monument but, indeed, a functioning astronomical observatory on the hallowed ground of Stellafane East, where the yearly Stellafane conventions are held. Built entirely through the efforts of amateurs, the McGregor Observatory houses the largest operating Schupmann medial telescope in the world—a 13-inch $f/10$ instrument.

The telescope sits atop a massive concrete pier that is pinned to the bedrock below. The chief optical designer, Scott Milligan, dreamed of building a large refractor back in 1987. With the help of friend and master optician Philip Rounseville, they have constructed a long focal length instrument that is particularly splendid for observing the planets. The telescope has a folded light path which utilizes both refracting and reflecting elements to yield a color-free image. The hand grinding and figuring of the glass took over 600 hours work, to say nothing of the design work, ray tracing and testing that intervened. The Mangin mirror, a double-pass component, was hand figured to an accuracy of $\frac{1}{30}$ of a wavelength of light.

Scott and Philip's employer, Mr. Harry Vandermeer of Optical Systems and Technology, provided workspace and test plates for making the optics. Through his influence, the glass blanks were generously donated by Schott Glass. Other things, such as a vibration isolation table for interferometric testing, could not be afforded. Scott Milligan was able to improvise one out of wood and inner tubes that did the trick at only $250. This type of ingenuity hearkens back to the days of club founder, Russell W. Porter, who worked on the 200-inch telescope at Palomar. Porter encouraged telescope makers to scour the junkyards for possible telescope components.

Douglas McGregor, a dynamic member of the Springfield Telescope Makers, would have been "proud to bursting" over this observatory, as his close friend George Scotten put it. Doug's first homemade telescope was fashioned from a hollowed out bowling ball. His enthusiasm for observing planets, comets, Messier objects, and even the Moon could not be overstated. He went on to build other telescopes and trained others to do so, always infecting them with his passion for observing the heavens.

Notably influenced by him was the chief architect, designer, and builder of this observatory, John Martin. McGregor had pinned the nickname "Bulldozer" on him in the early days of their relationship. Not long after McGregor's passing in 1988, John Martin used a real bulldozer to clear the way to the granite rise that would become the foundation of the observatory. Along with an ancient machine that looks like a Panama Canal steam shovel and other marginal equipment, the rugged new land of Stellafane East was tamed enough to continue to hold the yearly Stellafane Conventions. These conventions were endangered by the loss of the original campground several years back. John Martin, chief carpenter; Charles Thayer; and George Scotten mortgaged their homes in order to buy the adjoining raw land. The conventions continued.

At the Stellafane Convention, Doug McGregor was best known as the Master of Ceremonies of the evening program where awards were given for the most innovative and prize-worthy amateur telescopes. The booming voice and fiery enthusiasm that first caught the attention of John Martin also affected the Stellafane conventioneers. "Doug dearly loved Stellafane," says Martin, "and he wished that his mortal remains would be placed there." After his untimely death in a car accident, Doug McGregor's ashes were cast upon the hill on which the original Stellafane clubhouse is built. Said Martin softly, "From that vantage point, he can see everything that we're doing."

During the years of construction, other members of the club assisted John Martin and Charlie Thayer as the observatory changed from a piece of rock into a massive building with a rolling roof. In addition, other astronomy clubs such as the Amateur Astronomers, Inc. of New Jersey, the Vermont Astronomical Society, and other local clubs sent support in the form of free labor, and in the case of the Amateur Telescope Makers of Boston, precious seed money for the tube assembly. Says president, Bob Morse, of the Springfield Telescope Makers, "Doug had no small effect. Stellafane attracts very special people . . . like a magnet . . . something intangible brings them together." Inside the observatory, a large brass plaque sparkled with the names of volunteers and contributors.

President Morse called upon Doug's children, Robert and Tennile McGregor, to cut the gold ribbon encircling the observatory. He likened the young observatory to a child going into the future, to nurture future children with new generations of inspiration and discovery. With Doug's parents, brothers, sister, and widow looking on, the observatory roof rolled back majestically the instant the ribbon was snipped. Amidst multicolored balloons and the strains of bagpipes playing *Amazing Grace*, the computer controlled telescope swung slowly towards the heavens. Steve Scotten, Doug's nephew remarked, "Doug would be overwhelmed—this is truly marvelous." Later that night, the heavens were visible from Springfield, Vermont, despite the thunderstorms that raged early that morning.

At this year's Stellafane convention, the McGregor observatory was open for viewing of the gas giant planets Jupiter and Saturn. The skies were very clear.

"Look at that!" Doug might have said at the top of his lungs, "Isn't that gorgeous!"

9.3   Notes on Making a 12½-Inch Full-Thickness Mirror of Good Quality

by Chris Gulacsik

9.3.1   Introduction

I have been a telescope-making enthusiast since the age of twelve. During the course of my hobby, I have had the pleasure of grinding and polishing 4.5-, 6-, and 12½-inch mirrors, with the latter two exceeding my wildest expectations. The 12½-inch paraboloid was installed in a Porter/Pearson style, split-ring mounting system pictured in Figure 9.3.1, and it won a 1st place ribbon at the 1993 Table Mountain Star Party.

My reason for writing this article is to encourage those who might entertain the thought of fabricating their own large, full-thickness mirrors. The sections that follow give a clear perspective on the effort required. It is important to recognize that invaluable guidance is provided by the references listed at the end of this article, including back issues of *ATM Journal* and its forerunner, *Telescope Making*.

9.3.2   Curve Generation

The objective was an $f/5$ curvature, which for a 12½-inch mirror results in a sagitta (i.e., mirror depth) of approximately 0.16 inch. The mirror blank was fine-annealed Pyrex, 2 inches thick, and the tool was 1-inch thick plate-glass. Rough grinding was performed with the mirror on top (MOT), using #80 aluminum oxide— ½ heaping teaspoon per wet.

Because of the gross roughness of the Pyrex blank, 110 wets were required to bring the mirror and tool into good contact. I used a centered, diametral stroke of ⅚ the mirror diameter for this purpose. A tangential stroke was then applied for the next 200 wets in order to "hog out" the curve as quickly as possible. The work was finished with a centered, diametral stroke (less than ½ the mirror diameter), to minimize deviations from a sphere. A total of 40 wets, alternating MOT (mirror on top) and TOT (tool on top), were needed for this last step.

An interesting graphical view of my progress is shown in Figure 9.3.2. Here, the measured sagitta is plotted against the number of wets; the dashed line indicates the average slope, which is 0.52 mils/wet. Note that a total of 4½ lbs of #80 abrasive and 350 wets were required to complete the job.

Fig. 9.3.1 *Chris Gulacsik shows off his 12½-inch split-ring Newtonian.*

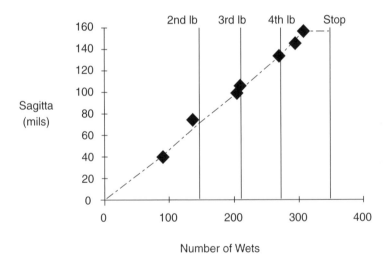

Fig. 9.3.2 *Plot of measured sagitta versus number of wets.*

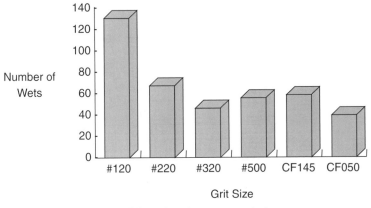

Fig. 9.3.3 *Number of wets versus grit size.*

9.3.3 Fine Grinding

At this stage of the process, I used #120 grade silicon-carbide, ½ teaspoon per wet, with a ⅓ normal stroke. The applied pressure was that of the hands and mirror only. The focal length was adjusted to 62 inches by biasing the number of wets: MOT decreases the focal length; TOT causes it to increase. For the finer abrasives, mirror and tool were alternated in position, and ¼ teaspoon of abrasive per wet was adequate. Figure 9.3.3 shows the number of wets required at any given stage to eliminate pits in the mirror surface produced by the previous grain size. Numbers 120 through 500 are all silicon carbide; CF145 is 14 μm aluminum oxide lapping powder; and CF050, 5 μm aluminum oxide. The last is extremely fine and does little to reduce the individual pit size.

The decision as to when a particular grit size should be discontinued was based upon close inspection of the mirror surface with a 10x magnifying loupe, assisted by an intense light source (a focused Maglite™). Larger-than-normal pits were identified and logged. Texereau says[2] that if these pits disappear after a further wet, then one can proceed to the next abrasive. For my mirror, I consistently found that, on average, three wets were required to eliminate the largest pits.

9.3.4 Polishing Lap Construction

The tried and true method of molding pitch bars, cutting them into small squares (⁵⁄₁₆ x ¾ x ¾ inches) and securing them in place via a thin layer of beeswax, works well. However, some patience is involved since a 12½-inch lap has a total of 112 squares. Fortunately, the lap only needs to be rebuilt perhaps three times. A simple modification of Texereau's wooden form[3] prevents unwanted adhesion and can

---

[2] Texereau, Jean. *How to Make a Telescope*. Richmond, VA: Willmann-Bell, Inc., 1984, p. 43.
[3] Ibid., p. 47.

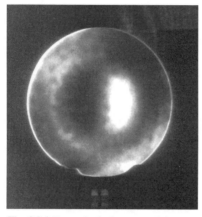

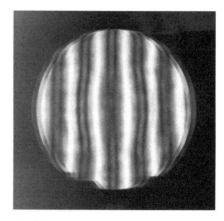

Fig. 9.3.4 *Foucault shadowgram of the central-depression defect.*

Fig. 9.3.5 *Ronchigram of the central-depression defect.*

mitigate stripping of the mold: carefully coat it with a layer of aluminum foil, shiny side up, before the molten pitch is poured.

One aspect of lap construction that impacts the ease of figuring is having some means of testing the hardness or temper of the pitch surface. Pitch will grow progressively harder with age and use. Too hard a surface results in loss of contact; on the other hand, if it is too soft, a turned edge is inevitable. In either case, the mirror can be polished, but the figuring process is recalcitrant. The tester featured in Appendix J of Texereau's book is not difficult to construct. There are no critical parts, with the exception of a small machinist's rule that can be purchased at most hardware stores.

Consider the pitch supply that I purchased from a well-known vendor and then put into storage for about a year before being used. Measurement trials at room temperature (72°F) yielded a fall rate of only 5 mils in 5 minutes, which is too slow. The storage pot contained 35 cubic inches of pitch. After it was remelted, spirits of turpentine were added, a teaspoon at a time, and samples extracted for tests. Upon introducing a total of 6 teaspoons, the fall rate was at an acceptable level of 25 mils in 5 minutes.

9.3.5 Polishing and Correction of Zonal Defects

The polishing operation was performed with cerium oxide and a ⅓ amplitude normal stroke. A total of 77 hours of labor were required to reach the point where only a hint of gray remained in the center of the mirror. The first 18 hours were spent with TOT to ensure that the edge was sufficiently polished; MOT was then used for the balance. On average, I was able to maintain a rate of one stroke per second with eight strokes made at each position around the post. A typical polishing session lasted 3½ hours.

The next 24 hours were spent experimenting with various methods of honing the surface to a nearly spherical shape. Figures 9.3.4 and 9.3.5 show the Foucault

Fig. 9.3.6 *Water-bath apparatus used for hot-pressing.*

and Ronchi shadowgrams part way through the process. The straight Ronchi bands are indicative of a sphere; however, a broad central-depression defect is clearly visible in Figure 9.3.4. The mirror edge is in good shape, as indicated by the bright diffraction ring. Measurements of the difference in radii of curvature of the inner and outer zones with the Foucault apparatus showed the offending feature to be only 0.2 μm deep. Texereau's stroke #12[4] applied for ½ hour (12 rotations of the mirror) was then used to smooth-out the ridge between the two zones. The end result was quite satisfactory: an undercorrected paraboloid.

9.3.6   Figuring

Successful figuring of any mirror surface depends upon securing good contact between the mirror and lap on a consistent basis. Otherwise, one is plagued by unpredictable effects. There are two accepted methods for accomplishing this: cold- and hot-pressing. Masters like Ritchey elected to use only the former, and routinely produced optics better than $\lambda/20$. My preference was to employ hot-pressing before each figuring spell, simply because it required less time to complete, especially when the lap began hardening with use.

My hot-pressing technique is a variation of Thompson's[5] and involves heating the lap with hot water in a large plastic tub. However, rather than submersion, it is suspended above the water surface; and the hot, humid air circulating within the tub provides the heating action. Figure 9.3.6 is a photograph of the setup. The lap is supported by an empty coffee can. Hot water, at a temperature of 150° F, fills the bottom third of the tub; the lid is closed, and the lap is heated for approximately 20 minutes. Afterwards, the mirror is painted with a creamy mixture of cerium oxide and water and is worked back and forth on the lap for 30 seconds. It is then centered, and a 34 pound weight placed on top for 30 minutes or more.

A full-thickness 12½-inch mirror weighs 21 lbs—nine times as much as a 6-inch mirror. The classical method of parabolizing, with MOT and a stroke length of ⅘ the mirror diameter, can be difficult to execute. Texereau's method of accented pressure[6] is much more tractable and involves placing the tool on top in

---

[4] Ibid., p. 92.
[5] Thompson, Allyn J. *Making Your Own Telescope.* Cambridge, Massachusetts: Sky Publishing Corp., 1980, p. 57.
[6] Texereau, Jean. *How to Make a Telescope.* Richmond, VA: Willmann-Bell, Inc., 1984, p. 97.

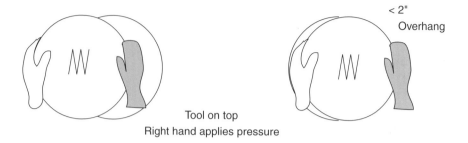

Tool on top
Right hand applies pressure

< 2"
Overhang

Deepening the Center   **Reducing the Edge**

Fig. 9.3.7 *Parabolizing Method: A Modification of Texereau's Method of Accented Pressure.*

a decentered position. To deepen the center, the right hand concentrates pressure on the tool edge over the area where material is to be removed, while the other hand merely supports the overhanging part of the tool (see left side of Figure 9.3.7).

In order to reduce the mirror edge, Texereau's technique was to work the center of the tool over the marginal zone of the mirror. After repeated attempts, I found that this invariably resulted in a badly turned edge. A good solution was found by experimentation and is illustrated on the right side of Figure 9.3.7. The tool was decentered, but only by a maximum of 2 inches. Pressure was then concentrated on the overhanging edge of the tool with the right hand. The action was rapid and had to be applied carefully. I limited each figuring session to eight walks or less around the post. The end product of my labor was a paraboloid with a $\frac{1}{6}$ wavefront error, 62.1-inch focal length, and a good-looking edge.

Figures 9.3.8, 9.3.9, and 9.3.10 are Foucault shadowgrams for the knife-edge at the intersection of light rays from the marginal, 70%, and central zones, respectively. The diffraction ring is visible to the naked eye, but does not clearly show in the photographs because of the chosen exposure. A slight microripple can be detected, and as expected, its spatial period roughly matches the spacing of the pitch squares on the polishing tool. Figure 9.3.11 is the Ronchi shadowgram at 144 lines/inch with the grating placed 0.18 inch inside of focus. It matches the theoretical image quite well.

9.3.7   Concluding Remarks

Fabricating a large mirror by hand has great rewards, especially when it is finally put to use out in the field. The images seen through a telescope built with a quality mirror are simply overwhelming. Stars appear as sharply defined pinpoints, and planetary details are clear and crisp. I hope that the discussion presented in this article has been of interest. If you have any questions, please feel free to contact me either through surface or electronic mail.

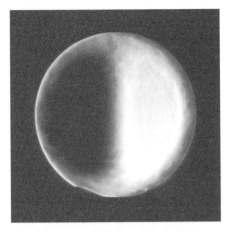

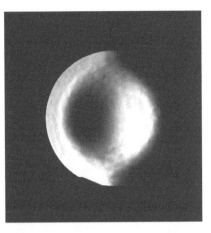

Fig. 9.3.8 *Foucault shadowgram: knife-edge at the intersection of marginal rays.*

Fig. 9.3.9 *Foucault shadowgram: knife-edge at the intersection of 70% rays.*

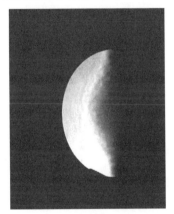

Fig. 9.3.10 *Foucault shadowgram: knife-edge at the intersection of central rays.*

Fig. 9.3.11 *Ronchigram at 144-lines/inch. The grating is 0.18-inch inside of focus*

9.3.8 Other References

1. Pearson, Joe. "An Equatorial Split-Ring Telescope," *Telescope Making*, Issue #27, Spring 1986. Waukesha, Wisconsin: Kalmbach Publishing.

2. Brown, Sam (ed). *Homebuilt Telescopes*. Barrington, New Jersey: Edmund Scientific Company, 1964.

3. Ingalls, Albert G. (ed). *Amateur Telescope Making*, Books I, II, and III. Scientific American, Inc., 1980.

4. Mackintosh, Allan (ed). *Advanced Telescope Making Techniques*, Vol. 1 & 2. Richmond, VA: Willmann-Bell, Inc., 1986.

5. Personal communication with Mr. Linke, Master Optician at Applied Optics, Inc., on December 17, 1991.

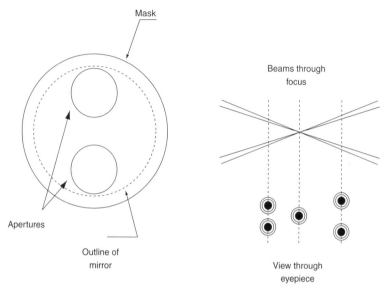

Mask

Beams through
focus

Apertures

Outline of
mirror

View through
eyepiece

Fig. 9.4.1 *The principle of the focusing mask.*

9.4 An Easy Way to Achieve Accurate Focus
By Steven Lee

There has been much written recently on the difficulties of focusing a telescope with a CCD attached. For astrophotography, a knife-edge focuser is the ultimate accessory achieving this aim. However, such devices are either difficult to make or expensive to buy—and they will not help you at all to focus a CCD. I present here a simple and reliable method of focusing a telescope that I have been using for many years and, until recently, had never seen used by other amateurs nor discussed in amateur journals. It is still little known, and some disbelieve that such a simple system can achieve accurate focus (until they try it). The method is suitable for visual, photographic, and especially CCD use; works on all types of telescopes; requires no fancy equipment, no exacting machine work; and yet is totally accurate.

It is basically a Hartmann test—a technique familiar to professional opticians and some advanced amateur telescope makers. The difference is that I use only one pair of holes in the mask. As can be seen in Figure 9.4.1, the two holes border the edge of the aperture and are a little less than half of its diameter. The mask splits the incoming light into two "beams" that, when the telescope is near focus, will appear as small spots (and will usually show an Airy disk and diffraction rings). However, until perfect focus is achieved, the two spots will appear separated as shown in the diagram. Moving closer to focus brings the spots closer; moving away from focus increases their separation. Only when a perfect focus is

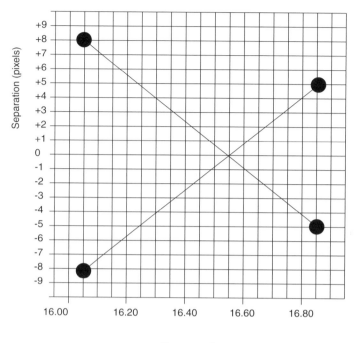

Focus reading

Fig. 9.4.2 *The two readings were taken at focus positions of 16.05 and 16.85 (arbitrary units), and the measured separations were 16 and 10 pixels. Reading from the graph, the desired focus is 16.54—exactly!*

achieved will the spots merge into one.

The mask is nothing special. Its only requirement is that it have more than one hole. There are no requirements for great accuracy in either the cutting or placement of the holes. Greatest sensitivity is achieved with the apertures farthest apart and not too large. Obviously, the larger the holes, the brighter the spots and the fainter a focus star can be. However, the sensitivity is lessened if the holes are too large. Experiment with different size holes until you find the best compromise between image brightness and sensitivity.

Figure 9.4.1 shows the mask on my 20 cm *f*/4.5 Newtonian. It has apertures of about 70 mm diameter, yielding each beam approximately *f*/13. It works perfectly for both photography and CCD work. I made my mask from the side of a cardboard box with the holes cut with a pair of scissors; it cost nothing and only took a few seconds to make. A more robust mask could be made from sheet metal, but a piece of paper will suffice.

Focusing is simply a matter of selecting the desired eyepiece or camera combination, finding a suitably bright star, and placing the mask over the front of the telescope. Then adjust the focus until the two spots merge into one. For CCD work, set the window closely around the spots and the readout to be as quick as possible. For photography just look through the viewfinder, although with short focal lengths the camera viewfinder may not provide sufficient magnification to resolve the spots when near focus. I use a small 4 x 10 finder telescope, focused on the camera's viewing screen, to magnify the image in this case.

If high magnifications are used, a reasonably bright star is needed to check focus. It is a good idea to check that there is no change in focus as the telescope is moved about the sky, should you need to move far to find a suitable star. Such changes are often caused by loosely mounted optics or flexure of either the tube or camera mount.

Focusing is made even easier if you have an encoder on your focuser. Simply place your Hartmann focusing mask over the telescope and adjust the focus so you see the two spots separated by a few pixels. Note the encoder reading, take an image, and measure the spacing between these two spots if you are using CCD; otherwise, you will need to adjust the focus so that the spots fall on repeatable marks, which you then measure. Drive through focus by roughly the same amount and repeat the measurement. On a piece of graph paper (or using your computer—the calculations are trivial) plot the result. Measure (or compute) the point of zero shift and set your focuser. Observe. With a motorized focuser, the entire operation could be carried out under computer control. (Here there is an opening for an enterprising CCD manufacture.)

The story does not stop there. Those who try the above method visually may notice something strange about the disk when the spots are in focus. Bands will appear across the spot. These are actually interference fringes caused by phase differences in the wavefront as it advances across the two apertures. This is the principle of the Michelson stellar interferometer.

One can also discover the state of the atmosphere by observing the two spots slightly out of focus. The two beams see the star through slightly different parts of the atmosphere and, should the seeing be poor, the two spots will dance about in a quite uncoordinated manner. Any motion due solely to the telescope will move both spots equally. The measure of this dancing coupled with the separation of the apertures gives you some idea of the scales involved with the turbulence in the air. This is the basis of a professional seeing monitor system called a DIMM (Differential Image Motion Monitor), and it provides an unambiguous, unbiased measure of the seeing.

One note of caution, however, is that if the optical system suffers from severe spherical aberration, focusing with the apertures at the periphery of the entrance pupil will not necessarily place one at the best focus. By making many apertures in the mask and carefully measuring the position of their respective images at different focus settings, you can determine the shape of the wavefront as it approaches focus, and hence the location of the best focus. This is the Hartmann test, from which I started this exercise.

(From a different perspective, quantitative mirror testing could easily be done with a CCD and a Hartmann mask. Software already exists to centroid stars from CCD frames, so all that is needed is some simple software to do the calculations from the 2 frames—one inside focus, the other a known distance away on the other side of focus. Professional opticians have historically only used the Hartmann test for final proof of the quality of their optics because making the measurements from photographic plates was too slow. However, with an amateur CCD—or just a video camera and frame grabber—and some simple software, the total time to do the test and reduce the results would be a matter of a few minutes.)

9.5 The Making of Voyager, a Binocular Chair

By Wayne Schmidt

Telescopes are great, as long as you do not have to use them. Flat, one-dimensional views, straining to reach eyepieces, and fighting the cold all conspire to challenge even the most stalwart astronomy enthusiast. Voyager is an attempt to overcome these drawbacks by mounting large binoculars on a motorized chair equipped with as many comforts as possible.

Large binoculars are used because they reduce eyestrain, increase light sensitivity, improve image sharpness, and provide enjoyable three-dimensional views. Optical design for the binoculars is dominated by a desire for easy mirror fabrication (since these are the author's first mirrors), undemanding collimation, a personal preference for low power, simple alignment of the two optical beams, and the highest possible contrast. These requirements indicate that the best design uses independent telescopes with 8-inch $f/8$ mirrors. To maximize contrast, the mirrors are mounted on cells without clips, and undersized diagonals are supported by wire spiders.

Eight-inch $f/8$ mirrors are not difficult to make. Purchasing a mirror kit with two Pyrex blanks, one intended to be the tool, provides two mirrors for the price of one. Mirror grinding is very satisfying. There is a quiet excitement to feeling grit break down as it cuts into glass. Polishing is tedious, but figuring and testing are thoroughly enjoyable as long as they are undertaken with patience. While written information implies there is one best way to make mirrors, some liberties can be taken without disaster. Varying the amount of grit or using a slightly different polishing slurry to accommodate the mirror maker's preference will not destroy the mirror and may make the overall project easier.

Both mirrors pass six-point Millies-Lacroix diffraction limit tests, and have smooth surfaces and good edges. The final focal lengths are 64 and 64.37 inches, well within the 1% allowable difference for binoculars. It takes 5 extra hours of work to get the second mirror's focal length to match the first's. Foam mounting tape holds the mirrors on plywood cells equipped with push-pull screw adjusters. Tape is used to eliminate contrast losses from mirror clips. Both mirrors and diagonals have enhanced aluminum coatings for an overall reflectivity of 86%. Normal aluminum coatings provide only 68% for three mirror systems. The telescope bodies are 11-inch diameter cardboard tubes for forming concrete pillars. They are 6 ft long and capped with 15-inch-diameter, 40-inch-long extensions. The insides

Fig. 9.5.1 *Wayne Schmidt and Voyager. Voyager's creature comforts include 5-inch thick contour padding in the seat, an electronically heated vest, a voice activated recorder for hands-off note taking, and an adjustable magnetic chart holder.*

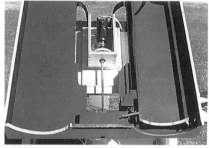

Fig. 9.5.2 *Removing the overhead cover and looking into the observing area shows the draw-tube focusers, 7 x 50 finder binoculars, and chart holder. Large and small wire circles are sized to match the fields of view of the finders and main binoculars. The wire hanging down is for the anti-dew system. Draw-tube focusers are used to provide the 7-inch focal length required to accommodate the inclusion of Barlow lenses in front of the tertiary mirrors.*

Fig. 9.5.3 *A threaded rod and captive nut drive provide the thrust, speed, and reliability needed to lift the 100 lb binocular assembly. A similar system (not shown) moves the binocular forward and backward.*

Fig. 9.5.4 *The chair breaks down into a dozen pieces that can be carried by a compact station wagon with a large roof rack. Voyager can be assembled in as little as one half hour by one person.*

of the telescopes are fully baffled, painted flat black, and covered with black velvet in critical areas. Dust covers with magnetic seals provide closure.

Spiders for Newtonian secondaries are typically four sheet-metal vanes 0.02-inches thick. Voyager's telescopes use three pairs of 0.008-inch wire in place of metal vanes to reduce diffraction losses. The combined effect of reducing the number and thickness of the secondary supports lowers the brightness of the diffraction spikes 65%. Wire spiders work well but installation is awkward.

The secondary mirrors are only 1.5 inches in diameter. This represents a 19% obstruction which is small for reflective binoculars. These secondaries provide 100% illumination of a 20-arcminute field of view. Surveying the size of 1,866 astronomical objects mentioned in *Burnham's Celestial Handbook* indicates that a 20-arcminute field is large enough for 95% of them. Visually, the drop off in illumination beyond the 20-arcminute field is not significant. Small secondaries are used to maximize contrast.

A secondary limits the field of view as a hole in a sheet of paper does. By moving the eye left and right, it is possible to look around the edges of the hole to see a wider field. Similarly, the field lens of an eyepiece can absorb more field of view from a small secondary than seems apparent. Adding a third mirror, the same size as the secondary is like replacing the hole with a long tube. Moving the eye left and right does not increase the view. Equally, the field available to the eyepiece is reduced. To eliminate this effect, Voyager uses tertiaries larger than the secondaries. Maximum size for the third mirrors is determined by the design of the mirror mount and focuser and by the minimum pupil separation. In this case, tertiaries that are 1.8 inches in diameter are allowable. They are mounted on non-moving pedestals inside drawtube focusers.

Fifty-millimeter Plössls provide a 1.4° field of view at 32.5 power. Exit pupils are 6.25 millimeters. Long focal length eyepieces provide the eye relief eyeglass wearers require and the low power desired. Using these binoculars discloses two valuable facts. First, eyepiece exit pupils should not match eye entrance pupils. If they do, any movement of the head causes a misalignment between exit and entrance pupils resulting in a sudden darkening of the image. This effect also makes the eyepiece separation adjustment more critical. Designing for five-millimeter exit pupils eliminates this problem. Second, the observer's heartbeat shakes the images. At 7 and 20 power this shaking is unnoticeable. At 32 power it is just barely detected. While the effect is insignificant at very low powers, it might be a problem for chair-mounted instruments with magnifications above 60.

Reflective binoculars with the observer looking in the same direction as the telescopes have images that rotate with the tubes. This means the simple procedure of rotating the telescopes to change pupil separation will not work. One scope must be mounted on a carriage with independent front and rear adjustments, enabling eyepiece separation changes and image fusion. The telescope's front pivots around the eyepiece and remains fixed once eyepiece separation is set. Remote controls move the rear of the telescope to merge the images while looking through the binoculars. The yoke holding both telescopes is mounted on a motorized elevator that moves up, down, back, and forth. This makes getting in and out of the chair easy. It also allows the binoculars to be moved to the best viewing position. Motor driven, $\frac{1}{2}$-inch-diameter, threaded rods with captive nuts provide enough thrust to move the 100 pound binocular assembly.

Safety is particularly critical in a system like this. If one of the telescopes breaks loose with the chair in a highly tilted position, the eyepiece could crush the rider's eye. For this reason all potentially hazardous structures have backups in case the primary members fail.

Straining to look through eyepieces is always annoying. By mounting the binoculars on a chair, the observer is always comfortable and relaxed. The chair uses a Dobsonian mount with metal casters for the tilt axis and a 12-inch diameter plate bearing for rotation. These provide almost torqueless pivots enabling small motors to tilt and rotate Voyager's 340 lbs and an additional 200 lbs of observer with ease. The body of the chair is made of $\frac{1}{2}$-inch plywood and 2 x 4-inch lumber. Marking the intersection of two vertical lines made when the chair hangs at two different angles determines the center of gravity. This is the location for the casters.

Testing several types of DC reversible motors showed that the best performance comes from rechargeable drills. They provide high torque, a chuck for easy attachment to the input shaft of a speed reducer, and low cost. Transmissions from old windshield wiper motors are used to further reduce speed and increase torque. Both tilt and rotation drives use 1-inch-diameter rubber-coated output shafts pressing against the edges of 3-foot-diameter plywood disks for final gearing. The chair has two drive speeds, 3° and 0.5° per second. Switching 1-ohm 20-watt resisters into the motor circuits provides the slow speed. Power is supplied by a 6 volt, 140 amp-hour deep discharge battery that stores enough energy for five nights of observing.

The chair is contour padded with up to 5 inches of foam. As the chair tilts, the padding and observer's back relax, pushing the rider upward. The motorized

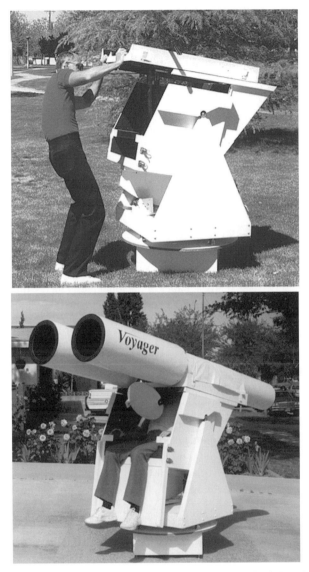

Fig. 9.5.5 *Looking like a NASA cameraman preparing to follow a space shuttle from the launchpad to orbit, Wayne gets comfortable and puts Voyager through its paces.*

elevator enables the eyepiece position to be raised, accommodating this change.

Trying to look through an eyepiece while shivering in the cold is not only unpleasant, it also reduces the detail that can be seen. The chair is made with protective sides and an overhead cover enclosing the observer and providing protection from wind and cold. Additionally, a homemade electric vest provides 25 watts of heat and guaranteeing the rider's warmth. Another advantage to enclosing the viewing space is that all external light is blocked, allowing maximum dark adaptation to be achieved and maintained.

An unexpected bonus of mounting binoculars on a chair is that the headrest fixes the position of the eyes relative to the eyepieces. Eliminating the unavoidable eye motions of conventional observing enables much finer detail than you would expect to obtain from 8-inch mirrors at low power. The effect is not subtle. Visible detail is increased almost as much as going from monocular to binocular vision.

Large binoculars can breathe life into astronomical views that must be seen to be appreciated; and riding a motorized chair—in complete comfort—increases the pleasure enormously. Keeping warm enables objects to be studied in greater detail. When riding in Voyager, all these effects combine to create a sense of identity with the night sky that is unattainable any other way.

9.6   The Perfect Telescope Is
By Randall Wehler

The perfect telescope is actually a matter of personal preference. Still, when I consider the many factors involved in the decision-making of buying or building a telescope, some are of more importance than others. For the telescope I own, the considerations were as follows:

1. Portability: My optical tube assembly (OTA) weighs only about 10 lbs with a compact tube length of only 14 inches, which would be considered portable by most amateur astronomers. This size lends itself to easy transport, set-up, take-down, and storage. It also fulfills the adage that "the best telescope is the one used most often."

2. Affordability: The purchase price of a new OTA of the system I chose was in the $750 and $900 range, with some used instruments selling for $600 or less. This cost seemed to fall at the lower end of the moderate-price range according to ads in the major astronomy magazines and product catalogs.

3. Sufficient Aperture: The optics of this system deliver more photons to the eye than does a 4-inch refractor or even 100 mm binoculars that are roughly in its weight and size class. Secondary mirror obstruction with this optical design may be slightly smaller than its Schmidt-Cassegrain cousins. By now, you have probably guessed that I am referring to a compound system.

4. Adequate Focal Length: The focal length of this instrument is compatible with both fairly wide-field (deep-sky) and more magnified (solar system), visual images. An $f/10$ system falls in the moderate range between fast and slow optical systems. Wide fields are not at the expensive of adequate image scale when needed, and higher powers are not at the cost of bright, larger area views. An $f/10$ speed is a nice compromise. With a

32 mm Plössl eyepiece of 52° apparent field, wide (1.1° actual field) low-power (47x) views are attainable. Using the old standby of 50x per inch of aperture, magnifications approaching 300x are possible on very steady nights.

5. Optical Performance: A number of telescope connoisseurs have stated that views through a Maksutov-Cassegrain—it is time I let the proverbial "cat out of bag" in identifying my "perfect telescope"—are of high definition with the optics in those commercially available today, which are often of premium quality and clarity. Some folks would say that the performance may approach that of an apochromatic refractor with an objective of similar size.

6. Collimation Stringency: Unlike some other telescopes, a Maksutov-Cassegrain does not require one to "line up" his mirrors correctly by hand and keep checking this collimation with repeated field use, as in a Newtonian telescope. Nor is there the necessity of critical alignment as in binoculars. Additionally, the closed tube helps to keep the optics clean and reduce air currents.

Well, there you have it—why I believe a 6-inch Maksutov-Cassegrain may be the ideal all-purpose telescope when one examines the above mentioned factors. With its relatively low weight and compactness, it lends itself well to being attached to the top of the yoke of my reclining chair binocular mount[7] with a pair of standard-size binoculars on the bottom of the yoke serving as a counter weight and second observing instrument. Retaining reclined, straight-through viewing positions with this OTA has made it a joy to use, and it may be one of the most ideal small-moderate aperture observing systems around.

9.7 Comments on Riverside 1995 and the Arrival of "The List"

By Randy Johnson

RTMC is fast approaching, and I wanted to share with all of you several aspects of last year's meet that I found enjoyable and educational. In cyberspace a number of telescope makers have been gathering for the last few months on the ATM mail list, an open forum for telescope-making enthusiasts of all levels of experience. A month or two prior to RTMC, several people mentioned that it might be fun to rendezvous with the members of "The List" bound for the conference. We planned to have everyone meet on the lawn in front of Combs Lodge at Camp Oakes on the Saturday of the Conference. It was great! More than 25 members of the list, including people from Seattle, New York, and even one fellow from France, gathered there to toast our alliance.

Another thing that happened, at least in part as a result of the computer net-

---

[7] *ATM Journal*, Issue #5, p. 18.

Fig. 9.7.1 *Dean Ketelsen of the Tucson Astronomical Society takes the first photos of a newly formed group of telescope makers who share their comments and questions via email.*

work, was a kind of impromptu telescope making workshop that evolved over the course of the weekend in Combs Lodge. Rich Combs, one of the list members, announced before the conference that he had obtained an OK to work on a 10-inch mirror in the small lodge and used the net to invite others to come by and take a whack at grinding and polishing. One of my friends, Peter Hirtle; his 8-year-old son, Tristan; and I had driven down from Seattle. Before leaving, we tossed a 6-inch blank, a grinding stand, a load of abrasives and polishing compound in the vehicle. We decided to participate in Rich's demo and work the 6-inch as a project around which Tristan could build his first scope. Francis O'Reily, another ATM list member, and his daughter, also 8 years old as I remember, flew in from New York with a couple of mirrors that they were having problems trying to figure. Master optician Bob Goff spent a good deal of time in Combs Lodge coaching the mirror-making activities. There were also numerous visits from a number of skilled optical workers over the course of the weekend including Gerry Logan and John Gregory. That, in itself, was neat. Here were novices mixing it up with the very best—exchanging information about stroke length, mirror on top vs. mirror on bottom, Foucault testing, Ronchi testing (and that's Ron-Key, not Raunchy testing—thank you, Bob Goff), and system design. You name it, it was probably discussed as we walked the barrel.

Before the end of the conference, Francis had made great strides with his figuring, Tristan's mirror was about half polished out, and Rich's 10-inch mirror was ready to star test. It was exciting for me to watch newcomers, who had never tried their hand at making optics before, go from thinking; "this is a fancy process best left to professional labs"; to "this doesn't look that hard" (it isn't) and "I'd like to try it myself."

Some of my old-timer RTMC friends talked of bygone days when the late Bob Cox would gather informally with telescope-making kin and discuss their projects and their problems in the outdoor amphitheater on the RTMC site. They said the discussions would go on for hours, and there was sense of accomplishment and camaraderie that would come with it. I think this is what the "Telescope Makers" in Riverside Telescope Makers Conference is all about. I hope that conference planners will see these benefits, too, and actively promote this component of the conference.

9.8 The Early History of Binoculars
By Peter Abrahams

The invention of the telescope was a sequence of events that cannot be assigned to an exact time or place. There are several written references to telescopic instruments in the centuries before Galileo, but there is no solid evidence as to their construction and use. However, it is known that the first patent application for a telescope was in October of 1608. Hans Lippershey, an eyeglass maker in what is now Holland, applied for a 30-year patent that would grant him exclusive manufacturing rights. After testing, it was requested that Lippershey produce an instrument that could be used by two eyes. On December 9, 1608, the inventor announced completion of a binocular instrument. On December 15, the binocular passed inspection; and two more, with optics of quartz crystal, were ordered. The patent was denied, based on the argument that the instrument was already known to other parties, but Lippershey was hired as telescope maker to the State of Zeeland. Louis Bell speculates that Lippershey's instrument was likely 3 or 4 power with an objective of 1½ inch or less in diameter. Henry King, whose *History of the Telescope* is more authoritative than Bell's work, agrees that Hans Lippershey applied for the patent and was requested to produce a binocular telescope with optics of quartz, but King is mute on whether the instrument was successfully completed. Quartz was known to be more difficult to work, and the request for crystal optics was dictated by the poor quality of optical glass of the era. The early desire for binocular instruments is not surprising to experienced observers of today, who are familiar with the problems that monocular instruments present to critical viewing. Lippershey's customers were the very first telescope buyers, and

Fig. 9.8.1

had no experience with viewing through an eyepiece to refer to. It is easy to imagine them peering through a primitive Galilean type eyepiece of poor quality glass and being overwhelmed with eyestrain and exhausted with fatigue from squinting. That they immediately desired a binocular instrument (without having seen or used one) is testimony to their imagination and the primacy of binocular perception in human telescopists.

Recent Galileo studies disagree concerning his construction of a binocular instrument. Giovanbattista de Nelli, in his 18th-century collections of Galileo letters and other works, wrote that in 1618 Galileo constructed a helmet with a twin telescope attachment to be used on board a ship. There are many references to the helmet in Galileo's writings, including construction, testing, presentation to sponsors and ambassadors, and the development of a gimbaled observer's chair to counteract the motion of the ship. Instrument historian Silvio Bedini summarizes the known documentation and accepts the notion that the helmet was binocular. However, telescope historian Albert van Helden notes an 1881 Italian history of "Cannocchiali Binoculari," by Antonio Favoro, in which it is claimed that the helmet had a single telescope. Van Helden believes that Galileo did not use a binocular instrument, and the idea must be regarded as speculative, pending further research.

There are many other references to very early binocular telescopes, including:

- 1613, by Ottavio Pinani, noted by Favoro.
- 1645, Antonius de Rheita, first published claim of invention.

- 1671, Cherubin d'Orleans, *La dioptrique oculaire*, a landmark in optical history, includes illustrations and details on binocular telescopes. The Museum of Science in Firenze, Italy, exhibited in 1988 a four draw, ornamented cardboard binocular attributed to d'Orleans.

- Pietro Patroni of Milan is mentioned by d'Orleans as another maker of binocular telescopes, circa 1700. Patroni seems to have been a prolific maker of telescopes and binoculars, since his instruments still appear at auctions.

- 1702, Johann Zahn, *Oculus artificialis*, illustrated what seems to be a hand-held binocular with a very flexible collimation linkage between tubes.

- Lorenzo Selva was part of a family of instrument makers in Venice through the 1700s. In his Dialoghi Ottici Teorico-Pratici (1787), many instruments are described and depicted, including telescopes and binoculars.

The optical designs of most of these instruments are not recorded. Most would have used Galilean optics, but some might have used a convex eyepiece to allow higher powers. These Keplerian optics would have been useful in an astronomical instrument; and their widespread use in telescopes insures that, at some point, a binocular telescope was constructed on that principle despite the difficulties in collimation that result from higher powers. By the nineteenth century, image erecting systems of two spaced lenses were in wide use in terrestrial telescopes, and no doubt some binoculars were made in that way as well. A twin Newtonian was proposed by M. Vallack and described by John Herschel in *The Telescope*, published in 1861. Vallack wished to view by reflecting one light path over the open end of the other tube. Herschel devotes 7 pages to the binocular telescope, noting that it has been used for viewing the Sun, Moon, and planets, "though without any very great practical advantage." Herschel also presages the modern battery commanders stereoscopic rangefinder by describing a stereo-telescope that had been built by A.S. Herschel in 1855, and another described by M. Helmholz in 1859. These instruments are two telescopes, with the objectives spaced perhaps 18″ apart, each directed to an eyepiece via mirrors or prisms. The widely spaced objectives give a greatly enhanced sense of depth perception.

Many of the instruments described in an early history of the binocular do not now exist, and some never existed. *The Lost Binoculars Most Worthy of Recovery* would seem to be the Galileans made for the U.S. Naval Observatory during the Civil War. Robert Tolles supplied a small quantity of field glasses, and Henry Fitz and Alvan Clark produced one binocular each. The Clark glass was tested in February of 1865; but the end of the war, soon after that, indicates that only one instrument was completed. None of these binoculars are currently known to exist. The Clarks made several binocular telescopes in later years, and the revised edition of *Alvan Clark & Sons, Artists in Optics* covers them on pp. 245–246.

The modern prism binocular began with Ignatio Porro's 1854 Italian patent for a prism erecting system. Throughout the 1860s, Porro worked with Hofmann in Paris to produce monoculars using the same prism configuration used in modern Porro prism binoculars. Other early makers of Porro prism optics were Boulanger (1859), Emil Busch (1865), and Nachet (1875). Some of these makers produced prism binoculars. A combination of poor glass and unrefined optical design and production techniques resulted in the failure of all of these ventures. These monoculars are very scarce today, and it is unknown if any of the binoculars survive.

The German optical designer Ernst Abbe displayed a prism telescope at the 1873 Vienna Trade Fair. Designed according to Porro's principles, but without knowledge of the earlier work, Abbe's new innovation was to cement the prisms. He then set aside the idea and went on to develop the theoretical basis of the modern microscope. His association with Otto Schott, glassmaker, and Carl Zeiss, instrument maker, resulted in a spectacular series of innovations by the German optical industry. The first high quality modern binoculars were sold in 1894, a product of the optical design of Ernst Abbe and the production techniques of Carl Zeiss. These antiques give very sharp views and are still one of the most attractive binoculars ever made. The documentation on the development of the binocular after this point has survived in much greater quantity, and the long story of the modern history of the binocular belongs in another chapter.

9.9   Alika K. Herring on Telescope Optics, a Telephone Interview

By Richard A. Buchroeder

An excellent description of the breadth and depth of Mr. Herring's work, both at Cave Optical Company and for Gerard P. Kuiper at the Lunar and Planetary Laboratory of the University of Arizona, will be found in O. Richard Norton's article in the Telescope Making section, May 1995, *Sky & Telescope*, pp. 81–86, "Master Optician, Master Observer." In brief summary, Mr. Herring figured over 3500 mirrors, typically 6-inch to 12.5-inch, mostly unsigned and found in Cave Astrola Newtonians—although almost 200 were made independently and bear his name and identifying code scribed on the back of the Pyrex blank.

Rumor has it that "those in the know" would occasionally buy a Cave mirror (which could have been made by any of several different good employees), and send it back with a specious complaint that it "wasn't good enough." These mirrors went to Herring, who refigured them, much to the delight of the devious owners! Unfortunately, cases of counterfeit-signed mirrors are known, so a prospective buyer is advised to test a mirror being considered for purchase. Although knowing how a master craftsman does his magic is not likely to turn the rest of us into great mirror makers, we can scarcely fail to profit from knowing what the maestro thinks about mirror making and the use of telescope optics. Therefore, I telephoned Mr. Herring several times and produced this composite in-

Fig. 9.9.1 *Alika K. Herring: Master Optician and Amateur Astronomer.*

terview.

Alika K. Herring was born on July 17, 1913 in Wailua, Oahu, Hawaii, to Jesse Douglas Herring and Liliani Kahanu Herring, who moved with Alika to the Mainland not long after. Sadly, his mother died while he was quite young. Alika recalls his father telling him that he started asking questions about astronomy when he was four years old. At age 9, he found a 3-inch lens and did his first lunar observing. With no mentor, and guided only by the thin first edition of *ATM*, he made his first mirror at age 13, a 6-inch $f/8$, which he remembers silvering by the Rochelle salts process in the school chemistry lab. Those were the depression years, but he said they were okay for him, because he was also a musician and could always earn a living. I asked him if he was a religious man, and he said "no"; he says his only regret in life is losing his vision due to macular degeneration in both eyes. He remains an avid musician and has made five Hawaiian recordings in recent years.

At Cave, Alika specialized in figuring mirrors, which were taken through three stages of grinding and then machine-polished. At first, coal tar, such as is used for roofing, was used for pitch laps, giving variable results depending on the seasonal and diurnal temperature of the Cave facility in Long Beach, California. Later, they went to Zobel pitch which came in various compositions, allowing the optician to stop fighting the lap. He cut approximately 1-inch squares in the pitch and figured with the mirror face up.

He pointed out that this minimizes the influence of heat distortion from your hands and allows testing more frequently. He would then trim the pitch squares frequently, reporting that this had a strong influence on the rate of polishing. However, he disliked and did not use the HCF (honeycomb foundation) lap, as it produced too rapid changes in the surface to suit his method.

Alika disdained cerium oxide, preferring to use optical-grade rouge such as

#309 American Optical Company. Final figuring would be done with decanted rouge, which is produced by first mixing some stock rouge in a tall jar, stirring and allowing it to stand for a few minutes. The solution near the top is poured into a cup, and squirted onto the mirror with an ear syringe. He preferred sub-diameter polishing tools. For example, he would use a 6-inch tool for an 8-inch mirror. He also preferred to work at a room temperature of around 70°F, and depending on the *f*-number, a zig-zag stroke overhanging the mirror by about an inch would have a 10-inch *f*/6 mirror ready for first test in about 20–30 minutes. No additional weight was placed on the tool during figuring, although the polishing machine would be loaded up during spherical polishing.

Production manufacture of mirrors is, of course, different than one-off fabrication, and so it was possible to finish Cave mirrors in only a few hours. Alika said that when making only one, you have to spend a lot of time understanding the rate at which the figure is changing. If you mess it up, the easiest way to recover is to put it back on the polishing machine and let it run until it becomes spherical again. Alika stated that he mixed small amounts of wax with pitch and put it on top of the lap to improve the lap. Cold pressing preceded a figuring spell, or followed it when the job was not finished by quitting time. When a mirror became seriously stuck to a lap, he would make a dam around the mirror and fill it with ice cubes; when things had chilled down, the mirror and lap could be rapped gently apart.

Testing was performed both at center of curvature, using the Foucault and Ronchi tests, and in auto-collimation against a flat, again with the Foucault and Ronchi tester. A 150-lines/inch ruling showing 4 or 5 lines was about right for an *f*/8 mirror. Back then, they had no interferometer; and Alika said that he remains suspicious of wave-based claims purporting extremely small fractional wave accuracy. Further, with Pyrex (a great improvement over plate glass in terms of thermal stability), best results were not always obtained with a null-figured mirror. For a 12.5-inch mirror, even hours of settling down would occur at night, during which the figure would bend into a state of over-correcting. It was not uncommon for opticians to refigure some mirrors, slightly under-corrected on the bench, to compensate the mirror for an anticipated observing environment. It is essential to remember, however, that a mirror for the mountain might not be a good one for the seashore, and vice versa. Be forewarned.

Alika went on to say that the optical polish and quality of figure that can be produced on plate glass or Pyrex are the same, but the degree of under-correction would naturally be different. Of course, the advantages of using costly zero-expansion material are evident but hardly consistent with a low-cost mirror. Alika charged about $150 to make a brand new, aluminized 8-inch *f*/8 mirror. However, mirror precision consists of many factors; and as owners of exceptional mirrors report, the work of the optician, more than the expansion coefficient of the substrate, is still what makes the big difference. I asked whether he knew of mirror figures ever changing with time; and Alika said he had seen it happen several times, especially with plate glass, after a bump. He said he was unaware of a mirror ever changing figure after aluminizing, but this assumes the coater took no aggressive steps to clean a mirror prior to applying the film. Cave mirrors, back then, were all coated by Pancro Mirrors.

Alika did not bevel the rounded edge of the cast Pyrex mirrors (nowadays, Pyrex is often cut from sheets and lacks the smooth finish all-around that the older blanks featured), and said he felt the effect of this upon the diffraction image was not noticeable. He preferred rounded mirror retaining clips, and straight-vane four-legged spiders to curved ones. As for optical windows, he doubted that one could ever be made accurately enough not to degrade the performance of the mirror.

Cave mounted its mirrors in cells that would raise eyebrows by today's more sophisticated standards, yet Alika reported no problems even with his personal 12.5-inch $f/6$ super-mirror. Cave cells, perhaps uniquely, use nylon-tipped lateral retaining screws and often put a layer of cork around the edge of the mirror. Some have suggested this reduces thermal problems with both mirror figure and tube effects. I asked Alika about the out-gassing, or odor, of the fiberglass Cave tubes, which was evident even on the 35-year-old telescope I own; he was not aware of any instance in which this had caused deterioration of a mirror or its coating.

Cave used to buy a lot of diagonal mirrors from a company called "Deep Sky," which were spec'd to ⅛ wave. Cave also made some by blocking them up; Alika said the diagonal in his 12.5-inch was taken from the middle of such a block. He always used Cave orthoscopic eyepieces, which were imported.

For shipping or storage of a mirror, he wrapped it in ¼-inch of tissue, taped sheets of cardboard on both top and bottom, and put it in a tight-fitting cardboard box taped tightly shut. Remember that he lived in comparatively dry, southern California; and trapped moisture in other climes may cause surface deterioration.

To clean an aluminized surface, Alika washed it first under the tap using a clean cotton swab and liquid detergent, but he advised against using Windex. He cautioned that soap must be avoided as it contains lye, which will corrode the aluminum film. Finally, he rinsed it with distilled water and set it on edge to dry by evaporation. Alika was active in the amateur movement, and he was the "Red Herring" mentioned by Albert Ingalls in the preface to *ATM3*. He has observed with many different kinds of telescope, including the Lowell Observatory 24-inch refractor, which he confirmed to be quite good. However, in regard to Lowell's canal observations, he suggested that Percival may have had eye trouble. Asked what makes a telescope "legendary," he said: (a) who owns it; (b) what was done with it; (c) its location.

To these I would add the requisite that the optical quality be far superior to what one commonly encounters in ordinary telescopes. Contrary to the idea that something on the order of quarter-wave is adequate, the discriminating observer knows that there is no clear limit at which superior surface polish and optical accuracy cease to reward the seasoned observer.

Asked whether mirror making is more of an art than a science, Alika said "yes"; you never finish learning that special talent is involved and that you must be serious-minded. He said that a small obstruction is important, and the diagonal in his personal 12.5-inch *f*/6 has a minor axis of 2.14 inches, 18% by diameter. He had to argue to persuade Gerard Kuiper to provide an especially small 10-inch Cassegrain secondary mirror for the 61-inch LPL planetary Cassegrain on Mt. Lemmon, Tucson, Arizona.

Regarding Schiefspieglers and "off-axis" paraboloids, he said, "trick telescopes are not worth the effort." As to observing with his own telescope, he was mainly an evening viewer and observed "everything in its season," but especially the Moon and planets. Although right-handed, he observed with his left eye because it was better. Both the eye and the mind become trained with time. He used color filters, especially on Mars where blue shows atmospheric detail and red the surface. He believed a clock drive is essential. Although eyepiece magnification depends on the subject, Alika felt that powers much in excess of 50x per inch produce no gain, and could result in lost image contrast. He was, of course, aware of locations that have the best seeing, one of which was Table Mountain near Los Angeles, northwest of Anaheim. He advised against observing on the desert floor and favored sites close to the ocean.

Alika reported that the best seeing in the world was on Haleakala and Mauna Kea on the islands of Maui and Hawaii, respectively. He used his personal 12.5-inch *f*/6 Newtonian extensively on both peaks! He was, as explained in *Sky & Telescope*, the person who performed the first site survey of Mauna Kea, among other now-famous observatory locations. Alika closed up his own optical shop in 1980, but his mirrors continue to bear witness to his exceptional talent!

Issue 10

10.1 Veiling Glare

By H. R. Suiter

10.1.1 What Makes a Good Image

The characteristics of a good point image (and therefore any image) can be stated in three conditions:

1. The central spot is as small as possible.

2. A minimum of the star's energy is found in a halo immediately around its base.

3. The unfocused glare in the field of view is as small as possible.

Condition #1 says that the image-forming system is as near to diffraction-limited as possible. The statement that the optics are "diffraction limited" is somewhat confusing because this usage does not imply the image is perfect. The image continues to improve right on down to zero aberration. What it says is that you have eliminated enough geometric aberration so that the spreading caused by diffraction dominates increases in the size of the image caused by anything else. Further lessening of aberrations will not decrease the spot size much further; i.e., you have reached the point of diminishing returns. Alignment errors, astigmatism caused by poor supports, and defocusing are among the curable sources of failure in this condition. Once these things have been corrected, only an increase in aperture can decrease the size of the central spot.

Condition #2 demands that the maximum contrast be delivered in fine details, such as cloud-bands on Jupiter, ring detail on Saturn, or spots on Mars. Failures in this condition are caused by large secondary mirror obstructions, the primary aberrations, and medium-scale roughness ("primary ripple").

Condition #3 refers to what is called *veiling glare* in the system. Veiling glare is a broad glow of light not obviously associated with something in the field of view. It can be caused even by a source outside the field of view, although it need not be. Its sources are poor baffling, poorly coated optics (causing internal reflections), small-scale surface roughness, and small obstructions such as dirt on the optics.

Even though spider diffraction has a known origin, this diffraction can be thought of as being a source of veiling glare because light is thrown far from its source. If you look at an extended object (such as the Moon), spider diffraction

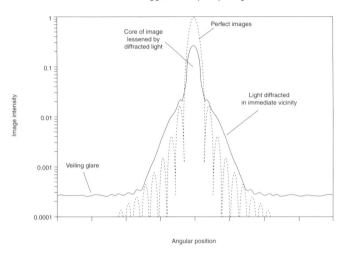

Stellar image with aberrations and veiling glare added Ñ point spreading

Fig. 10.1.1 *An image model is shown with two sorts of degradation. One is a glare or haze on the image that contributes to lessened contrast at all scales of image detail, and the other is an aura closely confined to the base of the image. This near halo would be destructive of narrow details of the type often encountered in planetary observation. The halo in the immediate vicinity contains most of the light diffracted by the secondary mirror and the primary aberrations. The glare, on the other hand, contains unbaffled light and light diffracted from dust, scratches, and spider vanes. The perfect image is depicted by a dotted line.*

spikes are originating on the entire lit area, crossing each other and becoming mixed with the general glow. For extended objects, the similarity with simple glare is striking. Thus, it is a convenient fiction to throw spider diffraction in with other sources of glare.

Looking at Figure 10.1.1, we can identify these various difficulties. A failure in Condition #1 is set by the width of the central peak over the limits imposed by diffraction (in the figure this condition is okay). Condition #2 is determined by the width and amplitude of the hump the central peak is sitting on, and the failures in Condition #3 originate with the plateau the image is sitting on. Of course, the causes of failure in Condition #1 fade into the causes of failure in Condition #2, and likewise, #2 merges into #3.

The difference between Conditions #2 and #3 is not often appreciated by amateurs. At times, people complain that their instruments do a poor job of showing dim nebulae or galaxies at low power and ascribe it to the central obstruction of the telescopes. "Obstruction reduces contrast" is a widely misunderstood, but often chanted, mantra of amateur observers. Central obstruction diffracts most light to within a few arc seconds of the stellar image; therefore, it cannot be interfering much with a fuzzy object many arc-minutes across. Secondary obstruction and other broad aberrations do reduce contrast, but their worst effects are seen in narrow planetary details, fine structure of globular clusters, or other small-scale objects. What these observers are complaining about is veiling glare, an affliction that can strike obstructed and unobstructed telescopes alike.

Glare is one of the most ignored features of amateur instruments (both home-made and professionally produced). Ironically, it is one of the least expensive and most straightforward problems to repair. It seldom requires that you replace the optics (although you may need to recoat them). It often demands that you baffle the instrument properly. Many amateurs blame their primary optics for sources of glare (accusing such mysterious things as micro-ripple); when, in fact, the problem is incomplete baffling, painting, or just plain inadequate instrument design. This is particularly true of Newtonians, partially contributing to their poor reputations. Well-baffled Newtonians, fitted with clean eyepieces, will amaze viewers with their black fields of view.

Baffling is the art and science of making sure that light reaching the focus is properly processed by the optics and does not originate elsewhere. Of course, no instrument may be perfectly baffled, but you can severely decrease the stray light.

Since baffling has been covered in other articles, I am going to discuss here is how to measure glare, and I will help you to know when those measurements indicate that your instrument is substandard.

10.1.2 Testing for Veiling Glare in Your Instrument

The test for veiling glare in a camera lens is to point the lens into a uniformly lit hemisphere of light with a single dark spot in the center. Because a spot of paint cannot be made totally non-reflective, this dark spot is usually a "light trap," or a hole into a large dark cavity.[1] The lens is focused on the entrance portal to this cavity, and the exposure is taken long enough so that the surrounding light area is nearly (but not quite) saturated. A ratio of the apparent average brightness within the hole to the average brightness of the surrounding area is called the "veiling glare index." You can estimate the brightness of the hole better by taking two exposures, one in which the white areas are contained within the dynamic range, and one which is deliberately overexposed by a known factor to favor the dark area.

The sensing system has to be carefully calibrated so that this number means something. Photographic film is only approximately linear, especially over the span of illuminations we wish to measure; so unless you possess a densitometer and make a good set of exposures straddling the dynamic range of the film, you can generate questionable results. The usual way amateurs would perform this test is by using CCD cameras. Of course, the CCD camera must be calibrated by subtracting dark images and flat-fielding. Another possible interfering factor in using CCDs is their strong red and near-infrared sensitivity. If you are interested only in the visual veiling glare, you may wish to use a filter.

Unfortunately, an amateur cannot cheaply construct a hemispherical apparatus large enough to do this measurement properly (such an apparatus for small ap-

---

[1] Geary, J.M. "Introduction to Optical Testing," *SPIE*, Vol TT15, 1993, pp. 1516.

ertures is described in Barton, 1981[2]). The amateur is forced to make tests in the field that are not so much "measurements", as "estimates." Still, making such estimates is a worthy pursuit because they can point out a severe difficulty with a telescope that did not appear until it was assembled. More importantly, before-and-after tests in the same geometry can measure relative differences between baffling schemes, even though the absolute glare is not well measured. These estimates of glare are especially important with the current popularity of open-tube structures.

The customary way that a telescope is used is pointing more-or-less into the center of the hemispherical dome of the night sky. This geometry is contrary to the way that a convenient test of veiling glare can be conducted, which is to image a light-trap target near the horizon. Half the usual dome of light is missing unless you happen to be observing from a beach or over snow. You may approximate the missing half-dome of light by laying the lip of the horizontal telescope on a large white card or table. The reflected light will simulate the other half of the sky. With big Newtonians this sky reflector might be one or more white cloth sheets on the ground.

Use your own judgment as to how to handle the unusual illumination of an open-tube telescope heeled over on its side. On one hand, having the whole open sky shining sideways into the tube may be unfairly harsh. On the other hand, if the scope is in actual use, you have direct skylight shining into the front-end of the tube from all directions, a situation that is lessened when looking at the horizon. You may have to arrange white or dark cards below the tube to simulate usage conditions. Close off the rear of the mirror cell if an unnatural amount of light is leaking into the system that way. Also, an open tube may require a shield extending sufficiently that direct skylight cannot be reflected onto the diagonal.

The light-trap target can be made inexpensively out of a cardboard box fronted by a white card (a corrugated box with a black painted interior is fine, but do not make it too shallow). A hole is then cut through the card and the card is attached to the box. The hole should be cut so that it is roughly $1/8$ to $1/10$ the linear dimension of the CCD's image. In the field setup, the box should be placed near enough so that the card is large in the CCD's field of view and far enough so that the focal plane is not greatly displaced from its usual position. Certainly the target should be at least 40 telescope focal lengths away, if not more. The hole is the light trap and, as long as the box is sufficiently nonreflective in the interior, its brightness can be considered a close approximation of zero (it is proportional to the ratio of the hole area to the inner box area). Even though no light trap can be dead black, it is far darker than required for the veiling glare estimates contemplated here.

The brightness of the card must closely approximate the average brightness of the whole scene in front of the telescope. You can adjust its apparent brightness somewhat by tilting the target's face toward or away from the sky. Also, the irradiance on the target can be increased by laying a white reflector in front of it, just

[2] Barton, N.P. "Measurement of transmittance and veiling glare index," *Proc. SPIE*, Vol. 274, 1981. Reprinted in *Optical Transfer Function: Measurement* ed. by L. Baker, Volume MS 60, *SPIE*, 1992, pp. 239–245.

as you did with the telescope.

This difficulty with accounting for card illumination points out the biggest inaccuracy in this estimate. Perhaps the best time to conduct a veiling-glare measurement would be the deepening twilight, when the Sun is gone and the Moon or streetlights do not yet contribute enough of the total illumination to skew the results. Cloudy skies simplify the problem of illumination non-uniformity.

Another difficulty is the way the CCD transfers the image off the chip. Many astronomical CCD cameras are not mechanically shuttered at all, but merely drag the collected image quickly off the chip face. The dark area could be transported across the image of the white card, unfairly giving an elevated veiling glare ratio. Exposures should be long enough to diminish the exposure during the transport time (again arguing for deepening twilight). Also, black stripes painted in the "drag" direction of the target would further diminish problems caused by this form of shuttering.

10.1.3 Expected Veiling Glares

In camera systems, one reference[3] gives these figures of merit for the veiling glare index (VG):

$$VG \leq 1.5\% \quad \text{very good}$$
$$1.5\% < VG \leq 3\% \text{good}$$
$$3\% < VG \leq 6\% \text{poor}$$
$$6\% < VG \quad \text{very poor}$$

Of course, we should expect better from a telescope because it has fewer internal surfaces in the optics. What value should we assign? A later article by the same authors[4] estimated the veiling glare that resulted from the camera body alone at about 0.5%. Thus, we can probably place the "very good" cut-off for a refractor or tilted component reflector at 0.5%. At the other end of the scale are instruments such as coronagraphs, which demand veiling-glare ratios of about nothing. Luckily, makers of coronagraphs know the incoming direction of most of the spurious light that must be baffled out.

A typical Newtonian or Cassegrain reflector has more irreducible sources of scattering or diffraction, so that its "very good" cut-off is probably nearer 1%. That is the sum total of dirt (0.1 to 0.2%), spider diffraction (0.3 to 1.0%), and imperfect painting and baffling in the tubes (0.1 to 0.5%),[5] so 1% will be a worthy

---

[3] Kondo, H., Y. Chiba, and T. Yoshida, *Proc. SPIE*, Vol. 274, 1981. Reprinted in *Optical Transfer Function: Measurement*, ed. by L. Baker, Volume MS 60, *SPIE*, 1992, pp. 309–313.

[4] Kondo, H., Y. Chiba, and T. Yoshida, "Simplified method for measuring and assessing the veiling glare of 35mm lenses," *Proc. SPIE*, Vol 467, 1983. Reprinted in *Optical Transfer Function: Measurement* ed. by L. Baker, Volume MS 60, *SPIE*, 1992, pp. 62–65.

[5] The way the higher number has been estimated is to assume a Newtonian with a fairly near tube opening beyond the spider. Let's guess that the limited cone of sky as viewed from the tube across from the focuser reduces the sky radiance by 2π. The squatty focus assemblies popular now among ATMs might reduce the viewed radiance only another factor of π. We are now down a factor of 20 merely from geometry alone. Now, if we have not been careful choosing our black paint, the reflectance might be as much as 10%, thus reducing the glare to $1/200$ of the sky brightness, or 0.5%. Of course, with proper baffling and choice of paints, this number can be reduced a further factor of 10.

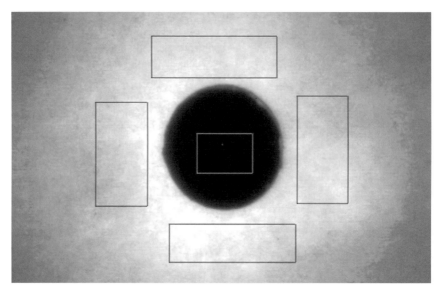

Fig. 10.1.2 *An example veiling-glare target*

target to achieve in these instruments. If we are to accept scaling of the lesser values, the 3% poor and 6% very poor grades become 2% and 4% in this constricted figure-of-merit. These numbers will need to be refined as people actually report doing this estimate.

I suspect a lot of telescopes out there will, at first, be closer to the "poor" rating than the owners realize. Using this estimation method and baffling improvements, particularly in the often neglected focuser tube, they will be able to achieve images with better contrast. Perhaps the best use of this estimation technique is not in measuring the absolute veiling glare index (it is not that accurate), but in getting a handle on which glare-reduction changes are working. You take a baseline measurement when you begin modification and another after you make the attempted baffling improvement. This will tell you what is working and what is a waste of time.

10.1.4 Addendum

We see the type of display you want to investigate in Figure 10.1.2. Indicated are the regions where you select the dark levels and the comparison levels. Average all four comparison levels to derive the veiling glare index. This photograph is taken with a video camera in the laboratory, but it demonstrates the same principle. In a measurement on a telescope that had not been carefully baffled, I got a veiling glare index of between 1.5 and 3.0 percent.[6]

---

[6] *Sky and Telescope*, Jan. 2001.

10.2 A Guide to Making Schmidt Correctors
By Robert Pfaff

A few notes on making a Schmidt corrector by the vacuum method. These were originally posted in 8 sections to the ATM listserver.

10.2.1 First—Why the Vacuum Method?

- It is possible to generate a zone-free plate with almost no testing.
- The process is somewhat like making a spherical mirror (no aspherizing to do).
- If the plate is a little under- or overcorrected, you can put the correction on the primary mirror.
- Almost any good glass can be used (I used Bk7 and plate glass).
- Any advanced ATM with access to a metal lathe, or who knows a machinist who could make the vacuum pan, should have no major problems.

10.2.2 Literature and Formulae

The formulae and math needed to make a corrector plate are a bit much to describe here. Besides, they can all be readily found.[7] Most of the information presented here is from Dr. Edgar Everhart.[8] I used the partial vacuum method with ½ correction each side of the plate. There is no need to calculate Young's modulus because the plate deflection will be measured with a dial indicator (see page 389 of the *S&T* article mentioned above). Also, you will not need to compensate for the weight of the tool as the pan is filled with water which will prevent further deflection. Be sure to do the example on page 392 to check your math.

10.2.3 Tools

- The vacuum pan as described in the next section.
- A 1-inch micrometer to measure the wedge of the corrector plate (0.001-inch). Enco sells imported micrometers for as little as $10.00.
- A dial indicator that reads to 0.0001 inch. The corrector plate will be deflected in the range of 0.002 inch to 0.007 inch. Enco has one listed for $39.95.
- A convex grinding tool to grind the plate. I used a tile and plaster tool. The tool was generated by grinding the back of my primary mirror blank against it. To generate my larger 16-inch tool, I glued a ½-inch plate glass disk to a plaster disk and ground it against the tool, see Figure 10.2.1. Remember, the tool will be convex in the range of 0.002 inch to 0.007-inch—almost flat. Use the spherometer to check this accurately.

---

[7] *Sky and Telescope*, "The Vacuum Method of Making Corrector Plates", June 1972, pp. 388–393.
[8] *Applied Optics*, "Making Corrector Plates by Schmidt's Vacuum Method" May, 1966.

Fig. 10.2.1

You can stop fine grinding at about 500 grit.

- A special spherometer has to be made to fit the plate diameter. The spherometer should be large enough to just set inside the edges of the plate. I built what I call a hybrid two-footed spherometer. A third leg at 90° from the center like a "T" shape is added so it will stand up leaving both hands free. Later, you can replace the dial indicator with a micrometer head and have a great two-footed spherometer to measure the sagitta of almost any mirror. For this, you will have to accurately measure the distance between the balls.[9]

- Your lungs. No, you do not need a vacuum pump to pull the slight vacuum needed.

10.2.4 The Vacuum Pan

A metal lathe must be used to make the vacuum pan. I machined my pan from 6061T6 aluminum plate stock, which is very common in the metal world. I purchased a square piece of plate (1 inch thick for my 11-inch pan and a $1\frac{1}{2}$ inches thick for my 16-inch pan). The corners were cut off with a wood table saw. A cheap, carbide-tipped saw blade covered with lots of motor oil was used on the saw to do this. Wear a safety face shield for this operation.

Three blind $\frac{1}{4}$ x 20 holes are drilled and tapped in the back of the plate for

---

[9] See *Advanced Telescope Making Techniques*, Vol. 2, p. 100).

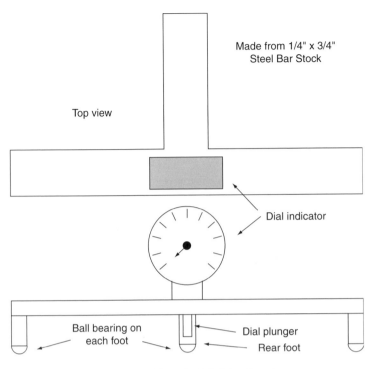

Made from 1/4" x 3/4"
Steel Bar Stock

Top view

Dial indicator

Ball bearing on
each foot

Dial plunger

Rear foot

Fig. 10.2.2 *Diagram of the author's ball-bearing spherometer.*

bolting to the lathe faceplate. A spacer, such as a washer, is placed on each bolt between the faceplate and the pan plate. I used ¹/₂-inch ball bearings instead of washers. Do not bolt directly to the faceplate as this will distort the pan plate. Machine the pan ¹/₂ inch to 1 inch larger than the optical diameter of the Schmidt plate. The front side of the pan is turned to about 0.100 inch to 0.150 inch in depth. The inside of the pan is the optical diameter of the Schmidt plate. In my case, I machined a ¹/₄-inch seal rim around the pan. This can be less, but I think the extra width of the rim gives a better seal. Also, the ¹/₄-inch rim is covered by the Schmidt corrector cell, which will cover any turned edge.

After the machine work on the pan is complete, drill an internal L-shaped hole in the pan that enters the side of the pan and exits into the pan. (The hole should be ¹/₈ inch or less in diameter—see Figure 10.2.3.)

A small valve, such as a good gas valve, is fitted to the entrance hole on the side of the pan. A valve that had ¹/₈-inch pipe thread was fitted to the pan by using a drill and pipe tap. I had problems with some valves leaking, causing loss of vacuum. To fix the problem, I disassembled the valve, lapped the parts and reassembled it with laboratory vacuum jar grease. A short length of small diameter hose is attached to the valve, which is used to draw the vacuum.

The last step in making the pan is to lap the seal rim. Lay a piece of plate

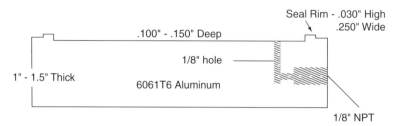

Fig. 10.2.3 *Diagram of the vacuum pan.*

Fig. 10.2.4 *The vacuum pan with corrector and spherometer in place.*

glass on some carpet. With some fine grit grind the pan rim against the plate glass. Check the rim at many points around it for flatness. Get the rim flat to 0.0001 inch. Use the spherometer that was made to do this. This is the only critical part in making the pan, and like mirror making, it is ground in. It is also important to check this at about the same temperature at which you will be grinding and polishing. I painted the inside of my pan with black lacquer. This helps when checking for pits as the plate is polished out.

10.2.5 The Glass

I have no knowledge of the glass or optical business. I can only relate my experiences with finding glass for my two projects.

For the 11-inch corrector, I was able to obtain a piece of Bk7 with a coating that acted as a filter. The wedge was only 0.0002 inch, and the coating ground away immediately. There were no problems here. With the 16-inch corrector, I was not that lucky. In this case, I needed a 16-inch x ½-inch plate. Plano-parallel Bk7 would do the job nicely; however, a quote of more than $1,000 pushed it out of my low-budget range. I talked with another glass supplier that had a glass called "water white." The dealer said it was plano-parallel and was for optical windows and filters. Yet, even though it sold for $100.00 per square foot, it only came in ¼-inch thick. Next, I located a glass dealer who makes glass tabletops. I ordered a 16-inch tabletop in what the glass dealer called "crystal white." Also ordered was a 16-inch disk of standard green tinted plate glass. Both were unedged. This was a big mistake on my part as the edge had ⅛-inch chips all around. The crystal white, although it was only lightly tinted, cost twice as much as plate glass and had 0.015-inch wedge! Not being up to grinding 0.015-inch wedge from a 16-inch disk, I ordered another 16-inch (edged and polished) disk. This one had only 0.0004-inch wedge. Yes, at a thickness of ½ inch the corrector does have a green tint, but in my Schmidt camera it is no problem.

I think any good glass can be used in a plate if the thickness is in the correct range and the wedge is 0.001 inch or less. The index of refraction only needs to be in the ballpark. Remember, the original 48-inch Schmidt had a plate glass corrector, and plate glass is still the glass of choice for some of the companies manufacturing SCTs.

10.2.6 Mounting the Corrector to the Pan

The first thing to do is zero the spherometer dial indicator. Adjust the dial indicator in the spherometer so that there is enough range to read the concave plate and the convex tool. Lay the plate on some carpet, then place the spherometer on top and zero the indicator. Next, lay the spherometer on a flat surface, such as a machinist's granite flat. Take a reading and record it. Any stable flat could be used if you mark it where the spherometer will sit for a reading. This is needed to calibrate the spherometer in case you bump it or calibration is lost some other way. This reading may be different from the one that was recorded on the glass plate, because the glass may not be flat.

Next, seal the corrector blank to the pan. I put a thin coat of laboratory vacuum jar grease on the rim of the pan. Other thick greases could be used. A friend had good luck with non-hardening Permatex. The pan is filled with distilled water, which should be preheated, and then cooled to drive the air from the water. Hard water can etch glass. Lay the plate on the water filled pan, trying to keep out as many bubbles as possible. Place the spherometer on the plate. Wearing safety goggles since plates sometimes break, suck on the small hose connected to the pan valve while watching the dial indicator. When the plate is deflected the correct amount, close the valve. I put RTV around the edge of the plate to prevent vacuum leaks.

After a few hours, the plate will lose some of its deflection due to a weakening of the grease seal. When this occurs, redraw the deflection. It should be stable for 24 hours before any grinding is done. A 0.0002-inch change from day to day

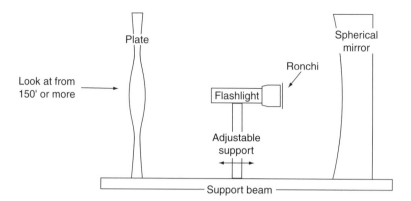

Fig. 10.2.5 *Diagram of the testing setup.*

is normal. Note the size of the bubbles under the plate. An increase in size will indicate a vacuum leak.

Once the plate deflection is stable, start grinding with the preformed tool. You may start with a grit size of about 500. The panplate is on the bottom. Use very light pressure in your grinding. Check the tool with the spherometer frequently as it will have to be reground to shape now and then. Also check the plate deflection to be sure it is holding. Grind through the grits, as if making a mirror, down to about 4 microns. Polish out with a standard pitch lap. At the beginning and end of each session, recheck the deflection in case it changes due to leaks.

Some ATMs have suggested rotating the plate a few times during grinding to minimize astigmatism. I prefer not to break a good vacuum seal and loose the calibration. Because the plate is to be ground on both sides, you should mark the plate and pan. When side one is done, release the vacuum, turn the plate over, and rotate it 90° with respect to side one. This should cancel most of the astigmatism.

Now reseal and draw a vacuum for work on the second side. Grind and polish as before. Release the vacuum, and the plate is finished!

10.2.7 Testing

A simple way to test a Schmidt corrector plate used in a classical camera is shown below:

A test rig is made of wood. The corrector is at the radius of curvature of the mirror. The flashlight is near the focus of the mirror, adjustable, and pointing at the mirror. The flashlight is covered with a Ronchi grating. Look at the corrector from a distance of 150 feet or more—the greater the distance, the better. You should see straight lines. Lines flaring at the top or bottom indicate that the setup is out of alignment. Unless the bands are extremely bad, you will want to make any necessary corrections on the primary. My two plates needed no correction. I could only see some aberration with one Ronchi line showing, which was probably due to the primary mirror. For correctors in SCTs or Wrights, which are cous-

Fig. 10.2.6 *The author's testing setup.*

ins of a true Schmidt, the corrector is tested in the complete system. A flat (water, mercury, or glass) would be ideal. An artificial star (as described in *Star Testing Astronomical Telescopes* by Suiter) with a Ronchi grating should work. As in the case of the true Schmidt, the correction, if needed, should be done on the primary.

10.2.8 A Few Extra Notes

- You might ask why the pan size is larger then the plate size. The extra lip provides an easy way to seal the plate with RTV. Also when you are done with the pan, the center can be cut out, and the outer ring machined into a cell for the corrector.

- Another tip, if you use a machine for grinding-polishing, is to cover the machine table with recycled pitch about ¼ inch to ⅜ inch deep. Groove about every 2½ inches. Warm the pitch with a lamp, place a 4 mil plastic sheet on top, and lay the pan on top of the plastic. Mark a spot on the table and pan so the pan can be replaced at the same location on the table each time. This will help prevent astigmatism. Unlike carpet, the pitch will float the pan but will not flex as the tool overhangs the table on each stroke. (This also works great for mirrors.)

- With the help of Tom Lum, a knife-edge test for the corrector was found in *ATM III*, p. 364. Place the corrector directly in front of the paraboloidal mirror, and use a knife-edge tester at the focus to read the zones. It is almost like testing a parabolic mirror, but the zone measures are double the distance. The formula is delta $\Delta R = y^2/R$ when both the knife-edge and light source move together. There is more interesting information about Schmidts, including an article about Bernhard Schmidt, in *ATM III.*

- Make the grinding tool a little larger than the corrector plate. This will give a better surface on which to place your spherometer when checking the tool's R.C.

- If you have polished glass, you can check it for striae by placing the glass in front of a spherical mirror and observing the glass with the knife edge test. The striae will appear as streaks or veins.

- I found another article on Schmidt plates using vacuum deformation in *Applied Optics*, January 1972. Here they used slow-setting plaster pads in the pan to provide stiffness to the plate.

I hope this will persuade someone that a willingness to work with a little math, metal, and glass can be fun and rewarding.

10.2.9 Other Articles of Interest:

1. "Construction of a 12-inch Schmidt-Cassegrain," *Sky & Telescope*, November 1976, pp. 382-386. (Some information on vacuum pan design.)

2. Wright-Schmidt Construction Notes, *RTMC 1990 Proceedings*. (Special data on plate thickness ($\frac{1}{30}$ to $\frac{1}{40}$ of plate diameter—$\frac{1}{34}$ is best.) Also includes a write up on the pan.)

3. Ingalls, A. (ed). *Amateur Telescope Making,* Book 3. Scientific American, pp. 371–372.

10.2.10 Related, But Not Using the Vacuum Method

1. "An Australian 11.5-inch Wright Telescope," *Sky & Telescope*, May 1972, pp. 320–326.

2. "A Diffraction-Limited Schmidt-Cassegrain Telescope," *Sky & Telescope*, April 1966, pp. 231–235.

3. "A Phoenix Amateur's 12.5-inch Schmidt-Cassegrain," *Sky & Telescope*, April 1970, pp. 254–260.

4. "A Schmidt-Cassegrain Optical System with Flat Field," Part 1, *Sky & Telescope*, May 1965, pp. 318–322.

5. "A Schmidt-Cassegrain Optical System with Flat Field," Part II, *Sky & Telescope*, June 1965, pp. 380–384.

6. Rutten, Harrie, and Martin van Venrooij. *Telescope Optics Evaluation and Design.* Richmond, VA: Willmann-Bell, Inc.

Fig. 10.3.1

10.3 The Perfect Telescope is a YOLO Reflector
By Beat Kuechler

Building a very good telescope of my own has always been a wish of mine. Aside from the possible financial advantage of building one's own telescope, the main fascination for me has always been that of being a "researcher," while using my own materials. Until a few years ago, all I had to work with were directions for building a Newtonian. I constructed several of these, the largest of which had a 250 mm aperture.

About two years ago, I ran across an article on the Yolo system in the *Journal of the Swiss Association of Astronomy* by H.G. Ziegler. He presented the system and announced the creation of an ATM group, which I immediately joined. Since then, I built a prototype with a 150 mm aperture and an instrument with a 200 mm aperture for observation on my roof balcony.

The first glance through my 200 mm f/12 Yolo shows the eminent advantage of the unobstructed reflector. The contrast transfer, which is highly praised in literature, lives up to its reputation, enabling recognition of the planetary details that are not visible with my Newtonian systems. It enables fine structure analysis of the lunar surface, and with adequate seeing, the double star Zeta Boötis can easily be resolved.

One can expect a Yolo telescope to be more difficult to build than a Newtonian. In issue Number 1 of the *ATM Journal*, José Sasián presented various TCTs that can be built by amateurs, including the Yolo by A. Leonard.

The secondary mirror of the Yolo is a toroid with two radii of curvature. The radius difference of my secondary is 144 mm and was made approximately by fine grinding a few hours. Polishing and figuring was then no great problem and seemed easier to me than parabolizing my Newtonian mirrors. The surface can be exactly controlled by null testing.

Performance of the Yolo could satisfy even with the primary mirror left spherical. Spherical aberration, however, can be totally corrected by hyperbolizing. Measuring the longitudinal aberration of the hyperboloid with the Foucault method requires some practice and patience, but the surface varies so little from spherical that failures are not to be expected during figuring. Tested on a star, the defocused point spread function on both sides of the focal plane of my Yolo were identical.

I was the first of our ATM group to finish an instrument; and because of the good results, I encouraged the others to continue their task. We owe much to our founder, H.G. Ziegler, who did the basic work.

I can recommend the construction of a Yolo to anyone who wants to build an excellent, but not exceptionally large, telescope.

10.4 An Improved Moving-Mirror Focuser For Compound Telescopes

By Thomas A. Dobbins

Changing the distance between the primary and secondary mirrors in a Cassegrain telescope changes the location of the instrument's focal plane by a factor that rather closely approximates the square of the amplification factor of its secondary mirror. The Cassegrains of 12- to 40-inches aperture that were produced for professionals and universities during the 1960s, by firms like Boller and Chivens, featured motorized lead screws ("linear actuators") that moved the secondary mirror back and forth for remote focusing. The wires supplying power to the motor were run along one of the vanes of the spider.

The manufacturers of the popular catadioptric derivatives of the Cassegrain (i.e., the Questar and Quantum Maksutov-Cassegrains) have also taken advantage of the "optical leverage" inherent in these systems to achieve an unprecedented degree of versatility. A displacement of the primary mirror in a typical Schmidt-Cassegrain (f/10 effective focal ratio with an f/2 primary and a 5x secondary amplification) by only one millimeter moves the focal plane by about 25 millimeters! Small wonder that these instruments can focus on subjects only a dozen feet away and accommodate a wide variety of accessories, such as off-axis guiders and erecting prisms.

Image quality can suffer if the primary mirror to secondary mirror spacing in any compound telescope is changed too much. The optically tolerable range of motion is determined solely by the focal ratio of the primary mirror, regardless of the amplification factor of the secondary or the aperture of the system. If on-axis diffraction-limited performance is to be preserved, the nominal spacing in a sys-

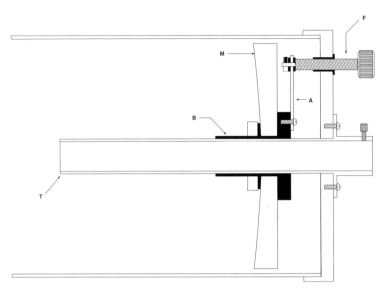

Fig. 10.4.1 *Conventional Moving-Mirror Focus Mechanism.*

tem with an *f*/3 primary should not be varied by more than 1.0 mm, while an *f*/3 primary allows a displacement of the focal plane corresponding to these changes, spacing is determined by the secondary mirror's amplification factor. Most commercial catadioptrics permit the user to vary spacing well beyond the limits of optimum optical performance.

The internal focusing mechanisms in commercial SCTs are, with minor variations, constructed like the system depicted in Figure 10.4.1. The primary mirror (M) is mounted on a bushing (B) that slides to and fro over the central baffle tube (T). An arm (A) attached to the rear of the bushing captivates the tip of a focusing screw (F). When the focusing screw is turned in one direction, the primary mirror and its supporting bushing are pushed forward toward the secondary. Turning the focusing screw in the opposite direction pulls the primary away from the secondary. Some manufacturers have machined a lengthwise internal groove in the sliding mirror support bushing that follows a guide pin in the baffle tube (or vice versa), in order to eliminate any radial motion that may be imparted to the primary mirror by the torsion of the focusing screw.

The alternate application of opposing (pushing and pulling) forces at a considerable distance from the primary mirror's axis of motion (i.e., the center line of the baffle tube) results in a phenomenon that is called "mirror shift," which manifests itself as an annoying displacement of the image when the focusing knob is turned. In better instruments, fabricated to tight mechanical tolerances, this apparent displacement may be an unobtrusive 10 arc seconds in amplitude—difficult to detect without the aid of an ocular equipped with a crosshair reticle. The typical commercial SCT, however, exhibits a displacement of 30 to 40 arc seconds; it is by no means unusual to encounter instruments in which it exceeds a minute of arc.

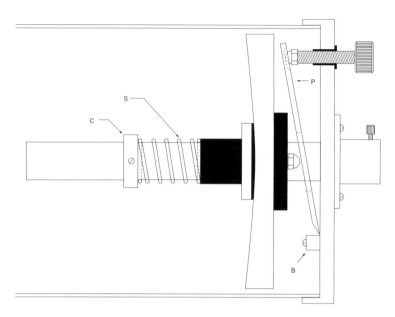

Fig. 10.4.2 *Improved Spring Loaded Focus Mechanism.*

While mirror shift may only be a source of chronic irritation to visual observers, it can be a real handicap for astrophotographers. After painstakingly achieving precise focus, the hapless astrophotographer all too often finds that the primary mirror of his telescope has settled backward due to the action of gravity during a long exposure, shifting the location of the telescope's focal plane and producing distended star images.

In recent years, the market has responded to the growing demand for a better way to focus these popular telescopes. University Optics (Ann Arbor, Michigan) offers a rack-and-pinion focuser, and Jim's Mobile, Inc. (Lakewood, Colorado) offers a Crayford focuser. Both screw onto the industry-standard accessory thread (2-inch pitch diameter, 24 threads per inch) of the Celestron and Meade SCTs, permitting the user to achieve coarse focus with the manufacturer's moving-mirror mechanism while avoiding image shift by utilizing the accessory focuser for final focusing.

The amateur, building a Cassegrain-type telescope from scratch, can avoid the pitfalls of mirror-shift while retaining the benefits of a moving-mirror system by incorporating a mechanism kinematically designed so that the forces transmitted by the focusing screw are applied directly along the mirror's axis of motion. The essential concepts of the following design were kindly related to the author several years ago by Dudly Fuller of the London firm of Broadhurst, Clarkston, and Fuller. He stated that it had been used successfully in a number of compound telescopes of British construction over the years.

As shown in Figure 10.4.2, a stout compression spring (S) and retaining collar (C) are placed over the baffle tube in front of a sliding mirror support bushing. Spring tension can be adjusted by varying the location of the retaining collar.

The baffle tube and the mirror support bushing can be fabricated from aluminum. The outside diameter of the baffle tube should be no less than 0.0010 inch and no more than 0.0025 inch smaller than the inside diameter of the bushing. If possible, make the mirror support bushing at least twice as long as the outside diameter of the baffle tube. Both of these components should be anodized to prevent galling.

After considerable experimentation, I found that the best lubricant between the sliding mirror support bushing and the baffle tube to be "STP Oil Treatment." Some readers may remember television advertisements for this product that featured unsuccessful attempts to suspend the tip of a flat-bladed screwdriver dipped in this material between a thumb and forefinger. It has a consistency similar to liquid dishwashing soap, enabling it to be retained by capillary action, and does not thicken excessively at low temperatures.

The force of the compression spring drives the mirror support bushing backwards against a pivoting plate (P), which is slotted to permit the baffle tube to pass through it. The rear of the mirror support bushing is drilled and tapped at two locations on its center line 180° apart to receive two headless set screws, which are allowed to protrude far enough to attach a pair of acorn nuts. Lubricated with a bit of viscous grease during assembly, the acorn nuts become points of sliding contact between the bushing and the pivoting plate, ensuring that the force imparted by turning the focusing screw is always transmitted directly along the axis of motion of the primary mirror.

The use of acorn nuts has proven very satisfactory with primary mirrors of 6 inches diameter; but the additional weight, and required spring tension of larger mirrors, requires the use of a pair of radial ball bearings as rolling, rather than sliding, contact members. The bearings are attached to the opposite sides of the mirror support bushing with shoulder screws that also serve as axles.

The point of contact between the pivoting plate and the focusing screw is a shallow, elongated depression milled into the center of the rear surface of the pivoting plate with a ball-end mill. Care should be taken to avoid any tool chatter during this machining operation, for it is essential that the surface be absolutely smooth. The end of the focusing screw is fitted with an acorn nut which rides in the depression, lubricated by a small amount of viscous grease. The focusing screw can be cut on a lathe from a piece of stainless steel round stock. A $\frac{1}{2}$-inch diameter shaft with a 24- to 32-pitch thread is recommended. For adequate focusing sensitivity, the thread should be no coarser than 24 threads per inch, while 40 threads per inch will prove too fine.

A threaded focusing screw bushing is machined from brass and either pressed or epoxied into a hole in the tailpiece. A recommended length is two to three times the diameter of the focusing screw. The focusing screw is lubricated with a medium-viscosity grease before being screwed into its bushing. The spring tension that is transmitted to this sub-assembly by the pivoting plate pre-loads the threads, providing instantaneous response when reversing direction and counteracting the effects of thread wear that may arise over time. The problem of mirror "settling" that plagues astrophotographers is also eliminated.

The opposite end of the pivoting plate is beveled and rests against a backstop bar (B), which is, in turn, attached with two screws to the interior surface of the tailpiece (or, preferably, to an intermediate collimation plate that is in turn attached to the tailpiece with three pairs of opposing screws) beside a strip of coarse (~80 grit) sandpaper that is epoxied in place to prevent any slippage of the pivoting plate against the interior surface of the tailpiece. The length of the backstop bar is equal to the width of the pivoting plate. Alternatively, a hinge may be employed to attach the pivoting plate to the tailpiece, but most of the examples I tested exhibited too much "slop" to give optimum results.

The critical angle of the backstop bar is adjusted until the beveled edge of the pivoting plate remains in contact along its entire length when the focusing knob is screwed in and out over its range of motion. If precise perpendicularity is achieved between the backstop plate and the axis of the elongated depression on the rear surface of the pivoting plate in which the tip of the focusing screw rides, all traces of image shift will be eliminated, even at high magnifications.

Following initial assembly, point the telescope at a distant terrestrial target and insert an ocular equipped with a crosshair reticle. If image shift is detected when the focusing knob is turned in opposite directions, remove the tailpiece and carefully adjust the backstop bar until image shift disappears.

Ed. Note: Tom is co-author, with Charles Capen and Donald Parker, of *Observing and Photographing the Solar System.*

10.5 Table Mountain Star Party 1996

By William J. Cook

This year's Table Mountain Star Party was held on July 12 and 13, with many individuals arriving earlier in the week. In the past, attendance has ranged between 400 and 500. This year, 629 people were on the mountain by mid-day Saturday, and by nightfall more than 700 astronomy and telescope making enthusiasts were there. While, for some, the Northwest maintains its reputation for clouds, rain, and generally poor seeing conditions, it was certainly not evident by the mix of amateur astronomers and telescope makers present. Some had driven more than a thousand miles to attend the event as verified by license plates from as far away as Texas and the Midwest. It would appear that the Table Mountain gathering has achieved a footing equal with events such as the Winter Star Party and the Texas Star Party.

Many attendees, some of whom having attended each year since it went regional in 1991, felt that this year's event was the best. It was definitely the most enjoyable for me. Perhaps it was because of the nice weather (mid 80s without a cloud in sight), or because I put the camera away and just tried to enjoy astronomy for a while. Sometimes I get so wrapped up in trying to get the *Journal* out that I totally forget why I wanted to get into optics in the first place. That did not happen this year. I had an observing plan! Early in the day, I went around the telescope fields to determine where instruments of special interest were being positioned. Then, I put out the word that I was considering getting a 12.5-inch to 16-inch Dob, and that I wanted to find some of the ugliest telescopes on the field (in that size range) to look through that evening—knowing that the snazzy telescopes would be collecting long lines of anxious observers. I received a number of gracious offers and, with twilight's last gleaming, I set out to have some fun.

Like basketball players, amateur-built telescopes just get bigger and bigger. This year there were at least 3 telescopes larger than 36 inches in aperture, including Steve Swayze's 40-inch, and a 41-inch (computer controlled) monster called "Hercules," attended by Dan Bakken. Considering that any instrument I get will have to be portable, my interest in this telescope was more casual than critical. But then, casual observing is more fun anyway. When I arrived for a view, the line was surprisingly short; and being vertically challenged myself, I felt comfortable with the fact that it was pointed to objects near the horizon—Jupiter or perhaps Hale-Bopp. However, just as I was getting ready to take my place at the eyepiece, a problem with the drive system popped up, and the eyepiece headed skyward without human assistance, ending up somewhere between where it would be at the right heights for Goliath and the Jolly Green Giant. I decided to move along to other instruments and come back later in the evening.

Another way in which this year's event was special was related to the atmosphere—or should I say, ambiance. Things seemed much more casual for attendees and administrators alike. I am certain that part of the subdued tone was dictated by the higher than usual temperatures. Gene Deitsen, one of the senior organizers, felt that it was because some changes had been made in the parking arrangements. It could have been the fact that organizers were getting accustomed to the routine of orchestrating such an event.

An additional area, perhaps as large as two football fields, was opened up for parking and observing. Many attendees had the option of simply setting up camp and their telescope right out of the back of their vehicles. This spaciousness added to the feeling of reverence that carried on into the evening. At 10:40 there was a slow-moving fireball in the east which lasted for at least 4 or 5 seconds, and when I called out for a "heads-up," I was heard by people 40 or 50 feet way—I was by no means shouting.

There were many beautiful and interesting ATM projects on the mountain this year. Unfortunately, most were not entered in the telescope judging competition. There were only 13 entries, and not all of those were telescopes. Perhaps this was because award winners from years past wanted to give others an opportunity to shine, or due to the fact that they were too busy preparing themselves for a few nights of cloudless skies and moderate evening temperatures.

While short focus Dobsonians dominated the scene, large aperture refractors made their mark as well. Kreig McBride of Bellingham, Washington, had the larg-

Fig. 10.5.1 *The telescope judging team gets a closer look at Hercules, the 41-inch (computer controlled) giant entered by Dan Bakken.*

Fig. 10.5.2 *Greg Jones describes his Laurie-Houghton telescope to Dennis di Cicco.*

est refractor again this year.

The 8-inch *f*/13.6 was painted bright yellow and green, and looked like it had been originally owned by an over-zealous John Deere dealer. There were also several Astro-Physics refractors present, including one owned by Tom Masterson (also of Bellingham) on a very well-made equatorial mount.

The most interesting ATM work was presented by Greg Jones of Lake Oswego, Oregon. He exhibited an 8-inch Lurie-Houghton taken directly from *Telescope Optics Evaluation and Design* by Rutten and van Venrooij. While I am decidedly biased toward Houghton derivatives, it was plain that I was not alone in my thinking. Greg not only made the optics himself, but the fork mount, convenient counter balance system, the electronic drive mechanism, and snap on eyepiece/camera mount as well. The instrument had a wide, relatively flat field. Even so, Greg was not totally pleased with its performance. Recognizing the photographic potential of such an instrument, he plans on removing the four-vane spider and touching up the corrector to mount the secondary directly to it.

It was a pleasure to meet old friends and be introduced to new ones. Table Mountain 1996 was the most enjoyable Star Party I have attended in recent years, and I look forward to next year's event.

10.6 Riverside 1996
By Kreig McBride

Camping for the weekend? One of the most important items to plan for is food. Take four hungry guys, a motor home, and a pre-planned menu; release them into a grocery store to pick up the goods; and good-bye menu, hello junk food! I gained 5 lbs on my second trip to Riverside.

Now the four of us were looking forward to a warm, sunny vacation after nine months of Washington clouds and rain, and we dressed accordingly with shorts, T-shirts, and a gallon of sun screen. Of course, it snowed! Luckily, we brought our snowshoes and had great times looking into telescopes with white dew caps.

The Riverside Telescope Makers Conference is a bit of a misnomer as most of the activities are related to vendor sales of new, used, and surplus government equipment, complemented with plenty of talks covering a wide range of subjects. CCDs are a hot topic, and it seemed that a majority of the talks were related to this subject. There were plenty of photo shots of the recent Hyakutake comet provided by the audience, and the variety provided a real insight into the many imaging techniques used. This was a great primer for those planning to photograph Hale-Bopp as it flies by this coming spring.

Another event, and the one I enjoy most, is looking at telescopes and related equipment on display by their owner/designers. I used to collect toy erector sets; therefore, when I saw Crystal Ackerman's 3-inch f/5 Erectonian, I had to stop for a look. The whole scope was constructed using erector set parts, and she frequently hand holds the scope using a 17 mm eyepiece for excellent views of Jupiter. I liked the simple concept of using parts that are easy to find and very adaptable to any design one might have for a small scope.

Fig. 10.6.1 *Crystal Ackerman's 3-inch f/5 Erectonian.*

Fig. 10.6.2 *David Pitou, of Boise, Idaho, brought the 20-inch Newtonian pictured here. The instrument featured many innovations.*

On the other extreme was a 20-inch *f*/4 high-tech Newtonian constructed by David Pitou of Boise, Idaho. What made this scope so interesting was not the single-arm hollow aluminum fork with constant tension spring drive, the ultra-light honeycomb carbon fiber/kevlar tube, or the double eyepiece holder located at the focus; it was the tube rotation system. Simple, elegant, and "easily" constructed, this system consisted of two nylon webbing belts, wrapped around both the tube and rollers, attached to the equatorial mount that allowed the tube to rotate while the tensioned belts held the tube securely to the mount. The tube rotated easily and maintained alignment while doing so. You just had to see it, and this design is one

Fig. 10.6.3 *Two flat bands are attached to the OTA and are the tracks for rollers that allow the tube to rotate. Clamps are used to tension the nylon straps which are wrapped around both the tube and independent rollers, holding the OTA in place while allowing tube rotation.*

Fig. 10.6.4

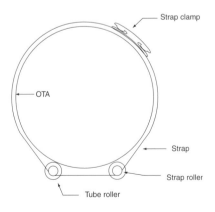

Fig. 10.6.5 *Drawing by Kreig McBride.*

of those "Why didn't I think of that?" type of ideas. In the drawing, you can see how the mechanism works and how easy it is to adjust tension, making the interchange of scopes an almost effortless maneuver. I watched as an owner of a major telescope manufacturer checked this scope out thoroughly. We may see many of Dave's ideas showing up on company scopes in the future.

Curtis Hruska showed off his very nicely made 5¼-inch *f*/30 refractor used for solar observing. The singlet lens is made of red Bk7 which acts as a pre-filter and is mounted in a titanium tube. Curtis uses a Daystar filter on this scope, and the results are impressive. So is the scope.

One of the highlights for me was meeting John Gregory of Gregory-Maksutov fame. John is a humble and wise man with plenty of excellent advice on just about anything related to optics. Even my simple and basic questions were answered with clarity, insight, and inspiration. John is currently working on a 12-inch Houghton of his own design, and I cannot wait to see this new scope when it sees "first light."

Although the weather probably reduced the attendance and size of the swap meet, the conference was still a success. Several amateur telescope makers met Sunday afternoon for an opportunity to exchange ideas and demonstrate optical fabrication techniques. The meeting was not part of the official agenda, but as interest in telescope making increases, it should (will?) be included in future conferences. I hope so. That is why I attend.

10.6.1 RTMC Awards '96

Merit Awards (in order of registration)

- Robert Pfaff, Torrance, CA, Unique Modification of Split-Ring Mount and Construction of a Folded Newtonian.
- Bill Lennartz, Riverside, CA, Excellent Wooden Construction in a Large Equatorial Newtonian.
- David Pitou, Meridian, ID, Outstanding Workmanship Using Advanced Materials.
- John Lightholder, South Lake Tahoe, CA, Outstanding Construction of a Lightweight Newtonian.
- Bruce Sayre, Applegate, CA, Excellent Design and Craftsmanship.
- James and Dan Ritts, La Canada, CA, Excellent First Telescope/Junior Division.

Honorable Mentions (in order of registration)

- James Stevens, Phoenix, AZ, Large Quick-Take-Down Dobsonian.
- Dan Schwabe, San Diego, CA, Effective Azimuth Bearing Sealing System.
- Stephen Collett, Bakersfield, CA, Craftsmanship.

Fig. 10.6.6 *Curtis Hruska shows off his 5.25-inch f/30 solar telescope.*

- Randy Johnson, Seattle, WA, Interesting Use of Materials.
- John A. Volk, Monrovia, CA, Well-Done First Telescope.

Warren Estes Memorial Award
- Don Pies, Redondo Beach, CA.

10.7 Stellafane 1996
By Matt Marulla

Well, in 1996 the latest Stellafane tradition continued—miserable weather on Friday and Saturday afternoon gave way to the best night anyone could ask for. As in each of the last four years, we had rain on Friday afternoon. It was not too bad this year; no one's tent got washed away.

Saturday dawned crisp and clear with activities getting underway quite early. The first swap tables were set up at 5:45 AM (!); and, as usual, it was a bustling mix of trash and treasure throughout the day.

This year's twilight talk was conducted by none other than John Dobson, who did not really "give a talk" so much as "talk with" the audience. Most people close a speech with "Any questions?" Well, that's how John started. When he had answered them all, he was done. The keynote speech was "The Search for ET" by Paul Horowitz of Harvard University on the subject of, you guessed it: SETI. I have to admit that the solid overcast giving way to clear skies right as the stars were coming out distracted quite a few listeners who trotted off to the observing areas in twos and threes during the talk, the author included... But enough of this— On to the scopes!

This year saw a dramatic increase in the number of telescopes entered in the mechanical and craftsmanship competition—from 16 last year to 33 this year! Maybe it is a result of interest generated by Shoemaker-Levy-9/Jupiter + Hyakutake + Life on Mars + Hale-Bopp. Astronomy has been getting more than its average share of the headlines for the last few years. But whatever the cause, let's hope the trend continues.

10.7.1 Dobs Galore

As usual, there were many Dobsonian scopes on Breezy Hill. Here are some of the highlights:

Scott Reeves's 8-inch Travel Scope collapses down to an 11-inch x 14-inch x 17-inch cube and weighs only 25 lbs. Also collapsible, Michael Hamilton's 6 inch f/9.5 Dobsonian, built by his father, Forrest, uses telescoping (no pun intended!) truss tubes.

Steve Eldridge is making quite a name for himself as a mirror maker. One of his scopes took a first place optical award last year, another received a first place optical award this year, and one of his mirrors is in John Desbiens's award winning 12.5-inch f/6.4 Dobsonian. Bill Tomaszewski's 6-inch f/8 Dob-Newt is optimized for planetary viewing. It features: (1) a unique focuser that lets the eyepiece actually drop below the level of the tube; (2) a finder bracket that incorporates a solar finder; (3) an offset primary that permits an even smaller diagonal; and (4) a spider employing a single vane held under tension with two sets of wires. Bill's scope received a special innovative component award and a 3rd place optical award.

10.7.2 Equatorial Scopes

Paul Valleli brought a beautifully restored and re-worked 10-inch f/15 Ritchey-Chrétien on a Springfield mount. The scope was originally built by Dino Argentini of Danvers, Massachusetts, in 1963. He showed it at a Stellafane tent talk in 1968. The scope was later put into storage, and the mount was damaged by battery acid sometime in the 1980s. Paul did the restoration during June and July of 1996.

Steve Durham once had a cherry tree in his front yard. After the local power company told him it had to come down, Steve harvested the wood and used it to build the tube, equatorial mount, and tripod for the scope pictured in Figure 10.7.4. He reports there are two larger trees that may come down in the near future!

Allon Wildgust of Brandon, Vermont, picked up three different awards for his 8-inch f/5.4 Newtonian. It sits on what I am taking this opportunity to christen, a "Jamieson Mount." It was first described in an article by Scott Jamieson in the May 1994 *Sky & Telescope*; this mount is a cross between a Dobsonian and an

Fig. **10.7.1** *Michael Hamilton's 6-inch f/9.5 Dobsonian.*

Fig. **10.7.2** *John Desbiens's award winning 12.5-inch f/6.4 Dobsonian.*

Fig. **10.7.3** *Bill Tomaszewski's 6-inch f/8 planetary Newtonian had a number of interesting features, including a focuser that allows the eyepiece to drop below the wall of the tube.*

Fig. **10.7.4** *Steve Durham's vision for what can be done with a cherry tree seems considerably more practical than that of a certain founding father.*

Fig. **10.7.5** *This 8-inch f/5.4 Newtonian won 3 awards for Allon Wildgust of Brandon, Vermont.*

Fig. **10.7.6** *Phil Owen's pipe mounting for his 8-inch f/6 Newtonian.*

Fig. 10.7.7 *The Arunah Hill group's 13-inch Fitz refractor.*

Fig. 10.7.8 *Doug Arion's 6-inch f/7.5 equatorial Newtonian. Instruments like this, and those of Steve Durham and Allon Wildgust, make me wish I could share the color photos with each of you.*

Fig. 10.7.9 *George Helmke's meteor camera.*

Fig. 10.7.10 *Two of Dennis O'Connell's PVC telescopes.*

Fig. 10.7.11 *Michael Hill's combination spectroscope, spectrohelioscope, and prominence scope.*

equatorial. Kinematically, it is a German equatorial, but it uses classical Dobsonian components and construction techniques. This scope is my personal choice for "Best of Show."

Phil Owen's first scope was built for under $250. Although it appears to be a basic pipe-mounting, the 8-inch *f*/6 Newtonian actually glides on Teflon bearings that are housed in the machined pipe fittings.

Doug Arion's 6-inch *f*/7.5 equatorial Newtonian has a completely open tube assembly, even for the finder. Unfortunately, Doug says his observing location in Kenosha, Wisconsin, has plenty of light pollution, which, of course, is not good for a scope this open. But . . . it is beautiful. Even the various knobs are made from oak and brass to match the rest of the scope.

The Arunah Hill group of Cummington, Massachusetts, all worked together to complete the restoration of a historical 13-inch Fitz refractor. That massive "guillotine" mounting is not overkill; the tube assembly alone weighs over 300 lbs.

George Helmke's meteor speed determination device cuts off the lens at a predetermined rate during long exposures, resulting in dashed meteor trails. The spacing of the dashes is then used to measure the meteor's speed.

Dennis O'Connell of Corning, New York, is some kind of "King of PVC Scopes." His collection includes a folded refractor and a Dobson-style solar telescope. I am not referring to the mounting here; John Dobson also invented this unique solar telescope concept where the filter and diagonal are the same piece of glass. If the filter ever breaks or falls out of place, there is no risk of viewing the unfiltered sun, because your diagonal is gone now as well.

Finally, Michael Hill of Marlboro, Massachusetts, showed his "solar observatory in-a-box." It is a combination spectroscope, spectrohelioscope, and prominence scope. There are two diffraction gratings, three concave mirrors (two 2.5-inch $f/10$ and a 3-inch $f/10$), two flats, and a slit. As I understand it (which is none too well!), the first grating images the sun on the slit in first order mode giving a classic solar spectrum with Fraunhofer lines. By adjusting the first grating, a second order spectrum can be imaged on the slit. Then, if the second grating is adjusted, a full disk image of the sun is seen that varies in wavelength. By isolating the H lines, one can observe solar prominences.

10.7.3 That's all folks!

This was my seventh Stellafane, having moved to New England from the west, uh… seven years ago! Like all Stellafanes before, it is not just the scopes, it is not just the people, it is not just the night skies; there's something magical about this place. Something about setting up your scope on the same ground that Porter and Hartness did. It is the closest thing to hallowed ground we telescope makers and astronomers have. I will be back next year!

10.7.4 Telescope Competition Summary—Stellafane '96

Mechanical Design

- 1st place: Allon Wildgust, Brandon, VT, 8-inch $f/5.4$ Newtonian.
- 2nd place: Arunah Hill Group, Cummingon, MA, 13-inch Fitz Refractor.
- 3rd place: Kelly Jons, Willoughby, OH, 60 mm folded refractor.
- 4th place: Scott Reeves, Gorham, ME, 8-inch Newtonian Travel Scope.

Craftsmanship

- 1st place: John Desbiens, Hillboro, NH, 12.5-inch $f/6.4$ Newtonian.
- 2nd place: Allon Wildgust, Brandon, VT, 8-inch $f/5.4$ Newtonian.
- 3rd place: Kelly Jons, Willoughby, OH, 600 mm folded refractor.
- 4th place: Doug Arion, Kenosha, WI, 6-inch $f/7.5$ Newtonian.

Junior (Mechancial)

- 1st place: Rachel Hubbard, Auburn, MA, 6-inch $f/4.2$ Newtonian.
- 2nd place: Chris Sousa, Townsend, MA, 8-inch $f/9$ Newtonian.
- 3rd place: Joan O'Reilly, Putnam Valley, NY, 6-inch $f/6$ Newtonian.
- 4th place: Elizabeth Ward, Ridgefield, CT, 6-inch $f/6$ Newtonian.

Special Award

- Nick Bulzacchelli, Pound Ridge, NJ, Equatorial Sundial (Helio-Chronometer).

Innovative Component Award

- Bill Tomaszewski, Coopersburg, PA, 6-inch $f/8$ Newtonian.

Optical Category (under 12½-inch aperture)

- 1st place: David Groski, Newark, DE, 8-inch $f/7$ Newtonian/Dobsonian.
- 2nd place: Allon Wildgust, Brandon, VT, 8-inch $f/5/4$ Newtonian.
- 3rd place: Bill Tomaszewski, Coopersburg, PA, 6-inch $f/8$ Newtonian.

Optical Category 2 (12½-inch aperture and over)

- 1st place: Steve Eldridge, Chester, NH, 16-inch $f/6.22$ Dobsonian.
- 2nd place: Robert Midiri, Coatsville, PA, 16-inch $f/6.9$ Dobsonian.

Junior (Optical)

- 1st place: Chris Sousa, Townsend, MA, 8-inch $f/9$ Newtonian.
- 2nd place Joan O'Reilly, Putnam Valley, NY, 6-inch $f/8.5$ Newtonian.

10.8 Telescopes for CCD Imaging, Part V
What We Learned from Comet Hyakutake
By Richard Berry

The passage of spectacular Comet Hyakutake was fortunate for those interested in CCD imaging, because it taught us how to plan for the upcoming apparition of Comet Hale-Bopp, which offers the potential to be an even more spectacular object. With the practice run afforded us by Hyakutake, we can plan more intelligently for Hale-Bopp.

Comet Hyakutake taught CCD imagers that bright comets are not deep-sky objects. Deep-sky objects and bright comets have very different properties. Comet Hyakutake was bright, large, fast-moving, and dynamic. These are not the characteristics of deep-sky objects, which are dim, small, unmoving, and unchanging. Imaging a bright comet calls for a new imaging strategy.

My images of Comet Hyakutake were carried out using a Cookbook 245 CCD camera with three different telescopes: my faithful 6-inch $f/5$ Newtonian, my 4-inch $f/5$ Tele Vue Genesis, and a 200 mm $f/3.5$ telephoto lens. As the comet came closer and grew larger and brighter, I switched from the 6-incher to the Genesis to the telephoto lens. The comet was so bright that blooming from the pseudonucleus and inner coma was a big problem. To capture the tail, my integrations were 60 seconds, but I lost the inner parts of the comet to saturation. Lesson #1 is: Think Bright. Fast focal ratios and long integration times are not necessary. Instead, consider longish focal ratios, short integrations, and large image scales to capture detail. Because it will be bright, techniques like eyepiece projection could enable you to record jets and near-nucleus features a few tens of arcseconds from the pseudonucleus that observers have traditionally relegated to visual observation. At $f/30$, integration times of 1 to 10 seconds should yield well-exposed images of the pseudonucleus, inner coma, and jets.

Comet Hyakutake was a large object. At the focus of my 6-inch telescope, I could record only part of the comet. Even with the 200 mm lens, I could get only $1.8°$ of comet in the frame. For wide-angle shots, I took photographs. I used my CCD to capture structure in the inner part of the tail. In this way, I was able to capitalize on the sheer size of the comet because my images showed part of the comet that is normally overexposed in photographs and seldom watched through an eyepiece. Lesson #2 from Comet Hyakutake is: Plan to Record Details. For whole-comet wide-angle imaging, stick with photography because it is better. With the CCD, zero in to capture unique small-scale features that photography and visual observation miss.

Next, Comet Hyakutake was fast-moving. This meant that long integration times produced blurred images as the comet moved against the background stars. On the night of closest approach, any integration longer than 5 seconds was blurred by the comet's motion against the stars—and that was using my 6-inch with a focal length of 750 mm. To get sharp images of the inner coma with long focal lengths, integrations would, of necessity, be shorter than a second or two. With deep-sky objects, the longer the integration the better; but with comets, motion blur can be seen as a significant problem. Thus Lesson #3 is: Shoot Short Integrations to Avoid Motion Blur; and shoot lots of images. You can always combine them later.

Combining them later was what I did. I found the motion a distinct benefit because it allowed me to eliminate the background stars. Here is how it worked. Suppose you shot a dozen images of the comet over 15-minute intervals; and further suppose that you used a program such as Multi245 to align the images so the nucleus fell in exactly the same pixel in each of those images. If you were to take the median value of each pixel in each of the 12 images, that value would represent the brightness of the comet. Although passing stars would change the average value of the pixel in the 12 frames, it would hardly change the median value. Thus the image made by making a median frame would be free of stars. Without a star background, you could use powerful image-processing routines to extract detail from the image, and see detail that was very faint or impossible to see in the original data.

I did this type of analysis on five nights, except that I shot 58, 71, 111, 137, and 200 frames. By dividing these sequences into groups of a dozen frames and making sequences of median-combined and processed frames, I constructed "movies" showing the comet's tail flowing away from the comet like a flag in the wind. Of course, there is a wind out there in space, the solar wind, and its effect on the comet was clearly visible. So rapid was the motion in the images I shot, that I suspect that any photographic exposure more than two or three minutes long must have blurring in the direction of the flow down the comet's tail.

Thus the Lesson #4: Expect Dynamic Activity. The movies show Comet Hyakutake as a living thing buffeted by the solar wind, throwing off shells of dust and gas as it spins across the sky. Hale-Bopp is going to be a big, bright comet, and quite possibly, even more active. Think in terms of capturing something that is continually changing, not some faint fuzzy that will look just the same tonight as it will a hundred years hence. Plan your observations to record events that take minutes to happen.

Comet Hale-Bopp will not be the same as Comet Hyakutake, of course. It should be intrinsically more luminous but pass us at a much greater distance. Therefore, the total brightness will be comparable, but the scale of the image will be smaller. Even so, it would be a mistake to ignore the Big Lesson we learned from Comet Hyakutake: Bright Comets are *not* Deep-Sky Objects. Plan your CCD imaging to record structural details in an object that is bright, fast-moving, and dynamic.

Note added in 2003: Comet Hale-Bopp turned out to be much less dynamic than Hyakutake because it was dominated by dust rather than gas.

Issue 11

11.1 A Dialogue on Spider Diffraction
By H. R. Suiter and William Zmek

[The following is written as a discussion between an imaginary novice (italics) and a knowledgeable TM concerning spider diffraction. It is not a transcription of a real conversation, but a recollection of scores of conversation fragments we have heard and in which we have participated.]

I realize that so-called "spider diffraction" is the cause of those faint streaks I see to either side of bright objects in my telescope, but what is the mechanism? I have a friend that says it is caused by light that reflects from the side of the spider vanes, but isn't reflected far enough to be scattered entirely out of the field-of-view. He has begun using spider vanes that are not elongated very far along the tube, and coating them with black felt (actually, he sticks the felt from Velcro tape on his spider). He claims that if spiders can be made completely non-reflective, the streaks will go away.

Your friend has made a very good guess about one origin of scattered light in the instrument; but in the case of spider diffraction, he is wrong. The source of those streaks of light is not as simple as ordinary reflection, but goes much deeper. It is the wave nature of light itself. Below we will model simple aperture pupils neglecting the spider's physical length. In the calculated images of a point source, however, we still see those streaks. The wave-optics mathematical model is completely innocent of the nuts-and-bolts details of the instrument it is modeling, but it does a good job of explaining the streaks.

But I can't see why the mechanism of mirror reflection doesn't operate. Could it be that the modeled effects are truly there, but they're vastly smaller than reflection from the sides of the vanes?

In a properly set-up instrument, you won't see reflections from the sides of the vanes in the same view as the bright object. That's not to say you'll never see them. We've noticed them occasionally with objects held off-axis (often out of the field-of-view), but they disappear when the bright object is brought to the center of the field. Vanes are in their own shadows when the telescope is directed right at a bright source. Your friend, in sticking thick pieces of felt to both sides of the

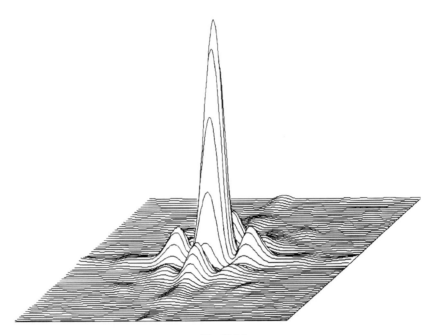

Fig. 11.1.1

spider, is actually making spider diffraction *much* worse, because he's making the vanes thicker.

So why does having a thin obstructing line across the pupil affect things? I would think that as long as it has a much smaller projected area than the secondary mirror it's supporting, it would cause much less damage to the view.

The effect of a spider is to modify different characteristics of the image than a secondary mirror does. It is quite true that for many purposes the effect of a spider *is* negligible. In fact, it can be positively helpful. Once, one of us had to measure star positions on a photographic plate. For dim stars this was easy; you could just set the crosshairs of the measuring engine on the centers of the stars. Bright stars were much too large for this method to give much accuracy. Yet, diffraction spikes poked out of the fuzzy, overexposed blob and allowed the measurer to accurately locate the star's position.

Unfortunately, visual observation is not one of the purposes for which spider diffraction can be ignored. Human beings, for better or worse, consider the aesthetic qualities of an image almost as important as the information it contains. We look for visual impact, and one of the ways of assessing this impact is the "snap" in the image. Preservation of high contrast is one of the ways we can create "snap" in an image (the other is resolution). In astronomy, contrast means showing velvety blackness where we don't expect light.

Perhaps this has something to do with maintaining the illusion that we are actually looking at the object instead of a degraded reproduction of the object. We don't like to have our illusions shattered, and two ugly streaks of light on either side of a bright planet is a sharp reminder of unreality.

Then spiders decrease contrast while the secondary doesn't?

Both degrade the reproduction, but by different amounts and in different ways. To understand this idea, we've got to discuss an odd physical concept. It is called Babinet's principle:

> "The wave disturbance of an aperture is equal to a uniform, infinite wavefront minus a blockage the same size as the aperture (except for a trivial phase shift)."

I don't understand. What you're saying is that if I place a dime between me and the sun, I get the same pattern as a hole the size of a dime. I don't believe it.

Babinet's principle doesn't work that way. Remember, Babinet's principle is only stated for the wave disturbance, called the "field." We are more interested in the transfer of energy, since our eyes are energy detectors. Energy or power in telescopes is proportional to the wave disturbance squared. The diffraction pattern is the pattern laid down by the deposition of power on the retina and averaged over many cycles. When you square the field, the two "dime-sized" situations are different. Also, the dime isn't sitting in a focused wavefront.

Okay. So what has this got to do with real telescopes?

Simply this: Vanes or obstructions can be approximated as their inverses. (Note that we didn't imply equality with their inverses.) While, at first, this seems non-helpful, it can be pointed out that, to a fair approximation, we can think of the aperture as very large and model any obstructions as slits and holes in a black background. Thus, we might approximate the diffraction pattern as a less-bright normal pattern surrounded by the hazy glow characteristic of slits (instead of vanes) and pinpricks (instead of dust). This approximation is especially good for processes that operate on greatly different size scales, such as the whole aperture versus a narrow vane. One constraint is that the energy of the whole pattern—the tight image plus the glow—must add up to the total energy that entered the telescope and successfully avoided the obstructions.

Look at Figure 11.1.2. What we've got there are three components of the point-spread function (i.e., "diffraction patterns"), originating in three separate processes and modeled as "Gaussian" functions. The narrowest, containing most of the energy of the image, is the point-spread function of the original unobstructed aperture, only lessened in intensity by the amount that is "stolen" from it by the other processes. The less-wide pattern is the point-spread function of a 33% obstruction. This blockage might be characteristic of a 6-inch f/5 Newtonian with a 2-inch minor-axis diagonal. The widest pattern is caused by diffraction from a narrow spider vane. It mimics veiling glare but is distinguished by its being directed

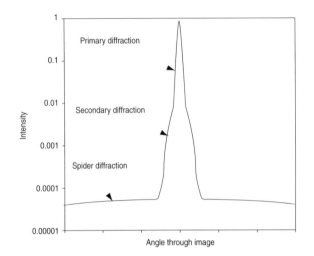

Fig. 11.1.2 *Diffraction Components: Three prime diffraction components are sketched in this figure of the intensity versus image angle. At the core of the image is the tight spot caused by the bounding of the aperture itself. Next down is the squatty glow caused by diffraction from the secondary. In actual practice, this diffraction pattern interferes strongly with the primary diffraction pattern and is unrecognizable as a separate artifact. Lowest down is the broad glow of a spider spike along its axis. Although the presence of the rest of the diffraction pattern modifies the appearance of this spike somewhat, its greatly different scale allows it to be still separately recognizable.*

in only a few directions.

Real diffraction patterns actually don't strongly resemble these Gaussian plots, but such a simple model does; in fact, they represent the broad-brush behavior of the system.

I'm getting confused. You say that there are three sources of diffraction considered here, but that the narrowest source gives the widest pattern?

We have a sort of topsy-turvy relation between the size of the diffracting region and the tightness of the diffraction pattern. The biggest region of all, the unobstructed aperture pupil, gives us the smallest diffraction pattern. (That's why big telescopes are preferable.) The medium-sized obstruction gives a bloated diffraction pattern, but it's barely recognizable because it interferes strongly with the main pattern. The vanes, because they are acting like narrow slits, result in the widest pattern. Because the vanes operate on such a narrow width scale, they interfere with the main pattern only weakly.

But how does Babinet's principle help us get rid of spider streaks?

It enables experimentation before you actually build a telescope. It also allows one to invert the appearance of a vane to a good approximation of a slit and makes it possible to investigate the appearance of a vane by actually placing a

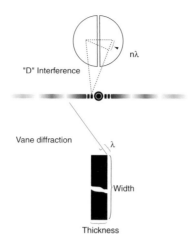

Fig. 11.1.3 *Sources of spider diffraction.*

mask with a slit shaped like the notional spider in front of the telescope. In doing this, you will strip away the primary diffraction pattern to investigate the effect of the spider vane by itself. You can merely cut holes in masks and place them in front of an existing instrument.

What you will see will exactly reproduce the spider vanes' component to the final image right?

Not exactly. Remember, Babinet's principle means an inversion of the entire diffracting aperture. That means the vanes invert to little slits, the obstruction inverts to a small hole, etc. The normally clear parts invert to an opaque disk in an infinite wavefront. Absolute inversion of the whole aperture is, of course, patently impossible. Once you block the wavefront with a mask, there is no convenient way of reintroducing it outside its previous maximum extent beyond the diameter of the mirror.

In fact, precisely inverting the aperture is not even a good idea. When you calculate the energy deposited on the retina, you would get the wrong answer because it includes a lot of previously untransmitted wavefront. Yet Babinet's principle is useful in the case of spider diffraction, which because a much different scale than the whole aperture. You will see the component of diffraction caused by the spider alone.

The only discernible difference between the "bare slit" diffraction pattern and the spider's diffraction pattern in the final image is a form of fast mottling across the streaks, having their origin in what might be called "D" interference (see Figure 11.1.3). The slit pattern has no such mottling because there is no supporting mechanism for it. However, this mottling is scarcely observable in white light images. Many colors add together to fill in the gaps once you're more than six or so rings outside the pattern.[1] Of course, the spikes are seldom bright enough to trigger true color vision. The typical perception of these spikes is "panchromatic" black and white.

I would imagine that this experiment has been done and the problem of spider spikes doesn't have too many right answers.

Right. The publication most accessible to amateurs on the effects of spiders was written by John L. Richter and appeared in the May 1985 issue of *Sky & Telescope* magazine.[2] In it he showed the effect of gradually curving the spider vane and spreading the streaks in a fan shape until they reach a very dim diffraction pattern that is circularly symmetric. Let's look at the effect of a single slit (Figure 11.1.3). The important thing to notice is that a single vane puts out two spikes at angles perpendicular to its own extent. If one views the spider as a series of short line segments, each one sending out its own spikes and each slightly tilted, then the effect of a curved spider doesn't start repeating until an angle of 180°—a semicircle—has been traversed.

Just because this experiment has been done, however, doesn't mean that you can't learn a lot by trying it again yourself. Center a bright star, and then look at the image through broad slits, narrow slits, and curved slits. Keep in mind, however, that the purpose of such imaging is not accurate reproduction of the star. It's making those spikes as dim as possible in any one direction.

So, all I have to do to prevent spider diffraction is curve my vanes?

Emphatically, no. Unless you get rid of the spider, you can't eliminate spider diffraction, any more than you can get rid of the diffraction caused by the limited area of the primary aperture. You can spread the light from spider diffraction around so it superficially looks like it is gone, but it is still there. If your aim is aesthetic, however, the ugly appearance of the spider spikes may justify such action.

Examples are easy to dream up. Say your club frequently presents public-viewing sessions, and you're tired of answering for the ten-thousandth time "Why does Venus have four bars sticking out of it?" Just don't delude yourself that this form of diffraction is removed. It still contributes to the overall glare of extended images.

Think about what is happening for a moment. At any one time a straight-vaned spider is throwing spikes out to the side of a star. These are apparent to casual inspection because they don't cover the whole image area. The curved vane is throwing light everywhere in the image. These are not apparent because at any one location, not enough light intensity is stacked up to be bright enough to see. Thus, for stellar viewing, curved vanes seem to be superior. How-

[1] Suiter, H. R. *Star Testing Astronomical Telescopes.* Richmond, VA: Willmann-Bell, Inc., 1994, p. 159.
[2] Richter, John L. "Rx for the Newtonian Reflector," *Sky & Telescope*, May 1985, pp. 456–460.

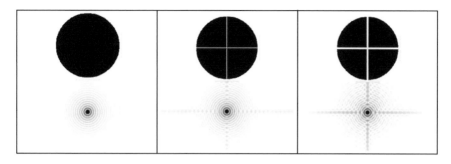

Fig. 11.1.4 *Simple Spiders: The effect of gradually increasing spider vane thickness is seen in this sequence. Please note that in all the stellar-image figures we have exaggerated the effects of spider diffraction by stretching the contrast to emphasize dim brightness levels. We had to do this to have a hope of placing the image in the dynamic range of the printing paper and ink. Therefore, you should not entirely trust the brightnesses you see. Comparisons can be made between frames, of course, if you keep in mind that we have seriously distorted the displayed brightnesses. On the left frame, we see the diffraction structure of a clear aperture. Note that too many rings are visible. This is caused by the unusual stretching function that brings out dim detail. In the center is a spider without a secondary mirror; and at the left, we see a spider with double the thickness of the one in the middle.*

In the center frame, we see the whispery-thin spikes emanating from the star and going through the first spider-diffraction minimum beyond the edge of the frame (see Figure 11.1.3). At right, the spike goes through minimum right at the edge of the frame. Keep in mind that twice the light is being packed inside half the area. Therefore, much brighter diffraction spikes from the thick vane are expected. Note also that the "edge" or "perimeter" length of the two spiders is about the same, but the energy diffracted by the thicker spider is greater.

ever, that is not the way most people use telescopes. They use them for large bright-field objects.

If the curved-vane makes the spikes dimmer, why doesn't it always help? What is different for extended objects?

Say that we are observing a tiny low-contrast craterlet on the floor of Plato on the Moon. The craterlet is surrounded by a crater floor of more-or-less equal brightness, but the spider diffraction washing over the craterlet originates from different areas in the two cases. Only those spots of light along the direction of the spike can contribute extraneous glare in the straight vane case, but *all* of the spots can contribute a little glare when you are using a curved vane. The net effect is that the glare originates strongly from a few directions, or weakly from everywhere. With minor differences, the net effect is the same. There is no difference between the two cases in extended-object imaging.

To increase the contrast, we must actually decrease the amount of energy diffracted, not just redistribute its intensity in the field-of-view.

*I want to get rid of the spikes, but I don't want to make a problem worse. Can anything **bad** happen if I curve my vanes?*

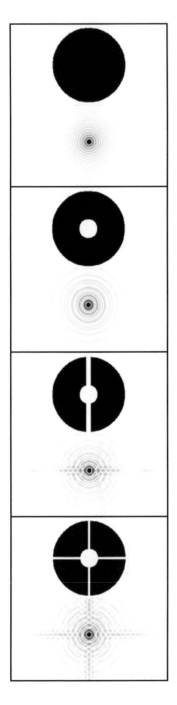

Fig. 11.1.5 *Spiders and Secondaries: The top frame repeats the clear aperture pattern and the next frame depicts the diffraction pattern of a spiderless annulus. Before anyone gets worked up about the much worse appearance of the annular pattern, let us reiterate that we have seriously distorted the contrast of these frames to emphasize dim details like spider diffraction. In actual use, the annular pattern would be little different from the clear aperture (see Ref. 1). The point of all of this emphasis is that spider diffraction is not that great an effect.*

We see the effect of a spider in the last two frames. Notice that the spike is not along a vane but perpendicular to it. The area covered by the spider is the same in both spider frames, but the last frame shows the beneficial effect of dividing the thickness and placing it into more vanes. The energy in the two stubby spikes in the third frame is about the same as the energy in all four spikes of the last frame.

Note also that the presence of the secondary mirror changes the crispness of the spike (compare with Figure 11.1.4). In the clear aperture, the spike emanates directly from the diffraction spot and has the same characteristic width as the diameter of the Airy disk. In the annular apertures, it seems to have acquired a companion spike on each side that makes the whole thing resemble a backbone. The secondary breaks the symmetry of the aperture and causes an interference that has an off-axis maximum. In realistic apertures, these side maxima and minima are difficult to see. Spider diffraction is so dim that it is hard to perceive any structure in it.

Not in principle, but in practice, bad things can happen. Curved-vane spiders have dispensed with one of the most fundamental advantages of straight-vane spiders. Curved spiders cannot achieve structural stability by tension. They must be self-supporting. Thus, they are usually made a little thicker to achieve structural strength. Straight-vane spiders can achieve excellent stability by being tensioned like wires, and in doing so can be made much thinner than a curved-vane spider ever can.

This is not to say that curved vanes cannot be made with the typical thickness of straight-vaned spiders. Curved vanes may achieve stability the same way that a thin soft-drink can does—by lengthening them along the tube—but a long curved surface must be very carefully mounted if it is to remain parallel to the axis of the telescope. In short, you may find that tilt in your spider has destroyed the beautiful symmetry and reintroduced diffraction spikes.

If a very thin vane is straight, will the spikes still occur?

Yes, but the amount of energy the spider diffracts from the peak is approximately proportional to the area of the pupil it covers; and if it is made extremely thin, it is forced to spread less light into a larger area. The most objectionable feature, the strong spike, can be attenuated almost to the vanishing point.

I don't see how making the vane thicker would diffract more light. I thought that because diffraction is a function of edges, we would make the core image dimmer but see the same spider diffraction because the edges haven't gotten any longer.

One of the hardest things to understand about diffraction is that it occurs because there isn't any wave originating from a section of the pupil. If we increase this shadowed area, everything changes. True, we can mathematically model the diffraction as a combination of the original wavefront and a wave originating at the edges, but it makes no difference.

For a generic obstruction of any shape, the normal "tight" diffraction pattern of the whole aperture seems to be lessened in brightness by about twice the fractional area. This factor of two seems puzzling, and has been called "the extinction paradox." It was discussed by H.C. van de Hulst[3] using the charming analogy of a flowerpot on a windowsill. If we measure the starlight passing the flowerpot, we get the energy passing through the window minus that blocked out by the area of the flowerpot. Seems simple enough.

Now, if we were to back off to the extreme distance where the obstructed window would act like a pinhole camera (or where Fraunhofer diffraction would be operating, forming an approximation of the Airy disk), we would find that disk diminished in intensity at the center by about twice the area of the flowerpot! That's the paradox. If the flowerpot blocked out 1% of the energy up close to the window, the far-away diffraction pattern is lessened 2%.

---

[3] van de Hulst, H.C. *Light Scattering by Small Particles.* New York: Dover Publications, 1981 (originally published in 1957).

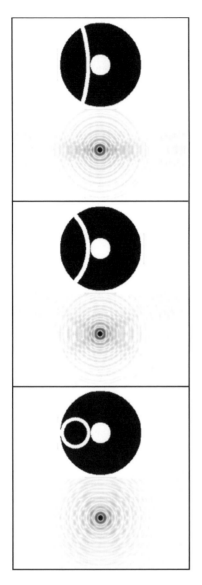

Fig. 11.1.6 *Single Curved-Vane Spiders: We will investigate curved vanes by using a single, beefy vane. We use such a thick vane in order to see even the spread-out diffraction spike. In the top figure, the total curvature is 30°. Its pattern covers (roughly) 30° to the right and the same to the left. It is this fore-and-aft "lighthouse" behavior that characterizes spider diffraction. When the curvature increases to about 80° in the middle frame, we see the brightness lessen as the fan is blown out to cover approximately 80° in both the left and right directions. Clearly, a 180° spider would be required to cover the full circle (shown in a later figure), but already curving through this smaller amount, the diffraction in any one direction is much lessened. The presence of additional structure is caused by interference with the secondary's shadow and by the fact that the spider is somewhat offset from the center.*

The bottom figure is a full-circle spider. Though this spider diffracts light through the circle twice, it is a tempting choice because it is easier to make (we have actually seen this spider implemented in a Cassegrain reflector). The pattern is mostly circular with the exception of a very indistinct broad fan to the right and left. Again, this is caused by interference with the off-axis main pattern. In practice, this pattern would be scarcely noticeable. You would have a hard time seeing anything but a circular image.

So where did the missing energy go? Up close to the window, we thought we had the energy accounted for. Somewhere between the window and the diffraction pattern something happened to that energy, or so it would seem.

The extra energy has been diffracted far out of the main pattern. Half the missing energy smacked into the flowerpot and was immediately lost, either absorbed or reflected completely out of the field-of-view. The other half—about the fractional portion obstructed by the flowerpot—again diffracted into the nether regions of the image. We have already seen by Babinet's principle where it diffracted to. If the ·flowerpot is big and squatty, the light resembles the secondary's pattern in Figure 11.1.2. If it is a long, thin vase, it resembles the spider pattern of the same figure.

Does the amount of light diffracted exactly equal the obstructed area?

Not precisely. A paper by J. E. Harvey and C. Ftaclas[4] does a more accurate calculation of the placement of light in the image. In it they define a fractional encircled energy caused by diffraction alone. They rightly ignore the light that hit the rear of the obstruction, because it doesn't degrade contrast or resolution. Their expression is:

$$EE(r) = \frac{(A_{ann} - 2A_{spiders})EE_{ann}(r) + EE_{rec}(r)}{A_{ann} - A_{spiders}} \tag{11.1}$$

$EE_{ann(r)}$ is an expression for the fractional encircled energy of a pupil with an unsupported secondary (i.e., an annulus) and $EE_{rect(r)}$ is an expression for the encircled energy of a rectangular vane. These are accompanied in the referenced work by elaborate tables of various values of both EEs. The tables will not be reproduced here, because they are bulky and subject to retyping error.

What is fractional encircled energy?

It's just the fraction of the total energy that makes it through the telescope and is found inside a circle of a certain diameter at the focal plane. In perfect, unobstructed telescopes the encircled-energy fraction is 84% inside a circle the size of the Airy disk. That means 16% of the energy of a star is still outside the Airy disk. (Note: In Suiter's *Star Testing Astronomical Telescopes*, an encircled energy ratio was defined by taking the fractional encircled energy of a real obstructed or aberrated aperture and dividing it by the fractional encircled energy of a perfect circular aperture—a ratio of ratios. It is not quite the same thing.)

This equation still looks pretty complicated. How does it compare with van de Hulst's simple fractional area for the dimming of the image caused by diffraction?

The simple fractional area of the obstruction is a good approximation of the intensity loss at the center of a point image caused only by diffraction. This is an especially good approximation for spiders. (Derivation in Appendix A.) Of course, the intensity that was diffracted from the main image reappears in the form of contrast-sapping light.

Okay. You've answered the question of how much light a spider vane diffracts— its own fractional area or thereabouts—but not the question of where it ends up. Isn't that simple? It ends up in the spike, doesn't it?

Yes and no. It ends up in the spike, sure enough, but the real point is that the

---

[4] Harvey, J. E. and C. Ftaclas, "Diffraction effects of telescope secondary mirror spiders on various image-quality criteria," *Applied Optics*, Vol. 34, No. 28, pp. 6337–6349.

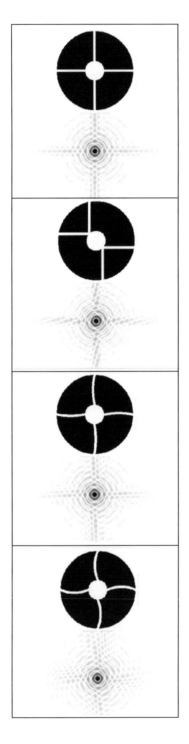

Fig. 11.1.7 *Four-Vane Spiders and Options:*

The top figure depicts the same spider that appeared in the bottom of Figure 11.1.5. The second frame shows what happens if we offset the vanes instead of curl them. A variation of this type of spider appeared in Jean Texereau's How to Make a Telescope, *second edition on page 128 and is there credited to older references (Morse and Hargreaves). In Texereau, the unnecessary stricture is imposed that the opposing blades should be made parallel to avoid additional diffraction spikes. However, we have already seen how two extra spikes are dimmer than one, if the area covered is the same. Indeed, the "breaking" of a single vane across the aperture can be seen to be a crude first step in curving it.*

The reason for this style of spider is structural, and we could probably make it thinner, but let's look at the diffraction structure if we leave it alone. The spike has been smeared out into a narrow fan by this offset alone!

If we actually do curve the vanes very slightly, we induce a very similar behavior. The third frame has four vanes curved through 30° each. The diffraction mapping, however, repeats itself twice; and we see four narrow fans. Curving the vanes through 45° each widens the fans, but still the diffraction patterns fall atop one another, and the diffraction pattern appears only on four 45° "windmill" fans last frame.

We could keep increasing the curvature of the four vanes, but we would not spread the diffraction over the whole circle until the vanes curve through 90°. This is wasteful, because the diffraction covers a full circle twice.

The point of this figure is to show how curved-vane spiders are inefficient with an even number of vanes diametrically opposed. A single-vane, 180° spider is more efficient. Three vanes curving through 60° is also good (not shown). If you do use four vanes curved through 45°, you should place them at odd angles for efficiency.

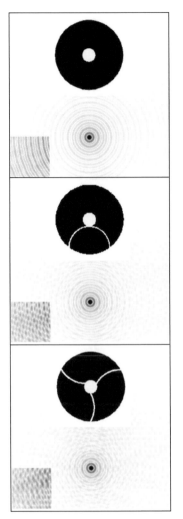

Fig. 11.1.8 *360-degree Pattern Curved Vanes:*

Some curved vanes that amateurs might actually want to try are depicted in this figure. On top is the annulus without a spider. This is a slightly different central obstruction as used in the other figures; so don't expect the appearance to be the same. In the middle is a 180° curved vane and at the bottom we show a triple 60° spider. In the corner, we do an identical amount of contrast stretching to show the distribution of light in distant portions of the image. Neither of the lower two spiders has a discernible spike pattern.

The decision about which of them to make should be based on practical considerations like the weight of the secondary and difficulty in fabrication. We have made neither, but we do have a bit of sheet-metal experience. We believe the 180° spider might be slightly less challenging to make because the secondary holder mount can be attached to the outside of the spider in a simple "bolt-on" arrangement. Please note that the angles are measured to the edge of the mirror, not to the edge of the tube; and keep in mind that many "wiggling" adjustments, such as placing a diagonal slightly farther away from the focuser in fast instruments, are no longer easily available by turning screws. They must be designed right into the spider.

shape of the spike changes with the thickness of the vane. In Babinet's principle, we treat the obstruction as a virtual slit. Slits, like aperture pupils, have diffraction patterns that depend on their width, corresponding to the vane thickness in Figure 11.1.3. A spike induced by a slit has a broad central lobe analogous to the Airy disk, but much bigger and only in the spike direction. This lobe reaches a minimum at an angle of $\lambda/\text{thickness}$ (λ is the wavelength), much like the angle at the edge of the Airy disk reaches a minimum around $1.22\lambda/D$. Thus, beefy spider vanes have narrower (and brighter!) diffraction spikes. A spider vane that can be thinned to $\frac{1}{2}$ its self-supporting thickness (because it is under tension) diffracts $\frac{1}{2}$ the amount of light into a diffraction spike twice as long. Thus, we don't expect the average brightness of such a spike at any one location to be more than a quarter as bright.

You can continue this thinning process. If you can somehow decrease the

Fig. 11.1.9 *Couder Vane-Masks:*

In ATM II, a suggestion appeared that was originally connected with a non-rotating tube telescope and an unfortunately-placed double star. A. Danjon found that if he covered the vanes with gently-curved masks, diffraction spikes were suppressed, and he could observe the companion star. In this figure, we calculate the prototype (on top), a notional variant, and a variant suggested by André Couder in that original article. While we notice a profound reduction of spiking in all but the bottom frame, we see a lot of extra light in the image. The bottom frame may have an imprecisely modeled mask, but we believe proper fabrication would be a problem with any realistic use of such a complicated mask, and the pattern appearing in Figure 11.1.9 may represent a typical attempt. In any case, the outside surfaces of the "rolling-pin handles" are uncompensated, so the performance of Couder's suggested spindle mask is questionable. We noticed better performance in simpler apertures, such as the four ellipses in the top figure, a "flower petal" version of the middle frame (not shown), or the middle frame's "diamond" figure. However, it is our firm opinion that these devices should only be employed in their original function—as a work-around. The important thing is minimizing the area covered, and in every case that area is large.

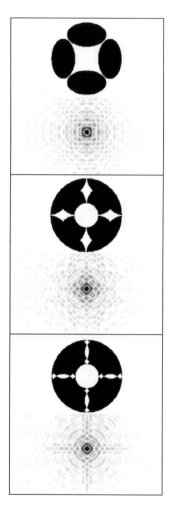

thickness by another factor of 2, you get a factor of 16, and so forth. It takes only the simplest analysis to reason that even though the spike is still there, it is *much* less noticeable, because it is so long and dim.

Do you lean toward super-thin vanes instead of curved-vane spiders?

It's not so much an inclination as a desire to point out that there are other options. Maybe the observing problems of the group putting on demonstrations for the public would be most simply served by a curved vane, but if you are interested in minimizing the effects of spider diffraction on the contrast of extended objects, you may well think about ultra-thin tensioned vanes.

How thin can the vanes be made? Is there a stopping place?

The thickness of the vanes is ultimately determined by the necessity of supporting the diagonal and the difficulty of setting the vane perpendicular to the mirror. The first condition depends on how much elasticity in the vanes will compromise the alignment when the telescope is directed at different points in the sky. You can obtain added strength without increasing cross section by using wider blades (i.e., longer along the tube), but these are more difficult to orient exactly perpendicular to the mirror and have greater mass, which may generate thermal problems. You can achieve better structural strength by making sure that the vanes are not radial in the tube (and are therefore not as subject to torsion or "twisting"). (See Texereau, p. 127.)

Current minimum thicknesses of commonly available commercial spider vanes are about ½ mm. There is a great deal of advantage yet to be gained by using vanes so thin that they are loose and floppy when untensioned. We believe that vane thicknesses approximating that of a razor blade (~0.13 mm or 0.005 in) are possible. Of course, we haven't tried these thin vanes because they haven't been recently available and would involve special construction.[5]

As far as the minimum thickness is concerned, the spikes keep broadening until the vanes are only about one wavelength thick. Then, as the thickness is reduced further, scattering processes different than diffraction take a strong hold. The material the spider is made of and how long it is become important.

Of course, we never have to worry about wavelength-thickness spider vanes. Even if they could be made strong enough to use, orienting the blades of such vanes perpendicular to the mirror to an accuracy of a half-micron would be practically impossible. A razor's thickness is hard enough to align.

Are there any effects other than misalignment that could render the thin vane ineffectual?

Yes. Most people coat their spiders with black paint because less light will be scattered that way. This is true, but must be balanced against another possible effect. Because they are near the end of the tube and radiate from both sides, black-painted vanes may be a good deal cooler than their environment (black things radiate heat better). Some suspicion exists that the layer of air next to them is chilled and becomes a weak refracting region. This refracting region can be much larger than the physical obstruction, giving the vane a greater virtual cross section.

Maybe the best way to coat your spider is to not coat the spider. Bare metal stays a tiny bit warmer. Unpainted metal has been used in professional-sized instruments to good effect, but more work needs to be done before we can positively say that using bare metal vanes helps in the more enclosed tubes of amateur instruments.

---

[5] Years ago, the firm of Telescopics sold a very thin spider with characteristics that seemed much like our ultra-thin spider. Sadly, neither of us ever inspected it, so we do not know if it qualified.

In a side issue, we can't help but believe that the recent anecdotal evidence for the superiority of tensioned piano wires over vanes isn't somehow related to thermal effects. The thickness of these wire spiders is in most cases about the same or even worse—so diffraction is similar—but the wires don't have as much mass. They're also less likely to be painted or to hold paint even if the telescope maker attempts to coat them.

Our point is that there is no reason wire spiders are inherently better from the point of view of diffraction theory alone. If they indeed work better, there must be secondary reasons to explain it.

I see that going to vanes that are ultra-thin might be more trouble than it is worth. Is there some other way I could get the same effect without incurring the same hassle?

It depends on what you think of as "hassle." If you are willing to accept some extra work during construction, you can do largely the same thing by using an optical window to support the secondary.

But I heard that optical windows have to be really precise, and they have stray reflections?

You heard correctly. The introduction of anything into the optical path must imply some degradation, but if we can force that degradation lower than another degradation, we have a net improvement. Optical windows can routinely be figured to less than $1/100$-wave RMS wavefront error (about $1/28$-wave peak-to-valley). We can invert the Strehl ratio formula (Suiter, p. 238) to find how much intensity is removed from the main diffraction pattern.

$$(2\pi/100)^2 \sim 0.004$$

If we equate this number to the fractional area covered by the spider vane, we can determine the thickness of the vane that the window "beats." For a 4-inch aperture with a $3/4$-inch diagonal and four vanes, any vane thickness greater than 0.2 mm diffracts more light.

Amateur astronomer and enthusiastic telescope maker Bob Bunge once made such a 4-inch long-focus Newtonian reflector. It compared favorably with 4-inch refractors in side-by-side tests.

In making this instrument, Bob had to deal with several difficulties. Chief among them was making the diagonal adjustable once it had been stuck to the window. It serves no point to eliminate spider diffraction on an instrument that cannot be aligned. There are several ways of handling this problem, but perhaps the easiest is to make the entire window tilt-adjustable with the diagonal fixed in position at approximately a 45° angle.

Another difficulty is internal reflections within the telescope. Any light reflected back from near the focal plane (the lenses of the eyepiece and the bottom of the chromed outer-barrel are strong candidates) is perceived as a half-focused glow. This problem is solved by using coatings and blackened eyepiece tubes, but also responds to deliberate tilts in the window. The window has no optical power, so it need not be squared-on. Thus, it is possible to suppress reflections by inducing a significant tilt in the plate. The worst situation is a very small tilt where the spurious reflection is still in the field-of-view. Since very slight tilts are almost impossible to eliminate (particularly if tilting the window is your method of secondary adjustment), it makes more sense to deliberately cant the window a degree or so and, in doing so, throw the reflection completely out of the field-of-view.

The final problem is obtaining an optical window with the required optical properties. Just any windowpane won't do; it requires superb, fine annealed optical glass figured to exceptional smoothness. You can coarsely test the window in the Foucault test, for example, by suspending it immediately in front of a spherical mirror (see Texereau for the technique). But for smoothness, you must be skilled in interferometric methods. It must be figured to high quality if it is to exceed the capabilities of an ultra-thin spider.

An even smoother flat, with wavefront disturbances of $1/500$ or $1/1000$ wavelength, would be better yet; but you have to figure out a way of testing such a thing. It helps that you are testing not for overall figure, but for smoothness. It hurts because so many familiar tests are unavailable. Of course, at some point, you've got to consider the roughness of your mirror optics. Perhaps any optical window much smoother than your primary would be acceptable.

Okay, we have three methods of reducing diffraction spikes. Can you review them so I can get my mind straight on them?

First, you can curve the vane. This is the easiest method, but it doesn't really get rid of spider diffraction—it just spreads it around. It will result in the elimination of preferred directions for spikes. Curved vanes will suppress the aesthetically unappealing nature of diffraction spikes. Curved vanes will need to remain thick, because they hold themselves up (with a few exceptions).

Second, you can use ultra-thin vanes, getting structural strength not from material thickness but from tension. This procedure doesn't get rid of the spikes but makes them dim and long. It doesn't attack the spikes directly but pulls the rug out from under them, so to speak. It results in a real lessening of spider diffraction.

Third, you can dispense with the spider altogether and support the secondary mirror on an optical window. This completely gets rid of spider diffraction but reintroduces the aberrations of the optical window. The aberration that most resembles spider diffraction is caused by medium to fine-scale roughness, because it blows the light removed from the image into a hazy glow. However, if the window can be made smooth enough, the effects of roughness can be made smaller than spider diffraction.

Well, all things being equal, I would just as soon go with the optical window in my 10-inch. Don't you agree?

Not necessarily. As optical glass increases in size, non-uniformities become more apparent, and smooth fabrication gets a lot more difficult. Windows are a lot easier to make or acquire for small telescopes. Also, an important thing to consider is that as the mirror gets bigger, the currently available spiders have a smaller relative cross-section. They're approaching "ultra-thin" status merely by staying the same thickness and allowing the telescope to grow around them. If your 10-inch were fitted with a $\frac{1}{2}$ mm thickness spider vane, it would divert the same amount of light as a $\frac{1}{100}$-wave RMS optical window.

Although we glossed over it before, roughness in optics typically causes a glow of much smaller radius than a thin spider does. If the average dimension of a diffracting facet in the rough surface is 6 mm across, then the $\frac{1}{2}$-width of the hazy glow is only about 18 arcseconds. A $\frac{1}{2}$ mm spider vane diverts the light an average distance of something like 100 arcseconds.

Thus, the hazy glow of the window is uncomfortably inside a radius smaller than Jupiter's, while the diffraction spike is largely outside the planet. The glow caused by the window looks "clean" but is still inside the planet doing damage. At the same time, the spider's spike looks "ugly" but is mostly outside the planet where it can do no harm.

In a way, spider diffraction is overt and honest, while other forms of optical degradation are more hidden and damaging. Spider diffraction attracts attention not because of the harm it causes, but because it is the most noticeable. Spiders commit their sins in public.

In small instruments, an optical window offers real advantages over thin-vaned spiders. In miniaturized, long-focus instruments such as Bob Bunge's, you have a greater chance of obtaining a perfect piece of glass, as well as the smooth primary that would justify such extraordinary measures. Also, narrow vanes in small instruments have to be tissue-thin to beat the window.

Of course, this recommendation ignores other possible benefits of optical windows, such as lessening tube currents.

So, which technique do you recommend?

We don't recommend doing anything if you are a planetary observer. If you have reasonably thin spider vanes already (~0.5% pupil area), you would have to perform miracles to notice any difference in the contrast of planetary details by changing your spider. The spider throws most of its diffracted light outside the planet anyway. More to the point, you possibly have larger problems in turned edge, mirror roughness, or overall figure. Cure more damaging difficulties first.

Deep-sky observers who look most of the time at dim objects don't have to worry about spider diffraction, because it is typically 5 magnitudes below the primary object, which itself may be on the edge of vision. We can think of exceptions, such as dim objects next to bright objects (double stars or galaxies uncomfortably close to bright stars), but even in these cases turned edge and substandard baffling are greater threats.

If you specialize in lunar observation, you have a potentially bigger problem with spider diffraction. The lunar surface is a bright-field object much larger than the diffraction angle of even a thin-vaned spider. Bright fields have a hazy glow extending from every point on the image, but much of the glow is still inside the Moon. For lunar observers, we would recommend an optical window or an ultra-thin spider. A curved vane will not help at all on the Moon.

In the case of casual observing, we recommend a curved-vane spider. Public observing sessions or quick-looks in your backyard often feature the bright planetary images subjectively disturbed by spider diffraction. If you're the type of person who reacts with distaste at seeing spider diffraction spikes, by all means try a curved-vane secondary support. Just keep in mind that using curved vanes more resembles sweeping dirt under the rug than truly eliminating the problem.

Okay. I'm willing to concede that the improvement is probably only cosmetic, but which type of curved spider is best?

The improvement is not cosmetic at all. A lot of visual observation has a psychological basis. The important thing is that you have improved the observation for your purposes.

Ideal is a thin spider that covers as little of the primary mirror as possible and at the same time adequately supports the secondary mirror. We spoke above about how the diffraction from a curved-vane support becomes evenly distributed when the support curves through 180°, but that angle is not required. If you can support the secondary by curving only through 90° or even lesser angles, go ahead and use your spider. As soon as the straight line is broken, the spike brightness in any one location decreases drastically. Spider diffraction won't be evenly balanced in such cases, but it will be much less strong than it was when it was stacked up in only one direction.

Perhaps the best curved-vane spider is a 3-vane support with equally-spaced vanes that curve through 60° each, and the easiest to make is a single vane curving through 180°.

How thin is "thin"?

A "thin" curved-vane spider would mean 0.020 inch for 4- to 6-inch telescopes, 0.030 inch for 6- to 10-inch instruments, and $\frac{1}{16}$-inch for 12- to 20-inch telescopes. One of us once saw a commercial 12-inch Dall-Kirkham-Cassegrain reflector that had a secondary supported by a $\frac{1}{4}$-inch-thick piece of round tubing extending to one wall (it was actually cast into the tube end-ring). Such a structure diverted light to an average radius of only 10 times the Airy disk. Surely, cleverer curved-vane spiders can be designed than this one!

Then you have only two conditions. Make a curved spider that covers as little area as possible and make it thin. Luckily, both of those conditions run the same direction.

It's three conditions, not two. The first job of a spider vane is to hold the secondary mirror without moving! This one veers away from the other two (thicker spiders are stiffer spiders). Nevertheless, it is the most important function of a secondary support. If we make a spider that moves depending on the position of the telescope, it will cause misalignment difficulties that dwarf any improvements we obtained by making it thin.

The way that you know if you've made your curved-vane spider too insubstantial is that alignment goes bad when you tilt the telescope another direction.

It is a little late at that point. Wouldn't I need to redesign the whole secondary support?

That's the first rule of odd spiders. Occasionally you fall flat on your face. However, a poor spider is easy to replace. Spiders are not usually fatal mistakes.

What about the Couder-style vane covers I read about in ATM II? Aren't they a good way of masking vanes?

They were a good idea for the purpose that elicited their design, but not as generic spider-diffraction reducers. Their original purpose was to allow observation of a poorly placed double star in an equatorial telescope that had a fixed spider. Other more prosaic solutions, such as rotating the tube, were not available in this large instrument, so this fix was a "work-around."

An article was reprinted in *ATM II*[6], with insufficient emphasis that this method was only a kludge. The article was accompanied by photos that didn't show the degradation very well. True, the diffraction spikes had lessened, but at what terrible cost in added light diffracted throughout the field of view?

A good way of thinking about this technique is to ask what fraction of the mirror such devices cover. As this fraction increases, so does the diffracted light.

I take it that you don't think much of it.

No. Unless you have a specific reason for it, as did the inventors, such a mask is an optical blunder.

Well, I must say this whole conversation hasn't been what I expected. I thought that advanced telescope makers really went crazy about anything that lessens the quality of the image, but here I've gotten answers ranging from "don't worry about it" to "get a life." Is spider diffraction really that negligible?

No. Spider diffraction is not negligible so far as it makes an image unacceptable. However, you must view spider diffraction in the context of other degradations. People often panic in response to imagined or perceived threats. They work themselves into a lather about one facet of telescope performance to the neglect, or worse, to the detriment of other facets.

---

[6] Ingalls, Albert G. (ed). *Amateur Telescope Making*, Book Two. Scientific American, Inc., 1978.

Nearly every reflector with an unmasked turned edge (and that is about all of them) has a much more serious problem than spider diffraction. Many people baffle their instruments half-heartedly, viewing it as a last-minute design issue. We have seen owners who tolerated direct skylight in open-tube Newtonians. Most TMs use unbaffled focuser tubes, even if they know better. In your last star-party outing, how many aligned fast Newtonian reflectors did you look through? Depressingly few, we'll bet.

Yet we have seen telescope owners who reacted indifferently to all of these problems start to foam at the mouth when they see spikes. Not because they are bad, but because they could *see* them. Perhaps a quote from Jean Texereau on the justification for optical windows will help to put things in perspective:

> *"On balance, is the effort justified? The answer is no if the telescope is not otherwise optically perfect in the sense already explained, and if the worker is not accustomed to judging stellar diffraction images . . . Finally, the enterprise is not really of value except for an amateur elite who are at the same time good opticians, good observers, and owners of rather large telescopes.*
> *Such an elite do, of course, exist."*[7]

Texereau was mostly speaking about the use of optical windows to suppress tube currents and other near-telescope atmospheric effects, but we can easily adapt these words to the justification for extra efforts on spider diffraction. At the very least, you should demand that the other, more damaging forms of degradation be minimized before you attack spider diffraction.

Any last words?

Only these. A lot of the perception of optical errors depends on training or predisposition. For an example of this, look at a Christmas card depicting the Star of Bethlehem. Nearly always, the Star is drawn with the optical effects of spider diffraction and photographic halation. Depicting these two optical defects has become the preferred artistic portrayal of this scene.

Of course, accurate optics is hardly the purpose of these drawings. No person, much less shepherds and Babylonian astrologers, ever saw a star with photographic haloes and spider diffraction before about 1870. It takes little effort to see the symbolism of painting the Star surrounded by a halo and with a superimposed cross—the message is purely religious, and needs no interpretation.

However, we can ask how the artists were prompted to depict a star in this way? We believe the answer is in the photographic archives of the major observatories, where overexposed stellar images on plates are frequently shown with a halation ring and with strong spider-diffraction spikes. These images are so beautiful and powerful that they've become a part of our culture.

One of us was showing the Pleiades to a four-year old girl at a public observ-

---

[7] Texereau, Jean. *How to Make a Telescope*, 2nd ed. Richmond, Virginia: Willmann-Bell, Inc., 1984, pp. 191–192.

ing session once, and upon climbing the ladder and reaching the eyepiece, she suddenly squealed and clapped her hands. He asked her what she had seen, thinking it had been a meteor crossing the field or some other reasonably rare event. She answered that the blue-white stars looked "so real." On further questioning, she said she knew the stars were real because of the spikes she saw emanating from the points of light. They looked "just like Christmas cards."

This is, indeed, a strange train of events. First, optical defects that we have spent many pages trying to eliminate were recorded on observatory plates. Further, these plates were reprinted in popular publications so many times that earthbound artists came to think of the defects as attributes of stars. Next, they drew these defects on Christmas cards, faithfully reproducing an optical property that no eye possesses. And finally, these defects were used as a measure of "rightness" by a little person unaware of their origin.

The final test of an image is that the eye sees what it expects to see, and if that includes spider diffraction, so be it.

(We would like to acknowledge Andreas Hayden for help in producing figures.)

11.1.1 Appendix A

Recall from the discussion of van de Hulst that the expression for the actual diminution of energy at the center of the image would include the light lost when it hits the rear side of the obstruction. Such diminution might be termed "darkening" because it is unrelated to the shape of the image. Including darkening in the energy accounting is excessively unfair to the telescope because of the percentages involved. Even a thick spider adds only about 1 percent to the obstruction, a number that could be easily compensated by having a fresh coating or an incrementally smaller diagonal. Such compensation would give an illusory improvement, however. Diffractive effects alone are modeled in Equation (11.1) as an effort to relieve this unfairness.

Please note that as the radius of the image goes to zero, $EE(r)$ also goes to zero, so we will be more interested in the ratio $EE(r)/EE_{ann}(r)$ than any absolute value. This ratio can be thought of as the degradation caused only by the spider. In other words, we think of an annular aperture as "perfect", and ask how much extra light is diffracted by the spider.

When we ask what form Equation (11.1) takes when r approaches zero, we note that $EE_{(r)}$ is much smaller than EE_{ann} near zero, so it can be ignored. (It is intuitively obvious that a narrow slit pumps much less light into the center of the image than a broad pupil.) Equation (11.1) becomes

$$EE_{(r)} = \frac{(A_{ann} - 2A_{spiders})EE_{ann\,(r)}}{A_{ann} - A_{spiders}} \qquad \textbf{(11.2)}$$

Rewriting the denominator as $A_{ann}\left(1 - \dfrac{A_{spiders}}{A_{ann}}\right)$ and expanding it in a series, we find:

$$EE_{(r)} = EE_{ann\,(r)}\left(1 - \frac{2A_{spiders}}{A_{ann}}\right)\left(1 + \frac{A_{spiders}}{A_{ann}} + \cdots\right). \qquad (11.3)$$

Keeping only the first-order terms of $\dfrac{A_{spiders}}{A_{ann}}$, we find

$$\frac{EE(r \to 0)}{EE_{ann}(r \to 0)} = 1 - \frac{A_{spiders}}{A_{ann}}. \qquad (11.4)$$

Therefore, the expression in Equation (11.1) reduces to the approximation of van de Hulst, or a simple ratio of areas for the degradation caused by spider diffraction alone.

11.1.2 Appendix B: Calculation of Diffraction Figures

The way these figures were calculated was to take a 256 x 256 model of the pupil with blockages in the required locations, blank-pad it to 1024 x 1024 and run it through a two-dimensional fast-Fourier transform (FFT) routine. The magnitude of the resulting FFT was then normalized to the unobstructed pupil's central intensity, and the quadrants reshuffled for display. A 128 x 128 thumbnail representation of the pupil that was used to calculate the diffraction pattern appears on the upper side of each frame.

Approximations used:

1. Monochromatic images—to test this approximation, we took a sample spider diffraction pattern and remapped it to a set of scales that mimicked the color response of the human eye. Then we added them together to simulate a color view as received by the sensitive non-color receptors of the eye. The spike maxima and rings were softened a little, but we didn't see much difference. The additional effort to make a color correction was deemed not worth the effort.

2. Sampled apertures—modeling the aperture as a set of 256 points presents the difficulty that "jaggies" or sharp discontinuities at the edges of the pupil, obstruction, and spider might induce unphysical artifacts into the calculation. These errors would not appear in the analog real world. We tried to ensure that such artifacts were small compared with the real spider light distributions by making the spiders unusually thick as compared with actual situations. That the pupil and obstruction were adequately sampled was demonstrated by the smooth, circularly-symmetric diffraction patterns when no spider was present. For a couple of curved-vane cases, we tested the assumption that our spider was thick enough to dominate "jaggy" effects by making the array somewhat smaller, thus making the "jaggies" bigger without changing the smooth structure. The results differed little. Sampling mostly degrades the dim backgrounds of the image.

3. Truncation of space—a necessary feature of the FFT is that space be modeled in periodically repeating regions. We have done this by truncating the array at 1024 x 1024 points. We limited the pupil, however, to a region of 256 x 256. This ensures that the Airy disk is adequately sampled in the final image. We avoided aliasing by calculating the diffraction effect of no obstructions narrower than 2 pixels, and we usually made the spiders much thicker than even this limit.

The FFT method, however, should not be used to calculate diffraction from realistic thin-vane spiders without using a huge number of points in the array. Our using thick spiders did not disturb the physics, but avoided the ill effects that would arise if we attempted to calculate an actual real-world situation.

11.2 A Springfield Mounted 9-Inch Refractor
By Michael Spooner

The refracting telescope has long had a reputation for providing high quality images of astronomical objects. However, the standard doublet lens design has been a source of debate due to its residual color error that gives a violet halo to brighter objects. As a result of a trade, I came into possession of a 9-inch unfinished lens and cell, and thus began a closer examination of the colorful world of achromatic lenses. The first problem I encountered was trying to determine the types of glass used in the lens. Both lens elements were semi-polished, and I started by making some educated guesses concerning the lens. First was that a lens of this size and cost would not be attempted without a reasonably accurate design. Second was that the glass would be standard choices. Third was that the curves would be worked fairly close to design. Using a spherometer, I determined the radius of each curve as closely as I could, and assuming a Bk7 and F4 glass combo, I headed for the computer to do some raytraces in three colors. Sure enough, the lens appeared to be of Littrow design with these glasses, and with a bit of figuring on one surface, the image could be nulled very nicely.

Armed with some basic idea of what I was working with, I headed to the workshop. I do not generally make detailed plans of what to build for a telescope, as I use recycled materials as much as possible and change the design as fortune (and available resources) dictate. Of course, the first thing was to make some pitch laps and work on the lens. I had bought a used 8-inch optical flat at RTMC a few years ago and setting up an autocollimation test showed a terrible figure on the lens. Now the problem was to determine which surface needed the most help. Since some of surfaces still had pits, I decided to polish and use a smooth stroke to try to help the figure while completing the polish. The concave surface of the flint was monitored with a Ronchi tester at the center of curvature and was worked to as good a sphere as I could manage. With the other three surfaces being convex, I decided to work the crown element as one tool that could be used on both sides of the lens; and I hoped that with three surfaces polished out—that the overall fig-

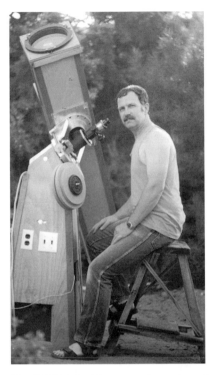

Fig. 11.2.1 *Michael Spooner at the eyepiece of his 9-inch f/15 Springfield mounted Refractor.*

Fig. 11.2.2 *Close-up of Michael Spooner's telescope mount showing eyepiece position and the "beaver tail" counter-weight arrangement. Note that even though the scope requires quite a bit of weight to do the trick, it is distributed in such a way as to cause the observer little or no discomfort or inconvenience.*

ure would be much improved. During this time I decided to use salad oil[8] between the lenses to reduce the figure effects, and I also reversed the crown from time to time to see if there was a benefit from one side to the other. This was very messy, and eventually I used clear Ivory dish soap, as it was easier to clean off prior to each polishing session. Evidently there were some errors still in the inner surfaces as the figure improved with the liquid between the lenses. However, it was still quite bad and after removing the pits and getting to the point where the figure did not change when reversing the crown, I started on the long radius convex surface number four.

Having some very deep zones, this almost flat surface proved to be very stubborn, and I spent a considerable amount of time smoothing it. There would be no pits in this surface when I finished! Eventually I managed to get an excellent

---

[8] Christen, Roland. "Gleaning's for ATMs—An Apochromatic Triplet Objective," *Sky & Telescope*, Oct. 1981, pp. 376–381.

null and was ready for the job of mounting the lens in a suitable tube. At RTMC 1995 I had talked with Gerry Logan about some of the problems of mounting large lenses and mentioned that I was considering a Springfield mounting for the 9-inch. Jerry is a professional optician who has brought some wonderful[9] scopes to RTMC and was kind enough to chat with floundering glass pushers like myself. He suggested an adaptation known as a beavertail mounting in which the counterweights are arranged in a wide flat configuration that slides between the observer and the pedestal. Since a Springfield directs the focus to an eyepiece aligned with the polar axle, the weights are bent at an angle to the plane of the declination axis to keep them from causing the user to reach an inordinate distance to the eyepiece. This angle also complicates the weight calculations for counterbalance as weight added or moved down the bar has an effect relative to the angle. The photos show this arrangement with more clarity. Since this requires the weights to be decoupled from the declination axle, the tube assembly needs to be balanced about the declination. This whole arrangement requires three flat mirrors to end up with a fixed eyepiece position that is the hallmark of the Springfield design. As I was not really certain how the lens would perform, I decided to fold the light path once and use an existing mount to field-test. I had recently obtained more flat mirrors (again at RTMC) and now had a 10-inch to test with and a 6-inch that I had picked up to fold the refractor. Knocking together a box and hunting up additional weights (this was going to seriously stress my 6-inch mounting), I set up in the back yard and got my first views with my own large refractor.

There are times when it is fun to do optics, but sometimes a project gives one a case of the blues. You guessed it! Way too much blue. The figure was really nice though, and it hurt to go back to the computer to come up with a solution. The easy way was to space the elements out and correct the color error that way, but I really did not want to spend hours smoothing the inner surfaces (they were still oiled). I really liked having the reflections cut down because of the oil interface. A recompute showed that regrinding the final surface from a radius of about 2500 inches to a radius of 1050 inches would leave the lenses in contact with a traditional color correction.

Here is where I digress for a moment about color correction. During the hours I spent initially determining the glass types by computing raytraces and looking up index values from glass tables, I became confused as to the use of the C, d, F correction and the C, e, F correction. The letters refer to different lines of color in the visible spectrum. My old college optics text used the d-line for the yellow index values while many more modern references use the green e-line in the calculations. It was interesting to see the different corrections that appeared by changing the e and d index in some designs. It would make an interesting article to study the results of such an experiment with different glass combinations.

Back to the telescope and the need for a new curve. I turned a tool from a scrap of aluminum to the required radius and ground the back surface with 500 grit and smoothed with 12-micron powder. The polishing went quickly, and the figure

[9] Berry, R. "Riverside 1982," *Telescoping Making* #16, Summer 1982, pp. 28.

Fig. 11.2.4 *The drive motor is mounted on a block of wood in an effort to dampen vibrations. The telescope is balanced in such a way as to allow it to be driven by the power of a 9-volt battery.*

Fig. 11.2.3 *Telescope saddle showing the cross-stitch plastic bearing.*

was not too far off when the pits were gone. Getting the figure to null again took some time and effort as my shop area (garage) lacks temperature control, and the pitch was a variable factor as the summer heat came early. Through persistence, the lens finally reached a stage I called finished; and I mounted it in the cell and set it back up on the bench for a final check. I nearly passed out when I put the tester in the light path and saw an immense turned edge! The cell was somewhat warmer than the lens and had transferred enough heat to totally bomb the figure. After cooling, the figure returned to normal, but I had not realized how much heat could be transferred to the lens from the cell. After having used the scope for a while, I found that the cell has not affected the figure during everyday use. The lens also shows a bit of wedge error that I could not completely eliminate by rotating the elements, but it does not seem to be too noticeable in the field. The lens elements themselves are 9.4 inches in diameter, but the flint has about a ¼-inch ding at the edge, and masking the crown to 9 inches keeps it out of the optical path. The defect would never be noticed in the final image. However, with the cell lip and edge effects from polishing, the 9-inch clear aperture is most reasonable. I used a 100-line Ronchi grating with the 10-inch flat to do the final testing. Considering that the focal ratio is 14.8, the bench tests indicate that the lens should do nicely.

Now with the lens taped and oiled (with a few drops of medical grade mineral oil) and in its cell, I was ready to start on the Springfield mounting. I was surprised to find that I had very little information available on this design. *ATM #1* has Russell Porter's info and *Advanced Telescope Making Techniques*, Volume 2 has a chapter by J. R. Bruman, and that is about it. My *S&Ts* from 1967 to date had no Gleanings article that I could find describing a Springfield. I understand that *ATM* #2 may have some additional information, but I have been unable to find a copy to purchase. So I used what was available and went to the scrap heap to see what would result. Previous experience had shown that plastic canvas used for crafts makes a smooth, stable thrust bearing when used between aluminum plates. For a radial bearing, I used a ball bearing fit through the R.A. gear into the polar plate. For a lighter scope, a simple bolt and locknut should work fine. I used a roller bearing to adjust the thrust friction at the back of the polar axle. The gear was hobbed on my lathe against a piece of ½-inch all-thread that I had ground spiral grooves in to form a cutting edge. I then used a piece of the same threaded rod to cut the worm to try to keep the periodic error as low as possible. Cutting teeth in the worm wheel was a slow process even though aluminum is relatively soft. A spiral fluted tap, as suggested by others,[10] would probably be quicker; but I used what I had on hand. The plastic canvas is used on both sides of the gear with one side being the motor clutch and the other being the slewing clutch. Some may be tempted to use Teflon for the bearings. The cold flow characteristics will cause the gear mesh alignment to change with time and possibly be a source of periodic error. The plastic canvas seemed stable and worked well between aluminum. As cheap as the canvas was, a yearly cleaning and replacement would not cause hardship. There seemed to be no problem for the motor to move the assembly if it was balanced properly. I mounted the motor and worm on the equatorial head instead of the base. This way I did not have to worry about the telescope tube running into the motor assembly. However, the motor vibrations had to be very low, or the image would be degraded at high powers. In an effort to minimize the vibration paths, I mounted the motor on a block of oak with slots cut between the motor mounting screws and the attachment plane of the block to the mounting. I used a small stepper motor to drive the scope and could supply enough power with a nine-volt battery. But I used an old car battery, as I was not sure how long a small transistor battery would last. I wondered about mounting the drive electronics on the plate with the motor and setting up a corrector using a TV remote to adjust the tracking. By using the same arrangement for the declination, a wireless control system could be achieved with no power or control cords for observers to get tangled in! At that point, I had R.A. tracking, and it worked fine for visual use or camcorder shots of the planets.

The declination axis was a similar set of plates except there was no radial bearing or motor drive. Because the light path coincided with this axis, a hollow pipe was used as the axle. An upgrade in this area would be to include a radial bearing and drive setup. The short declination axle was a piece of aluminum conduit threaded on one end to fit a conduit coupler, and welded to the saddle on the other end. This mounting could be built using hand tools, and tapping and bolting the pieces together. An outstanding example of a hand-crafted mounting would have to be the RTMC annual favorite 8-inch Newtonian Springfield built by Tim Parker and reported on by Richard Berry.[11] I did not have the issue of *TM* with the

---

[10] Mitton, John N. "Make a Telescope for $500: Model 3," *Sky & Telescope*, May 1989, pp. 488-491.

Fig. 11.2.6 *The old adage that states "the best telescope is the one that gets used" is very true. Still, how many of us have the luxury of being able to comfortably sit in one place and enjoy the performance of large aperture refractor.*

Fig. 11.2.5 *Final diagonal and focuser base illustrating the rigidity of the assembly.*

description, but Tim explained the details of the project when I saw him at RTMC. A metal cutting bandsaw was really useful to shape the parts, and a good set of files helped smooth things to final dimensions. A lot depended on the builder's ability to scrounge the pieces needed for construction. A wooden version was perfectly adequate for a lighter scope.

The eyepiece diagonal and focuser were held in pipe machined to fit in a base ring centered on the polar axis. The base ring had the possibility of accepting a polar alignment scope for initial setup at an observing site. I liked being able to transport the focuser and diagonal assembly separately from the pedestal. The base pedestal was a light, but very stable, wooden pyramid that I borrowed from the design used on Larry Myer's brass refractor.[12] I welded some flat aluminum stock to match the inner dimensions; this provided extra stiffness to the upper portion of the base. The mounting that was bolted to this assembly pivoted for latitude adjustment. Considering the length and weight of the tube assembly, the mounting. did a good job of providing a stable view.

One of the more difficult aspects of building this telescope involved setting up the mirrors for good collimation. In addition to having the lens square with respect to the final focal surface, the second and third mirrors needed to be aligned with each other and their respective optical axis. If they were not closely aligned with each other through the declination axis, moving from east to west would result in the image being somewhat off-axis when centered in the focus tube. Since

[11] Berry, Richard. "Riverside 1981," *Telescoping Making* #12, Summer 1981, p. 24.

[12] Myers, Larry. "Constructing a Large Brass Refractor," *Telescoping Making* #44, Spring 1991, pp. 6–13.

the Littrow lens design corrected coma only for certain glass combinations,[13] and as I further compromised the ability to correct aberrations by oiling the lenses in contact, I spent extra effort to provide the mirrors with enough adjustments to allow adequate alignment. I also tried to design the mounts to stay in adjustment as collimating was anything but intuitive—at least for me. Upon completion of the scope I journeyed to the Grand Canyon Star Party for 1996, and the scope required only a tweak to align. I have since replaced the two diagonal mirrors (they were back ordered at the time of the star party) with enhanced coated mirrors from Orion.[14] One of the older diagonals was curved and induced astigmatism at some orientations. This was not very noticeable, as only a small part of the mirror was illuminated by the light cone; but the coating was scratched, and I felt the objective deserved a sporting chance. The new mirrors cleared most of the astigmatism, and the clean coatings really helped the contrast. I am not sure if the astigmatism was a result of the residual wedge in the lens, or if the collimation was still doing something I did not fully understand. It may also be that the oil coupling between the lenses caused some problems during cool down. The mineral oil may not have been the best oil to use, as the lenses seemed to be difficult to separate. Perhaps I will try some of the oil suggested by Roland Christen in his article in *TM* 28,[15] if I ever find it.

I have read several articles about homebuilt telescopes that ended at this point, only to leave me wondering if the telescopes performed in the field as well as in the workshop. I have indicated a few of the problems, which I can see in this telescope when testing, either on the bench or in use. With the star test, and using Dick Suiter's book[16] as a guideline, I saw a bit of correction error. It corresponded to his images depicting about $\frac{1}{8}^{th}$ wavefront error. The color matched what he wrote concerning Taylor's description of a doublet's color error when a star is tested through focus. A 250-line Ronchi screen showed slightly bowed lines at focus. This indicated a very slight hump in the center, but the curve was so smooth and the images so good that I did not think it warranted risking the overall figure for such a small area. Most of the time, the astigmatism was absent, and the Airy disks were round and had 3 to 5 rings visible on the brighter stars. Around the double double Epsilon Lyrae, one faint ring was visible around each star. Yes, there was color; but I had not used many refractors, and so I did not really know if it was excessive. Of the few close double stars I viewed, Zeta Cancri and Zeta Boötes were quite easy as was Eta Coronae Borealis. I believed these stars were separated from 0.7 to 0.9 arcsec. More difficult to split was Gamma Coronae Borealis, which I believe was about 0.6 arcsec separated but had about 3 magnitudes difference between the components. The first time I split this, the seeing was not super; and I could not see the faint companion at 460 magnification. Slipping a 4.7 mm

[13] Sorenson, Andrew. "Building a 6-inch *f*/21 Littrow Refractor," *Telescoping Making* #36, Spring 1989, pp. 12–19.

[14] Orion Telescopes & Binoculars, P.O. Box 1815, Santa Cruz, CA. 95061.

[15] Christen, Roland. "Design and Construction of a Super Planetary Telescope Objective," *Telescope Making* #28, Fall 1986, pp. 20–23.

[16] Suiter, Richard. *Star Testing Astronomical Telescopes*. Richmond, VA: Willmann-Bell, Inc.

Fig. 11.2.7 *There are hundreds of types and configurations of telescopes. Yet, few can offer the same pleasure in viewing the Moon and planets as a long-focus refractor.*

in place for 720x brought it into view; and during instances of good seeing, a dark sliver of space could be seen between the Airy disks. Recently, Jupiter presented some fairly good views considering how low it was on the horizon (compared to where it was in 1993). There was some shading in the Red Spot and several white spots on either side in the SEB. Festoons and loops were seen, though the atmosphere seldom allowed steady images of these fainter details. At times, the image hinted at a multitude of fine detail on the planet; but by viewing stars at the same altitude, I found that the seeing suffers a great deal compared to views more directly overhead. I was fortunate to see the first quarter Moon nearly at zenith with excellent seeing, and the images were quite nice. The details of the multiple peaks comprising the center of the crater Tycho took some time to accurately sketch. Venus was the first available planet to examine and was placed (conveniently) high in the west at sunset. Of course it had a purple glow surrounding its white face, but by finding the planet as early as possible, the purple haze mostly faded into the background sky. Using 720x, the planet's edge was still razor sharp, and the cusps ended in needle points. I could hardly wait for more favorable conditions for Jupiter! However, the most satisfying aspect of these observations was not stretching on tiptoes or bending like a pretzel to reach an eyepiece. So come on over and sit on that stool, and perhaps a little later you will wonder how it came to be 2 A.M.

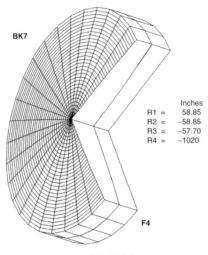

BK7

	Inches
R1 =	58.85
R2 =	−58.85
R3 =	−57.70
R4 =	−1020

F4

Fig. 11.2.8

so quickly.

11.3 Telescope Making Renaissance in Bellingham
By William J. Cook

March 23, 1997, was a special day for astronomy and telescope making in the northwest. We had relatively good seeing for viewing Comet Hale-Bopp and the lunar eclipse, and many of us had just completed our second great day of the annual Optics Workshop in Bellingham, Washington.

This conference was a real treat for me, for while it was the 6th event of its kind to be held in Bellingham (home of S.P.I.E), it eclipsed previous events in attendance, format, facilities, smoothness of operation, and productivity. Speakers included noted telescope maker and optician, John Gregory; Dan Bakken, who built the world's largest (41-inch) "portable telescope"; and Mel Bartels, who is facilitator of the ATM List.

The event was sprawled throughout the new Mathematics, Science, and Technology building of Western Washington University, which seemed to have been constructed with catering to such conferences in mind. Not only were participants surrounded by all manner of gizmos, gadgets, and displays to stimulate the imagination; but every conceivable form of audio/visual equipment as well. During previous conferences, those interested in lens design fundamentals had to huddle around a single computer screen and take turns getting close enough to see what was going on. This year, the presenters stood before a 17-inch monitor with the display being projected onto an 8-foot screen for all to see.

So many important things were going on at all times that it is difficult to describe any given demonstration or lecture as stealing the show. Mirror making and optical testing were going on throughout the event, and participants ranged in ages from 12 to 70, with some coming from as far away as Southern California and New York City. In the past, the mirror making portion of the event had largely consisted of one or two grinding platforms with each interested attendee getting only a few minutes of glass pushing to his or her credit. With the amount of room available this year and the conference being a two-day event, mirror makers had a good opportunity to make great strides toward coming away with a mirror ready to test or have coated. At one point, 8 mirrors were being ground or polished!

Efforts in the optical testing area were no less dramatic. The Foucault, Ronchi, null, and wire tests were explained and demonstrated; and some individuals had an opportunity to test complete telescopes. John Gregory even showed up with a 12.5-inch (aluminized) optical flat to help facilitate tests involving auto-collimation.

Sunday afternoon, the senior organizer Kreig McBride showed a series of spherometers he had made and demonstrated how measurements of relatively great precision could be made with the correct assemblage of items that are readily available at low cost.

I would think the second most important aspect of the entire event was the focus—no pun intended. Participants knew what they were there for and went about their business with a precision one would expect of a group of engineers putting the final touches on a project and who had been working together for months. There were teachers; there were students—and 15 minutes could change a person's role completely.

If this kind of camaraderie is the second most important aspect of the meet, then it is surely only surpassed by the sacrifice of the small but loyal team of selfless planners who spent weeks in planning and facilitating the event. While every avocation these days has its fair share of individuals with their own agendas, at no time during the conference did I detect a hint that any of the organizers had anything on his or her mind other than disseminating knowledge and strengthening our hobby.

Some might find fault in the fact that there was no star party associated with the event such as at Stellafane and Riverside. Those I spoke with felt that having a workshop venue was a plus. They felt that a clean comfortable environment with controlled lighting was much more conducive to serious optical fabrication and testing and that the attendance of those truly interested in telescope making made the gathering seem "more like an academic conference, than a sideshow."

The 7th annual event is slated for March 21st and 22nd, 1998, in Bellingham, Washington. While I cannot envision a great improvement over this year's event, Kreig McBride assures me that it will be the case.

Fig. 11.3.2 *Dan Bakken's 41-inch Dobsonian was the center of attention in the main classroom.*

Fig. 11.3.1 *Noted telescope maker, John Gregory, was the gathering's keynote speaker. John spoke on his Houghton telescope design and on the topic, "Telescopes I Have Known."*

Fig. 11.3.4 *Here, senior workshop coordinator Kreig McBride shows off some of the very fine spherometers he has made and discusses the relative accuracy of each.*

Fig. 11.3.3 *Above one of the storage cabinets we see the remains of one of the workshop participants (from the 1995 workshop) who did not even come close to discovering how long it takes to polish out an 80 grit pit.*

Fig. 11.3.5 *(Left) Now a resident of the northwest (Lyons, Oregon, to be exact), Richard Berry had his first opportunity to attend the workshop.*

Fig. 11.3.6 *Here we see Carl Fielding looking very serious as he sits at one of the workshop's many testing stations.*

Fig. 11.3.7 *SAS Board Member Loren Busch has read many books on telescopes, and at the workshop he was able to see one up close. Just seconds after this photo was taken, Loren was heard to exclaim, "That's not snow on Olympus Mons, it's...it's...lint?!*

11.4 Annular Baffles for Barlow Lenses
By Philip Moniot

Barlow lenses are very useful in obtaining high powers without the need for short focus eyepieces. However, during Barlow lens assisted lunar observation, the field of view in the eyepiece is subject to serious glare from reflections inside the Barlow lens tube. The glare occurs at intervals of arc of telescope position. As the image of the Moon approaches the optical axis, a series of reflections begins,

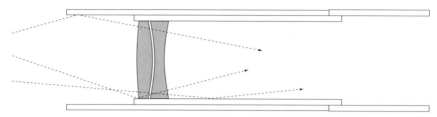

Fig. 11.4.1

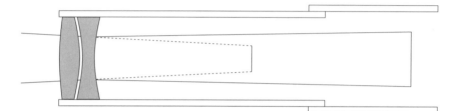

Fig. 11.4.2

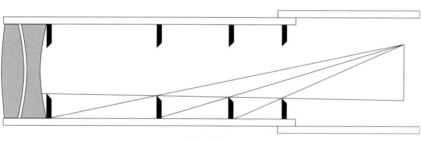

Fig. 11.4.3

climbing up the standard focuser's interior tube walls, past the Barlow lens, and into the Barlow lens tube.

Fortunately for us, the Barlow lens, being negative, allows its mounting to be 100% glare-proof. The reason is this: being negative, the lens can be smaller than the eyepiece field lens. If the Barlow lens is oversize, as is usually the case, free space is directly behind it for the first of a series of annular baffles.

The most serious glare spot in a Barlow lens mounting is the lens cell. By installing the first annular baffle immediately behind the cell, this glare spot may be completely blocked. The actual working area of the oversize Barlow lens will remain unobstructed (and preserve all of the telescope's resolution and light gathering power) across x as long as the aperture of this baffle equals

$$\text{Aperture} = \left(D - \frac{x}{m}\right)\left(\frac{L}{F}\right) + \left(\frac{x}{m}\right)$$

where

D is the telescope's aperture,

x is the linear diameter of the eyepiece field,

m is the magnifying power of the Barlow lens,

L is the distance of the Barlow lens inside the telescope's prime focus, and

F is the telescope's focal length.

Completion of the baffle system design is accomplished graphically by drawing straight lines from the edge of the eyepiece field across the optical axis to the base of each preceding baffle and installing annular baffles to intercept them. The interior sides of the mounting are now invisible from the eyepiece field and will always be dark. In practice, only two or three annular baffles are necessary.

The focuser can, nonetheless, defeat the Barlow lens baffle system. This happens if the Barlow lens tube is short and the focuser drawtube long. The glare will be seen when the bright portion of the Moon's image is well outside the eyepiece field. Light is hitting the focuser drawtube interior at a low angle, well ahead of the Barlow lens, and is being reflected directly into the eyepiece. The remedy is expansion of the baffle system into the focuser drawtube.

11.4.1 References:

1. Brown, Sam. *All About Telescopes*. New Jersey: Edmund Scientific Co., 1976. pp. 15, 67–68, 158–159, 204.

2. Hartshorn, C. R. "The Barlow Lens": *Amateur Telescope Making*, Ingalls, ed.: Vol 3, Scientific American edition, 1953, p. 215; Vol 3, Richmond, VA: Willmann-Bell, Inc. edition, 1996, 177–181.

3. Rutten, Harrie and Martin Van Venrooij, *Telescope Optics, Evaluation and Design*, Richmond, VA: Willmann-Bell, Inc., 1988. pp. 130, 149–152, 221–222. Although pages 221–222 are about baffle design for refractors and Newtonians, they reveal the basic principle of annular baffle design.

11.5 Rangefinders and Stereo Telescopes

By Peter Abrahams

In 1893, Ernst Abbe, working for Carl Zeiss, applied for a patent on their new prism binocular, but it was denied because of the earlier Porro prism glasses from several European makers. A revised patent was submitted for a prism binocular with enlarged objective distance, with the increased separation between the objectives being the protected feature. This was approved, and for 15 years no other optician could make a Porro prism binocular with objectives more widely spaced than the oculars. The rapid development of prism glasses by other quality makers caused the energetic Zeiss publicity works to seize their unique characteristic and proclaim its advantages in advertising. There is a real, if minor, increase in sense of depth that follows this increase in inter-objective distance. This is probably perceptible at close focus with standard hand-held binoculars, although there is wide variation in individual ability in stereopsis. Zeiss used the term "plasticity" to describe the enhanced sense of depth, and it is a very apt term, since nearby objects appear modeled or sculpted. This characteristic was quantified, with "specific plasticity" being defined as objective distance divided by ocular distance, and "total plasticity" as magnification times specific plasticity. Higher magnification adds to the effect. Increased perception of depth does allow the observer to distinguish between objects that might otherwise be of very low contrast. This advantage was the subject of many studies, papers, advertisements, and brochures around the turn of the century.

Zeiss also made theater glasses with closely spaced objectives for portability, and they were not shy about publicizing the advantages of this configuration. They claimed that in the theater, diminished depth perception is useful because the spectator will see the live actor as part of the painted backdrop. While these concerns are of minimal import today, the effects are real and were a very important part of the introduction of binoculars to the public.

The Zeiss prism binoculars of 1894 were the first commercially successful, the first mass produced, and the first high quality binoculars. At the same time, Zeiss offered 2 prism binoculars with objectives 12 inches apart (8 power), and 16 inches apart (10 power). A hinge between the oculars allowed them to fold in half, leading to the generic term *Scherenfernrohr* or scissors telescope. These were called by Zeiss, "Relieffernrohre," and were not successful. The 8 x 20 model was offered from 1894 to 1906, and the 10 x 25 from 1895 to 1908, and through 1918 for military use. They gave spectacular views of terrestrial objects, greatly magnifying the perception of depth in a scene and the appearance of modeled relief in an object. Here there is no exaggerating the effect. They were used as rangefinders in both World Wars, by several service branches of most of the participants in the conflict. Hand-held instruments were about 6 x 30, with objectives 18 inches apart and a folding hinge to reduce the length for transport. Tripod mounted instruments could have 50 mm objectives, for use at dawn and dusk. These were used by artillery forces to approximately judge distances. The smaller sizes were needed for quick judgments on shell bursts, when a large instrument or more complicated rangefinder could not work quickly enough. These "battery commander's rangefinders" can occasionally be found at gun shows or military collectors' meetings, and there are a few optical repair shops remaining that can correct their typical out-of-collimation condition.

Truly remarkable instruments were used by the U.S. Navy (among others), from prior to WWI through the 1980s, for controlling the large guns of their ships. Some of these rangefinders used coincidence sighting, where two images were brought together in the viewfinder and the distance read off of a scale. Others were stereoscopic rangefinders that gave a true stereo image of the target. A reticle for each eye was fixed in the tube and formed a stereo image that appeared to move towards and away from the observer when optical wedges were rotated. When the image of the reticles (an arrangement of diamond shapes) seemed to be at the distance of the target, the actual distance to the target could be estimated.

There was extensive research and development on these fire control instruments during the 1920s, and they were the primary tool used to aim naval guns through most of this century. The longest recorded distance for optical rangefinder controlled gunfire, successfully firing on a moving target from a moving battleship, is 26,400 yards, achieved in 1940 by the British. These rangefinders were designed around a particular gun, and the distances at which they were accurate were determined by the range of the gun. In the U.S. Navy, the Mark 41 (1930s) and Mark 75 (1950s) had objectives 11 ft apart, a near focus of 1,200 yards, and maximum useful range of 20,000 yards. These were made by Keuffel & Esser, weighed about 1,200 lbs, and had 147 glass elements including lenses, prisms, wedges, reticles, mirrors, and frosted elements. There were 15 ft models, weighing about 1,500 lbs, in a motorized mount that was connected with servos to a gyroscope to maintain the horizon at a level. The 11 and 15 ft models could be targeted on aircraft, and longer instruments were used to range ships and targets on shore. Larger models were made by Bausch and Lomb, including the 26.5 ft used with the common 16-inch guns. The Mark 52 consisted of a 25 power system with objectives 46 ft apart, weighing 10,500 lbs, and costing about $100,000 during World War II. Near focus was 5,000 yards, maximum use at 45,000 yards.

One interesting aspect of later rangefinders is that they were gas charged with helium, since it is the only gas with an index of refraction that does not change in the temperature range encountered by these instruments and the extreme length of the rangefinders mandated this stability. The use of helium necessitates yet another level of maintenance for personnel; one source notes that it can leak through steel—no doubt all seals and joints are somewhat porous to helium.

These instruments were closely held secrets during their era (still used in foreign fleets), and their size and weight ensured their dismantling on retirement. Very few persons have had the privilege of viewing through one, and the effect can only be imagined.

Bernard Merems of Patagonia, Arizona, is an ambitious ATM who is constructing a binocular refractor with a widened base between the objectives. At Riverside '96, Bernie described his half-finished project. Two B & L telephoto lenses, 5 inches in diameter and 40 inches in focal length, are mounted onto a prism housing so that the objectives are 16 inches apart. Light from the objectives enters the housing and strikes first surface mirrors, mounted at 45° from the lens's optical axes, to converge the light into prisms at the center. The first prisms are standard 90° reflecting prisms—used to direct the light back towards the oculars. These prisms serve as collimators by rotating around a vertical axis and are an unfinished aspect of the instrument design at this point. Collimation of twin telescopes is quite difficult, and it is likely that adjustment about a single axis will not suffice to correct all collimation errors.

The light exits these prisms in the same horizontal plane that it entered the

Fig. 11.5.1 *The view of the forward Mark7 16-inch/50 caliber gun turret of the USS Missouri (BB63). The ship was armed with three of these turrets, each with three guns. During World War II, Iowa class battleships like the Missouri used the Mark 38 Director System which was based on the 26.5 foot long Mark 48 stereoscopic rangefinder, with two Mark 69 telescopes, one for the pointer and one for the trainer, a Mark 56 telescope for the crossleveler, and a Mark 29 telescope used by the spotter to scan for other targets. These guns could fire a 2,700 pound armor piercing round 36,900 yards (21 miles) and a 1,900 bombardment round 40,180 yards (22.8 miles). U.S. Navy Photograph taken in San Franscio Bay, May 1986.*

instrument, having been reflected twice and giving a correctly oriented image. However, it was desired to direct the oculars downwards, at 60° to the horizontal, to allow comfortable viewing of the sky. This created many complications in anticipating final image orientation. Bernie finally consulted R. Buchroeder of Tucson on the subject and was advised to purchase two toy periscopes, which would allow rotation between the two mirrors. When held vertically, in using position, and the upper half rotated 360° to scan the entire horizon, the image rotated to upside down when pointing backwards and back to right side up when pointed forwards again. Continued perusal of this phenomenon is thought to give an intuitive grasp of the complexities of designing image erecting systems. The final image erecting design, if first light does not force any revisions, adds a deflecting prism with two silvered surfaces, with the light exiting at 60° to the horizontal. A total of four reflecting surfaces in this orientation should give an inverted and reversed image. For terrestrial use, part of the assembly will be replaced with a single prism or flat.

  Interocular distance will be adjusted by mounting all four prisms and the oculars, as a pair of assemblies, onto a sliding track with right- and left-handed threaded rods to change separation. Two-inch oculars will be used. The 16-inch objective separation places a distant limit on the close focus of the instrument, for viewing nearby objects would require the telescopes to swivel inwards towards each other. This reduces the required eyepiece travel, for they will not have to rack out for focusing on nearby objects. Many such details must await final assembly of the instrument.

Telescopes are typically used to increase resolution, contrast, and light gathering ability. Their potential for enhancing stereoscopic perception of depth is a fascinating and overlooked subject.

11.6 Optimizing Secondary Size for Newtonian Telescopes
By Charles R. Genovese, Jr. M.D.

Newtonian telescopes can offer performance rivaling any other design for image quality, as demonstrated in recent articles by Zmek[17] and Suiter.[18] However, selecting the proper size secondary mirror is critical. If optimized for planetary work, the central obstruction should measure less than 20% of the primary so as to maintain as much contrast as possible, and for prime focus photography a much larger diagonal is needed to prevent undue vignetting at the edge of the field. Most telescopes employ a compromise between these two extremes. Although Francis,[19] Texereau,[20] and Rutten and van Venrooij[21] provide formulas and discussions of the issues, rearranging the formulas and graphing the results gives a far better understanding of all the variables for any telescope.

Figure 11.6.1 illustrates that as we increase the primary to secondary separation, the diameter of the converging cone of light and the required secondary size to intercept it, N decreases. Simultaneously, the secondary to focal plane distance, D, decreases—as does the distance from the side of the tube to the focal plane. We shall call this parameter the "working distance." The focuser takes up part of the space, with the remainder called the "back focus" as seen in Figure 11.6.2. Usually a minimum of about 0.75 inch of back focus is required to focus high power eyepieces (unless some lens arrangement is employed to project the image further outside of the tube).

The next parameter is called the "field of 100% illumination." As we move outside this circle towards the edge of the field of view, the intensity of the image

---

[17] Zmek, William P. 1993. "Rules of Thumb for Planetary Scopes—I", In: Telescope Making section, *Sky and Telescope*, July, 1993, pp. 91–95; and Zmek, William P. 1993. "Rules of Thumb for Planetary Scopes—II," In: Telescope Making section, *Sky and Telescope*, Sept., 1993, pp. 83–87.

[18] Suiter, Richard and William P. Zmek. "A Dialogue on Spider Diffraction." See Section 11.1 on page 385.

[19] Francis, Peter. Newtonian Notes, 2nd ed.: *Designing and Building a High Performance Telescope from Commercial Components*, 1979, Kenneth Novak & Co., Ladysmith, WI, p. 59.

[20] Texereau, Jean. *How to Make a Telescope*, 2nd ed., 1984, Willmann-Bell, Inc., Richmond, VA, p. 424.

[21] Rutten, Harrie and Martin van Venrooij. *Telescope Optics: Evaluation and Design*, Richard Berry, ed., 1988, Willmann-Bell, Inc., Richmond, VA, p. 29.

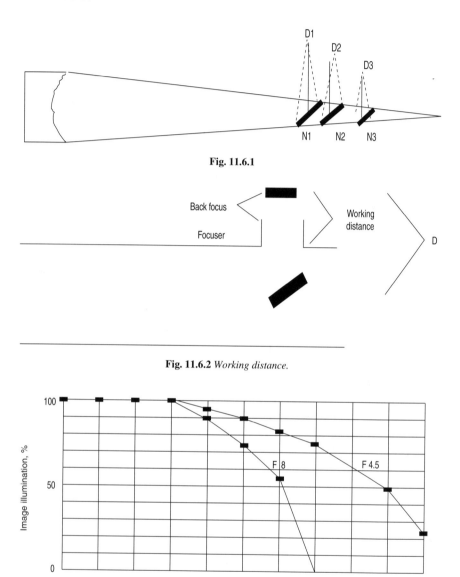

Fig. 11.6.1

Fig. 11.6.2 *Working distance.*

Fig. 11.6.3 *Relative falloff in illumination with distance from optical axis for different f/number telescopes.*

decreases (with low focal ratio primaries having a more gradual slope). For any given diagonal dimension, as the working distance is increased (by decreasing the separation between the primary and secondary), the field of 100% illumination decreases. This is illustrated in Figure 11.6.3.

Although long focal length eyepieces can technically benefit from fully illuminated fields of 0.75 inch to 1 inch or even more, the eye is relatively insensitive to mild decreases in illumination. For example, my 10-inch *f*/6 is usually configured with a 1.86-inch diagonal and 2-inch working distance. This gives only 0.6 inches of 100% field illumination (38 minutes of arc), but no vignetting can be detected visually, even with a 40 mm wide-field eyepiece giving a 1.8° true field. The field size requirements decrease as magnification increases. For CCD imaging, even the largest chip available for amateurs measures only 0.64 inch diagonally (most are half that or less), so again there is little benefit to a larger fully-illuminated field.

This, however, is not so with film! A 35mm frame measures 1.65 inches (42 mm) diagonally, and with a coma corrector, the entire field is usable. In order to fully illuminate a field this size, the diagonal must be larger than in the examples above; the working distance must be increased to accommodate a 35 mm SLR (55 mm of back focus). If an off-axis guider is employed (an additional ¾ inch to 2 inches), further increases in secondary size are needed. In addition, the faster the primary, the steeper the cone of light, and the secondary must increase even more. (As the telescope's aperture increases, this effect is less pronounced, as the increases in working distance are proportionally less.) With "fast" low-focal-ratio instruments, short 2-inch focusers are required to avoid vignetting at the edges of the light cone from the bottom edge of the focuser's drawtube.

The third parameter is the minor axis of the diagonal, which is commercially available in the following sizes: 1, 1.25, 1.55, . . . , 3.5, and 4.0 inches. The formula in Francis's *Newtonian Notes* for taking all these variables into account is:

$$N = \frac{MD + IF - ID}{F}$$

Where: D = distance from diagonal to focal plane
              N = minor axis size
              I   = field of 100% illumination
              F = focal length of primary
              M = primary diameter

A better arrangement, however, is to solve the equation for I:

$$I = \frac{NF - MD}{F - D}$$

By choosing an available diagonal size and several values of D, calculating the result and graphing I versus D gives a nearly straight line. Since only two points are necessary to define the line, it is convenient to choose the first value as the distance to the O.D. of the tube and the second to be 6 inches higher. On the *x*-axis one can then mark the minimum focuser height and the back focus required for each accessory; i.e., 2.2 inches higher for a 35 mm camera, and an additional inch for a Lumicon off-axis guider. A series of parallel lines are generated if this is repeated with several diagonal sizes, as shown in Figure 11.6.4. If one subtracts the tube radius from D, the "working" distance can also be plotted.

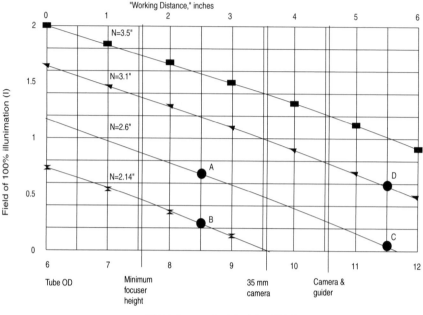

Fig. 11.6.4 *A 10-inch f/4.5 Newtonian with a 12-inch O.D. tube.*

By using the modified equation and graph, the interplay of all these parameters for any given Newtonian telescope can be easily visualized, the requirements
for various applications evaluated, and the best compromise chosen. More importantly, if the primary can be moved back and forth over only 3 to 4 inches and the
secondary easily changed to different sizes (such as with Novak's "Index-arm"
System), compromise is not necessary as a single telescope can be easily optimized for high-power planetary observation or for wide-field deep-sky photography. For instance, a 10-inch f/4.5 telescope with a 1.6-inch low-profile focuser
and a 2.6-inch diagonal could be used with a working distance of 2 inches, giving
¾ inch of 100% illumination for visual and CCD work. (Figure 11.6.4, point A).
This same arrangement could also be used with a Taurus camera (which requires
only about 1-inch of back focus, a 1° field at this image scale). A 2.14-inch diagonal could even be used for high power planetary viewing (0.3 inches of 100% illumination, Figure 11.6.4, point B). However, if the focus must be moved out 3 to
4 additional inches to accommodate an off-axis guider with a 35 mm SLR camera,
the field of 100% illumination drops to virtually zero (Figure 11.6.4, point C).
Amazingly, one prominent manufacturer provides 10-inch f/4.5 Newtonians with
exactly these parameters, giving no field that is fully illuminated! It is obvious
from the graph that a 3.1-inch or 3.5-inch diagonal is necessary for these circumstances (Figure 11.6.4, point D).

In summary, solving the rearranged equation with specific diagonal sizes

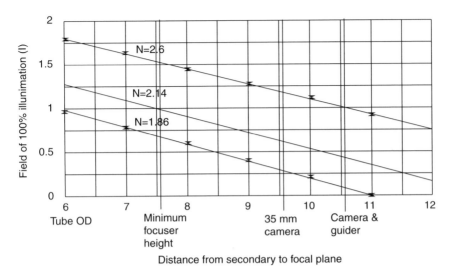

Fig. 11.6.5 *A 10-inch f/6.0 Newtonian: Compare with Figure 11.6.4 for diagonal sizes for a given 100% illumination and slope of lines.*

and various diagonal to focal plane distances generates a series of parallel lines that allows one to exactly visualize the interaction of the parameters for any Newtonian. A fixed compromise or a design allowing shifting the primary and changing secondary size to optimize for specific tasks can be evaluated.

11.7 Rebuilding a 60 mm Spectrohelioscope

By Fredrick N. Veio

My interest in making a spectrohelioscope began in 1962, and my first instrument was completed in 1964. Construction involved using simple woodworking techniques, nuts, bolts, and tools; and it was easily portable. The focal length of the telescope was 108 inches (2.7 meters), while the spectroscope itself had a focal length of 75 inches (1.9 meters). The grating was 32 x 30 mm ruled area, 1200 gr/mm. Linear dispersion was 4 Ångstroms per millimeter creating a 40-inch (1-meter) long solar spectrum from the violet to the red. The solar image synthesizer was a rotating glass disk painted black and cut with 24 slits of 0.005 inch (125 microns), producing 0.5Å half bandwidth. Center-to-center separation of the slits was 3 inches on the 4-inch diameter glass disk. As the glass disk rotated, the H-alpha line would lag at the exit slits by 0.004 inch (100 microns). All of the middle field of view was H-alpha. The extreme top and bottom was at a slightly different wavelength, but this caused no problems.

Now a new spectrohelioscope design is made so that the yellow He, Fe, Mg, and Na metallic lines can be employed at a necessary 0.1Å half bandwidth. The

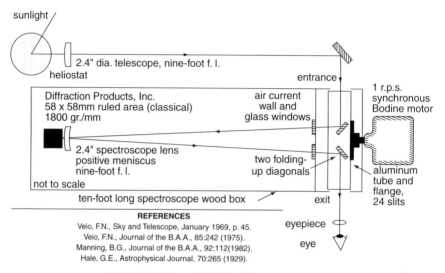

Fig. 11.7.1 *Universal Wavelength Spectrohelioscope.*

solar image synthesizer is redesigned. If the separation of the projected slits is re-duced to about 1.25 inches (32 mm.), the H-alpha lag is much less. Using a longer focal length for the spectroscope lens further reduces the H-alpha lag to about 0.001 inch (25 microns).

Instead of a glass disk, a short piece of aluminum tube 5 inches (125 mm) in diameter is attached to a wide flange, and the later is put on the output axle of a 1 r.p.s. synchronous motor. A thin slitting saw cuts 24 slits of 0.008 inch around the aluminum tube. Blades from pencil sharpeners decrease the slits down to 0.002 inch or 0.1Å half bandwidth. Two small ¹⁄₁₀-wave diagonals are fixed inside the aluminum tube. They fold up the entrance and exit sunlight beams of the spectroscope lens. Brian Manning of England was the first to use this folding-up technique with two fixed diagonals in conjunction with vibrating slits.

The telescope is still a 2.4-inch diameter of 108 inches focal length. The new spectroscope is a 2.4-inch lens of 108 inches focal length. The new grating is 58 x 58 mm ruled area, 1800 grooves/mm, 5000Å blazed wavelength. Linear dispersion is 1.8Å/mm, yielding about an 80-inch-long (2-meter) solar spectrum. An air current wall is near the two diagonals, preventing air currents in the spectroscope box. Two eyepieces were used: a 4.5-inch (112 mm) f.l. for the solar disk in H-alpha and other wavelengths, and a 2-inch (50 mm) for the solar spectrum. Except for young people, the human eye has low sensitivity to violet light. The violet Ca, H, and K lines cannot be utilized. Total costs are about $800.

11.8 An Enhanced Foucault Test For Astronomical Mirrors

By Dave Armstrong, Boeing Employees Astronomical Society

Anyone who has spent time discerning the subtleties of shading during a Foucault knife-edge test knows the fatigue that can occur. Another big factor in obtaining reliable, repeatable readings in this test is subjectivity. It is not always easy to decide if the extreme outer radius "cat eyes" of a zonal mask are graying equally.

I have worked on an enhancement of the knife-edge test, which has helped in the areas of fatigue and subjectivity while improving reliability and repeatability. The modification consists of using a video camcorder to pick up the reflected image from the mirror, putting it on a television screen and measuring the light variations on the image with Cadmium Sulfide Photocells.

Figure 11.8.1 shows my Foucault knife-edge tester, which has some changes from what you normally see in the publications. However, these modifications are not required to make this system work. The micrometer (mike) on my tester, used to move the carriage, is a "puller" rather than a "pusher." A spring or strong rubber band pulls the carriage against the mike. This makes the mike accessible and out in the open away from the camcorder to reduce the chances of bumping it. Normally the carriage slides on a single rail or rod and is then tilted by a vertical screw to cause the knife-edge to swing an arc to cut the returning light beam. I have mounted the carriage on two parallel chrome rods so the carriage cannot be tilted. The knife-edge is mounted on two 4-inch pieces of hacksaw blade and has a screw mounted horizontally on the right side (left side of Figure 11.8.1) to move the knife into the light beam.

The camcorder is placed where your eye would normally be with the returning light beam entering the center of the lens. The camcorder runs off the 110 v transformer power supply that came with it, as a battery would not last long. The image from the camcorder is sent to a portable colored 5-inch-diagonal television set. Removing the clear plastic cover from the front of the screen allows my light sensors to touch the screen. I suppose a larger screen television would work, but I am not sure what problems may be introduced by the curvature of the screen. The small 5-inch screen is fairly flat. The maximum mirror image diameter for this set was 4 inches. A wooden stand with a horizontal bar in line with the center of the TV image had a ruler, 4 inches + long, with $\frac{1}{10}$-inch divisions, taped to it. The ruler was placed so the 2-inch mark was at the center of the screen image. I found that a Cadmium Sulfide photocell from Radio Shack (cat. no. 276–1657) attached to an analog volt-ohm meter (VOM) was sensitive enough to register light intensity changes on the screen image when the knife cut the returning light beam. Two photocells, attached to separate VOMs, were used at equal opposite radii from the mirror's center. The photocells are approximately $\frac{1}{8}$ x $\frac{3}{16}$ inch in size and are taped to separate wood pieces so they can be moved along the 4-inch ruler.

Fig. 11.8.1 *Front view of Foucault knife-edge tester, with micrometer in front. The wooden knob on the left side pushes the knife-edge into the beam.*

Clothespins clamp the photocell/wood pieces to the ruler. A 1-inch length of plastic soda straw is placed over each photocell to make a tunnel and narrow the spot size measured on the T.V. image.

11.8.1 Taking the Readings

To start, the photocell stand is aligned in front of the screen with the 2-inch mark at the center of the mirror's image. Remembering that the 2-inch mark is actually "zero," slide the two photocells the same number of $\frac{1}{10}$-inch marks left and right of center (Figure 11.8.2). You can start measuring zones either from the center and work to the outside, or from the outside and work to the center. Clamp the photocells to the bar so they will not move. *Turn off all room lights and pull the shades.* You will now work only by the light of the TV set. This is why everything is so close together, as in Figure 11.8.3, so I could see what I was grabbing in the semi-darkness. Actually, I stacked the smaller VOM on top of the larger so both meter needles were visible, and used a handheld magnifier lens to read the meter scales more closely. The 1x kilo-ohm range on the meters worked best. With the knife-edge backed out of the light beam, read and remember each meter's initial ohm position. They may or may not have the same initial ohm reading—no problem as long as it is not too big a difference. Begin turning the knob that pushes the knife into the light beam and watch for movement on the meters. If one begins

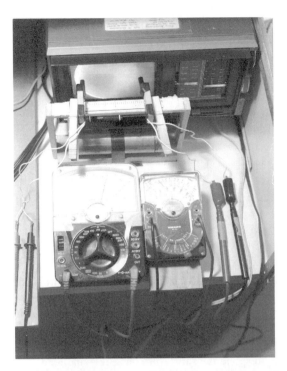

Fig. 11.8.2 *Photocells are in the black tubes touching the screen image of a paraboloidal mirror. Note the ruler bar in front of the screen.*

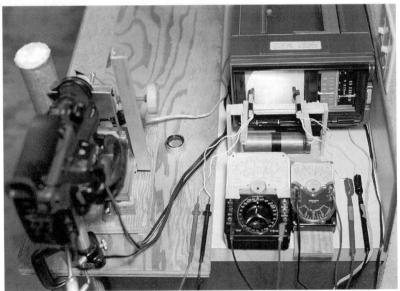

Fig. 11.8.3 *The compactness of the work area is shown with the camcorder on the left, behind the knife.*

Table 11.8.1 *Foucault Readings for a 10-inch f/4.8 Mirror*

Screen Image Radius (inches)	Micrometer Reading (x inches)	Actual Mirror Radius (inches)	Tex Program (1-x)
0.2	0.4840	0.5	0.5160
0.4	0.4716	1.0	0.5284
0.6	0.4650	1.5	0.5350
0.8	0.4521	2.0	0.5479
1.0	0.4398	2.5	0.5602
1.2	0.4279	3.0	0.5710
1.4	0.4074	3.5	0.5926
1.6	0.3937	4.0	0.6063
1.8	0.3669	4.5	0.6331

Note: This should be $1/16.4$ on the wave front.

moving first, turn the micrometer to change the carriage position. You will soon learn the direction to move the carriage in response to which meter moves first. Keep jockeying the carriage and knife-edge until both meters respond the same. I would run both meters up the scale an equal number of kilo-ohms. For instance, if meter #1 started at 18K and meter #2 at 20K, I might run them up 6K each to 24K and 26K, respectively. Once they respond the same, turn on a light and read and record the micrometer setting. Make a list with photocell radius in one column and the corresponding micrometer reading opposite (Table 11.8.1). Even though my mike only read to 0.001 inch directly, I estimated to 0.0001 inch (some mikes have a 0.0001-inch vernier). Repeat the above process for all the radii or zones you want to test. I typically tested 9 zones and have done as many as 18 zones. A 9-zone test, with practice, would take 25 to 30 minutes.

11.8.2 *F/Ratios, Image Size and Radius Conversions*

The *f*/ratio (mirror focal length divided by mirror diameter) of the test mirror will affect the image diameter on the TV screen. Use the zoom on your camcorder to enlarge the image to the maximum it will go or the largest that will fit the screen. Shorter *f*/ratios will give larger images, and longer *f*/ratios, smaller. My 10" *f*/4.8 gave a 4" image, but a 6" *f*/9 gave only 2.7". The minimum radius I could measure on the screen image, due to soda straw interference, was 0.2" on either side of center. How does 0.2" radius on the screen image relate to the actual radius on the mirror? Well, the actual mirror radius equals the mirror diameter divided by the TV image diameter times the screen radius in question. For example: (10" mirror)/(4" screen) image diameter = 2.5. Now multiply 2.5 x 0.2" screen radius to get the actual mirror radius of 0.5". Therefore, 0.2" radius on the screen actually equals 0.5" radius on the mirror. Four-tenths inch on the screen would be: 2.5 x 0.4" = 1.0" on the mirror, etc. The ratio, 2.5, is not true for all mirrors. An 8" *f*/6 mirror gave a 3.9" image and would have a ratio of 2.05. Thus the 9 zones measured on my 10"

Table 11.8.2 *Foucault Test Data for Dave's 10-inch Mirror.*
Note: Offset +0.0126 inches

Zone	100%	Dave's	1/4 Min	1/4 Max	1/8 Min	1/8 Max
0.5	0.00130	−0.01130	−0.02436	0.02436	−0.01218	0.01218
1.0	0.00520	−0.00280	−0.01218	0.01218	−0.00609	0.00609
1.5	0.01170	−0.00270	−0.00812	0.00812	−0.00406	0.00406
2.0	0.02080	0.00110	−0.00609	0.00609	−0.00304	0.00304
2.5	0.03251	0.00169	−0.00487	0.00487	−0.00244	0.00244
3.0	0.04681	−0.00071	−0.00406	0.00406	−0.00203	0.00203
3.5	0.06371	0.00289	−0.00348	0.00348	−0.00174	0.00174
4.0	0.08321	−0.00291	−0.00304	0.00304	−0.00152	0.00152
4.5	0.10532	0.00178	−0.00271	0.00271	−0.00135	0.00135

mirror went from 0.5" to 4.5" radius in steps of 0.5". With 18 zones, it went from 0.5" to 4.75" in 0.25" steps. The outer edge of a mirror is still a challenge to measure because of the bright edge diffraction present in a knife-edge test. A "good" problem to have in that department is a bright edge ring that is full circle, not just on the half opposite your knife's entering side—you do not have a turned edge either up or down.

11.8.3 Data Reduction

Once you finish a test and have your two columns of figures, image radius vs. carriage micrometer readings, what do you do with them? I had two programs on my trusty Compaq 286, which reduced the data to either tables or graphs comparing my mirror with a parabola. The first program was a relatively simple one written in BASIC by my brother. It produced tabular data listing the actual mirror radius, micrometer readings for a perfect parabola (center = 0), how my readings compared to the parabola (displayed as a +/− decimal inch) and four columns giving the maximum and minimum limits from a parabola for a ¼ or ⅛ wave mirror (Table 11.8.2). I could also display the data graphically (Figure 11.8.4).

Another program used was a Shareware program called *Tex—Mirror Number Crunching Made Easy,* Version 2.1, by Larry Phillips and Dale Eason. Their program was an upgraded version based on the program in the back of Jean Texereau's book, *How To Make A Telescope,* second edition. It produces several graphical representations of the data. Figure 11.8.5 (top), the "Relative Surface error" is very useful to mirror makers. This shows where the glass is high or low. The second, Figure 11.8.5 (bottom), is called the Millies-Lacroix Tolerance and gives a graph that matches what my brother developed. A third choice, when two or more micrometer readings have been entered, called a Monte Carlo simulation (Figure 11.8.6) is a statistical standard deviation program showing the most probable wave front accuracy of your mirror.

One thing I had to watch with this number crunching program was that it was

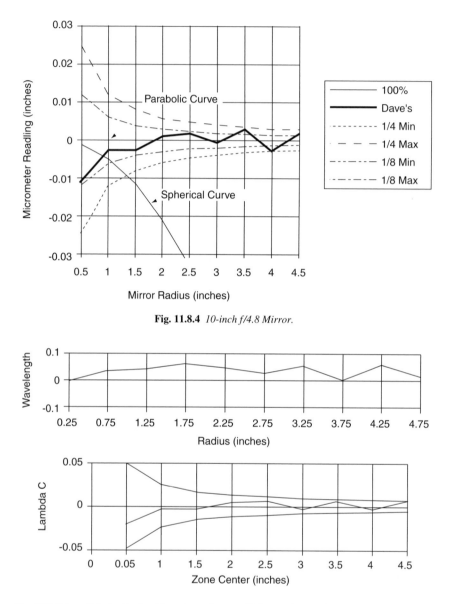

Fig. 11.8.4 *10-inch f/4.8 Mirror.*

Fig. 11.8.5 Top: *Relative Surface Error (+ surface too high; - surface too low).* **Bottom:** *Millies-Lacroix Tolerance (Parabola Removed).*

designed for knife-edge testers with a "pusher" micrometer and mine is a "puller." I had to input 1 − x as my micrometer readings. This program will also calculate the dimensions of a standard Couder mask given the mirror diameter and the number of zones. It was limited to a maximum of ten zones, and I often tested more. Also, my zones never matched the standard Couder zone radii. A slight difference

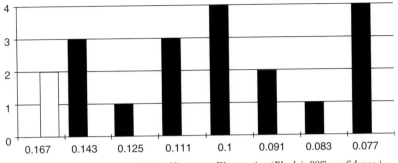

Fig. 11.8.6 *Monte-Carlo Simulation Histogram. Wave rating (Black is 80% confidence.)*

in zone radius made a big difference in how good the program said my mirror was. I had to adjust the radius limit numbers to make the program work correctly.

11.8.4 Helpful Hints

As in all knife-edge testing, the distance between mirror and tester knife is twice the mirror's focal length, which is called the radius of curvature.

When the screen image is evenly illuminated across the diameter, the initial reading on the two VOMs is nearly equal. This even lighting is greatly influenced by having the optical axis of the mirror and camcorder in a line. Work at this to get it the best you can.

My knife-edge tester had the light source projected through a narrow slit made from utility knife blades (see *Build Your Own Telescope* by Richard Berry). If the mirror under test was unaluminized, a slit width from 0.005" to 0.01" could be used; but when an aluminized mirror was tested, a slit width of 0.005" worked better. I set the width with a mechanic's feeler gauge.

This test is sensitive to air currents in the room that cause the VOM needles to constantly sway back and forth. I worked in a basement with a fireplace at the far end; and when the wind blew, I had to cover the front of the fireplace to keep air currents down. It helped when I built a light wooden framework draped with a plastic sheet to make a horizontal tunnel with the mirror in the far end.

I also found that I could do qualitative testing by placing a Ronchi grating (I used a piece of window screen) just ahead of the knife-edge. This can show irregularities in the mirror's surface, such as a turned edge.

11.9 Imaging in CCD Color
By Richard Berry

Unless you have kept your head buried in mud for the last three years, you will have noticed that CCDers have embraced color imaging. But if you are not paying close attention, you may have concluded that color CCD imaging is dauntingly technical and incredibly expensive. The ads in the glossies certainly make it appear that way. The purpose of this article is to show you that CCD color imaging is not only reasonably easy, but as a result of recent breakthroughs in thinking about color imaging, it is getting easier and better for everyone.

Color is a construct of the human eye and brain. The universe itself does not have color, but only light at different wavelengths. In the human retina, there are two types of cells, rods and cones; and among the cones, there are three subtypes. Rods are extremely sensitive to light, and respond most strongly to blue-green light centered on wavelengths of 500 nm (500 nanometers wavelength, or 5000 Ångstrom units). The rods give astronomers night vision, but without color. The three types of cone cells differ in containing pigments that break down in response to different wavelengths of light. One group of cones (beta cones) responds most strongly to blue light centered on 440 nm, another group to green (gamma cones) light centered on 540 nm, and the third group (rho cones) to orangish-red light peaking at 570 nm and extending redward to about 670 nm.

Nerve networks in the retina compare signals from the three types of cone sensors to derive color information which is sent to the brain and interpreted as color. Even though there are only three basic color stimuli, the eye and brain construct the millions of colors and shades from the differences and ratios among the three primary color sensations of red, green, and blue.

All color reproduction processes—photography, printing, television—recreate the sensation of color by mimicking the human visual system. The spectrum of light striking the sensor is separated into three components: red, green, and blue. The color information from the three bands is stored or transmitted separately and reproduced for the eye by dyes, inks, or glowing phosphor dots. The cone cells in the retina do not distinguish between the full-spectrum color of natural light and the reproduced color information conveyed to it in three recreated bands of color because the cones themselves are sensitive to only three bands of color. Hence, for all practical purposes, trichromatic (i.e., three-color) technology is capable of excellent color reproduction.

Until very recently, to make color CCD images, the standard method was to make images through three dyed-glass tricolor filters. The long-accepted filter set used in photography (Wratten #25 red, Wratten #58 green, and Wratten #47 blue) have broad transmission bands peaking at 440 nm, 545 nm, and 580 nm. When combined with typical emulsion sensitivities, these filters match the trichromatic sensitivities of the cones quite well. (For CCD tricolor imaging, an additional infrared rejection filter must be used to block wavelengths longer than 680 nm from reaching the detector, since this additional—and invisible—energy distorts proper color rendition.) Because the classic tricolor filters suffer from peak transmissions of only about 50%, recent CCD practice has been to use dichroic filters with 90% transmission over the filter's bandpass. Edmund Scientific's stock numbers A52,529; A52,532; and A52,535 are attractively priced at $59.95 for a set of three 50 mm diameter filters.

Despite its success, tricolor CCD imaging suffers from two significant difficulties. The first is an instrumental problem, because most front illuminated CCDs have low quantum efficiency in the green and blue region of the spectrum. To obtain adequately exposed images, exposures of 10 to 60 minutes are required, some three to five times longer than needed for the exposure through a red filter, and up to ten times longer than needed with no filter. The second difficulty occurs because gaseous celestial objects emit the bulk of their energy at single wavelengths, making the recreated color critically dependent on the exact transmission curves of the tricolor set.

This article describes two breakthrough techniques that allow amateur astronomers to overcome these difficulties. The first of these is quad-color imaging, in which we separate color information from brightness information by taking short red, green, and blue filtered CCD exposures to derive color, and a deep unfiltered CCD exposure to collect brightness information. A high-quality color image is then reconstructed in software. The second technique is to capture color information by taking exposures through cyan, magenta, and yellow filters and from them deriving the strengths of the red, green, and blue color components to recreate a color image. One potential advantage is that cyan, magenta, and yellow filters transmit more light than red, green, and blue filters; and another is that the spectral bandpasses of these filters do not block critical nebular wavelengths.

11.9.1   Quadcolor Imaging[22]

To make tricolor images, you must filter the light reaching the CCD, decreasing the signal generated by the CCD without changing the intrinsic noise level of the CCD. As a result, the signal-to-noise ratio of color-filtered images is much lower than it is for unfiltered images, often by a factor of ten. Unless the exposures are greatly lengthened, the components of tricolor images are coarse and grainy compared to unfiltered images. As a result, when amateur astronomers first attempt color imaging, their results are poor and they are often discouraged. Until recently, the only cure was running exposures long enough to attain a good signal to noise ratio despite the faint signal.

However, the quadcolor technique offers both high signal-to-noise ratio and good color quality. Quadcolor works by capitalizing on the fact that in viewing an image, the human eye-brain system places greater emphasis on accurate brightness information, or luminosity, than it does on precise color, or chroma. Shooting three noisy tricolor images produces a coarse, grainy color image; that is, an image in which the luminosity information is corrupted by noise. CCD tricolor differs from photographic tricolor methods because in the computer, it is possible to transform red, green, and blue images into luminosity and chroma components. In quad-color imaging, we simply make an additional exposure with no color filter, exposing long enough that the signal-to-noise ratio of the unfiltered image is very high. In the computer, we combine a set of red, green, and blue images and from

---

[22] Quadcolor later became known as LRGB or LCYM imaging. The "L" stands for luminance, the unfiltered exposure.

them extract the chroma components. The next step is to combine the luminosity information contained in the unfiltered image with the chroma components to produce a color image with luminosity information. The resulting images look very good to the eye-brain system.

Broadcast television has long exploited this technique. The luminosity information displayed on televisions in the US occupies a broadcast bandwidth of 3 megahertz, while the color information occupies only 1 megahertz of video bandwidth. The color on a TV screen is less sharp than the black-and-white information because color information has been sacrificed in order to present a sharp luminosity image, an engineering tradeoff.

In CCD imaging, quadcolor works amazingly well. If you shoot a tricolor image using 10 minutes in red, 25 minutes in green, and 35 minutes in blue, you will get a less pleasing result than shooting 5 minutes in red, 5 minutes in green, 5 minutes in blue, and 10 minutes with no filter. The total exposure time is cut in half, and the resulting image looks markedly better. In addition, because the unfiltered luminosity image has a high signal-to-noise ratio, it can be non-linearly scaled and its sharpness enhanced.

11.9.2   Color Accuracy

Celestial objects such as planets, stars, and galaxies have spectra that are rather well behaved, in the sense that their energy changes slowly with wavelength. Although an extremely red star can appear as much as four magnitudes brighter through a red filter than through a blue filter, and the very bluest stars appear about one magnitude brighter in blue light than in red light, the normal variation is only about ½ magnitude between the apparent brightness of stars in the red and blue light. Almost any set of filters will do a good job recording the smooth, predictable colors of stars, galaxies, and planets, so high-transmission dichroic red, green, and blue filters do a fine job.

In sharp contrast, the spectra of planetary nebulae and HII regions consist of discrete lines, so the amount of light reaching the CCD depends on the transmission curve of the color filters. As a case in point, the 500.7 nm line of doubly ionized oxygen (OIII) lies between the blue and green transmission peaks of most tricolor filter sets, and it dominates the spectrum of many planetaries. Visually, light at 500.7 nm appears blue-green in color, so the true visual color of the planetary nebulae should be blue-green; or perhaps, as the blue-green line combines with the second most intense spectral lines, which are the deep crimson H-alpha line at 656.3 nm and ionized nitrogen (NII) at 667.8 nm, they will appear a yellowish shade of green.

The classic Wratten tricolor filters have almost exactly equal transmission at 500.7 nm, so tricolor images of planetary nebulae taken with them appear bluish-green. However, the transmission curves of both the green and blue dichroic filters fall off sharply just short of 500.7 nm, so neither the green nor blue image properly records the strongest spectral line coming from planetaries. Instead, the blue filter picks up the fairly strong H-beta line at 495.9 nm, giving the image an unnatural blue cast.

Complicating the question of color accuracy is, of course, the fact that the familiar photographs of these objects that we use as our standard of "true" color suffer from the same types of color distortion. The sensitivity curves of the different color films interact with the line spectra of the nebulae in the same way that color filters do, so the photographs familiar from decades of reproduction form an unreliable standard for judging color, since they are often quite inaccurate. Chances are that when we succeed in making truly true-color images, they will look all wrong to us.

To address the filter problem, recall that red, green, and blue are additive color primaries, but the subtractive primaries of cyan, magenta, and yellow (called CMY) also exist. A red filter transmits red light, but the subtractive cyan filter blocks red light. The complement of green is the green-blocking magenta filter, and the complement of blue is the blue-blocking yellow filter. Each of these subtractive filters passes the other two additive primaries, so that cyan equals blue plus green, magenta equals red plus blue, and yellow equals red plus green. Given CCD images taken through cyan, magenta, and yellow filters, a computer can solve for the red, green, and blue color components and from those generate either a tricolor or quadcolor image.

One advantage of shooting through cyan, magenta, and yellow filters is that each filter transmits two spectral bands at once, so twice as many photons reach the CCD, resulting in a signal-to-noise ratio 1.4 times better than an equivalent set of images taken through red, green, and blue filters. In addition, all of the critical spectral lines lie in a region of high transmission for at least one filter. Cyan, magenta, and yellow tricolor, therefore, promises more accurate color rendition of difficult nebular objects and higher optical throughput for all types of color CCD imaging.

11.9.3   Current Status

At present, quadcolor imaging has proven its value over and over in the hands of CCD imagers Al Kelly, Chuck Shaw, and Rob West. My Color245 color synthesis software has been totally rewritten to accommodate the needs of quadcolor imaging and accordingly renamed QColor. At the telescope, you shoot sets of images through red, green, and blue filters, and another set with no filter. After calibrating the images, you load them into QColor. The three-filtered images are precisely registered to the unfiltered image, and the unfiltered image can be scaled and sharpened as much as you want using functions built into QColor. Qcolor produces a color image by extracting information about chroma from the three filtered images and information about luminosity from the unfiltered image.

CMY imaging is under test as you read this article. An experimental version of QColor can already load synthesized CMY images and recreate accurate color from them, so the software is ready and waiting until the imaging crews to have enough clear nights to acquire test images.

Note added in October, 2001: The color techniques described above flowered in the *AIP for Windows* software that accompanies the *Handbook of Astronomical Image Processing* (Willmann-Bell, Inc., 2000).

Index